Fuzzy Discrete Structures

Studies in Fuzziness and Soft Computing

Editor-in-chief
Prof. Janusz Kacprzyk
Systems Research Institute
Polish Academy of Sciences
ul. Newelska 6
01-447 Warsaw, Poland
E-mail: kacprzyk@ibspan.waw.pl
http://www.springer.de/cgi-bin/search_book.pl?series=2941

Vol. 3. A. Geyer-Schulz
Fuzzy Rule-Based Expert Systems and Genetic Machine Learning, 2nd ed. 1996
ISBN 3-7908-0964-0

Vol. 4. T. Onisawa and J. Kacprzyk (Eds.)
Reliability and Safety Analyses under Fuzziness, 1995
ISBN 3-7908-0837-7

Vol. 5. P. Bosc and J. Kacprzyk (Eds.)
Fuzziness in Database Management Systems, 1995
ISBN 3-7908-0858-X

Vol. 6. E.S. Lee and Q. Zhu
Fuzzy and Evidence Reasoning, 1995
ISBN 3-7908-0880-6

Vol. 7. B.A. Juliano and W. Bandler
Tracing Chains-of-Thought, 1996
ISBN 3-7908-0922-5

Vol. 8. F. Herrera and J.L. Verdegay (Eds.)
Genetic Algorithms and Soft Computing, 1996
ISBN 3-7908-0956-X

Vol. 9. M. Sato et al.
Fuzzy Clustering Models and Applications, 1997
ISBN 3-7908-1026-6

Vol. 10. L.C. Jain (Ed.)
Soft Computing Techniques in Knowledge-based Intelligent Engineering Systems, 1997
ISBN 3-7908-1035-5

Vol. 11. W. Mielczarski (Ed.)
Fuzzy Logic Techniques in Power Systems, 1998,
ISBN 3-7908-1044-4

Vol. 12. B. Bouchon-Meunier (Ed.)
Aggregation and Fusion of Imperfect Information, 1998
ISBN 3-7908-1048-7

Vol. 13. E. Orłowska (Ed.)
Incomplete Information: Rough Set Analysis, 1998
ISBN 3-7908-1049-5

Vol. 14. E. Hisdal
Logical Structures for Representation of Knowledge and Uncertainty, 1998
ISBN 3-7908-1056-8

Vol. 15. G.J. Klir and M.J. Wierman
Uncertainty-Based Information, 2nd ed., 1999
ISBN 3-7908-1242-0

Vol. 16. D. Driankov and R. Palm (Eds.)
Advances in Fuzzy Control, 1998
ISBN 3-7908-1090-8

Vol. 17. L. Reznik, V. Dimitrov and J. Kacprzyk (Eds.)
Fuzzy Systems Design, 1998
ISBN 3-7908-1118-1

Vol. 18. L. Polkowski and A. Skowron (Eds.)
Rough Sets in Knowledge Discovery 1, 1998
ISBN 3-7908-1119-X

Vol. 19. L. Polkowski and A. Skowron (Eds.)
Rough Sets in Knowledge Discovery 2, 1998
ISBN 3-7908-1120-3

Vol. 20. J.N. Mordeson and P.S. Nair
Fuzzy Mathematics, 1998
ISBN 3-7908-1121-1

Vol. 21. L.C. Jain and T. Fukuda (Eds.)
Soft Computing for Intelligent Robotic Systems, 1998
ISBN 3-7908-1147-5

Vol. 22. J. Cardoso and H. Camargo (Eds.)
Fuzziness in Petri Nets, 1999
ISBN 3-7908-1158-0

Vol. 23. P.S. Szczepaniak (Ed.)
Computational Intelligence and Applications, 1999
ISBN 3-7908-1161-0

Vol. 24. E. Orłowska (Ed.)
Logic at Work, 1999
ISBN 3-7908-1164-5

Vol. 25. J. Buckley and Th. Feuring
Fuzzy and Neural: Interactions and Applications, 1999
ISBN 3-7908-1170-X

continued on page 264

Davender S. Malik
John N. Mordeson

Fuzzy Discrete Structures

With 37 Figures
and 7 Tables

Springer-Verlag Berlin Heidelberg GmbH

Prof. Davender S. Malik
Prof. John N. Mordeson
Creighton University
Omaha, Nebraska 68178
USA
E-mail: malik@creighton.edu
mordes@creighton.edu

DOI 10.1007/978-3-7908-1838-3

Cataloging-in-Publication Data applied for
Die Deutsche Bibliothek – CIP-Einheitsaufnahme
Malik, Davender S.: Fuzzy discrete structures: with 7 tables / Davender S. Malik; John N. Mordeson. – Heidelberg; New York: Physica-Verl., 2000
(Studies in fuzziness and soft computing; Vol. 58)

Originally published by Physica-Verlag Heidelberg New York in 2000
MyCopy version of the original edition 2000

Hardcover Design: Erich Kirchner, Heidelberg
www.springer.com/mycopy

To
Lotfi A. Zadeh

Foreword

This ambitious exposition by Malik and Mordeson on the fuzzification of discrete structures not only supplies a solid basic text on this key topic, but also serves as a viable tool for learning basic fuzzy set concepts "from the ground up" due to its unusual lucidity of exposition. While the entire presentation of this book is in a completely traditional setting, with all propositions and theorems provided totally rigorous proofs, the readability of the presentation is not compromised in any way; in fact, the many excellently chosen examples illustrate the often tricky concepts the authors address. The book's specific topics - including fuzzy versions of decision trees, networks, graphs, automata, etc. - are so well presented, that it is clear that even those researchers not primarily interested in these topics will, after a cursory reading, choose to return to a more in-depth viewing of its pages.

Naturally, when I come across such a well-written book, I not only think of how much better I could have written my co-authored monographs, but naturally, how this work, as distant as it seems to be from my own area of interest, could nevertheless connect with such. Before presenting the briefest of some ideas in this direction, let me state that my interest in fuzzy set theory (FST) has been, since about 1975, in connecting aspects of FST directly with corresponding probability concepts. One chief vehicle in carrying this out involves the concept of *random sets*. At the outset, a basic interesting point (no proverbial pun intended!) connected with random sets - or rather subsets of a given domain - is that while all of the ordinary laws of probability are obeyed and formally all one is doing is replacing in "ordinary" probability evaluations "points" (or vectors) by sets of points (or vectors) in the randomization process, something very profound has taken place at even the simple finite domain cases: While the probabilities of the distinct outcomes of the random set in question must add up to unity, unlike our normal image of a probability function over that domain representing an ordinary random variable over same domain, the distinct outcomes, i.e., subsets of the domain, *need not be disjoint,* unlike our image of points. Of course, if we were to interpret such point as really representing possibly overlapping concepts as random sets do in general, without violating probability constraints, the same idea would unfold. But, the human side of this traditionally has been that when one sees a probability function - and equivalently, a discrete-valued random variable (or random vector) - one almost invariably thinks of complete disjointness. (Parenthetically, it would be interesting to see if any associated psychological studies have ever been initiated in this direction concerning our intuitive concepts of probability for random sets.) In turn, the very overlapping of the outcomes of a random set allow for a richer expression of probability concepts than just a point-limited random variable over the same domain.

In a related vein, certain types of possible overlap of random sets with fixed (but arbitrary) given sets from the same domain serves as a basis for the basic functions of Dempster-Shafer Theory - belief (superset relation), plausibility (incidence relation), doubt, and a fourth not usually named (each representing also the four basic Choquet infinite capacities - see Nguyen's now classic exposition of 1978, pp. 531-542 in the *Journal of Mathematical Analysis & Applications*). In particular, for the fixed sets restricted to be singletons from the domain, when overlaps occur they represent *one-point coverage events* generated by the effect of the random set covering such given points. Then, somewhat analogous to the way Dempster-Shafer Theory can be explained via random sets acting upon fixed sets, one can show that a fuzzy set (in the first order sense with membership function having range only in the unit interval) can be likewise interpreted as the action of a random set covering individual specified points of the fuzzy set's domain. In general, unlike the Dempster-Shafer situation, there may be many such (even an infinity of truly different) random sets providing such action resulting in the same fuzzy set representation. On the other hand, when the fuzzy set is actually a crisp set, the random set is uniquely determined as the constant crisp set itself. In fact, in general, each possible truly distinct random set representing a given fuzzy set in the above one-point coverage way is completely determine by both the given fuzzy set and a particular (but arbitrary) choice of copula, co-copula pair. Moreover, while even the algebraic structure of Zadeh's classical min-max-1-(.) logical operators over fuzzy sets is not equivalent isomorphically to that of the ordinary (boolean set) algebraic structure of sets - and hence random sets and their associated ordinary events, including one-point coverage events - certain localized equivalences can be obtained, including "isomorphic-like" correspondences between a number of fuzzy logic concepts and one-point coverage events. For example, Zadeh's extension principle couched in general co-copula form (extending max) applied to some crisp function g with arguments described by fuzzy set f can be interpreted as the one-point coverage action of the functional image under g of the random set determined by f and the copula-cocopula pair chosen on singletons of the domain of f.

For further information on these ideas, see, e.g., the articles by Goodman & Nguyen, "Applications of conditional and relational event algebra to the defining of fuzzy logic concepts", *Proc. S.P.I.E. Signal Processing, Sensor Fusion & Target Recognition VIII* (vol. 3720), Orlando, FL, April, 1999, pp. 25-36, or Goodman & Kramer, "Extension of rational and conditional event algebra to random sets with applications to data fusion", in *Random Sets: Theory & Applications* (J. Goutsias, R.P. Mahler & H.T. Nguyen, eds.) Springer-Verlag, New York, 1997, pp. 209-242.)

Finally, with some of the basic ideas outlined above relating fuzzy set concepts with corresponding random set ones, one may ask whether any natural connections can be established between the various fuzzified crisp concepts analyzed in *Fuzzy Discrete Structures* and naturally corresponding random set concepts. For example, section (4) on fuzzy graphs and shortest path depends obviously on how one carries out the fuzzification of graphs and subgraphs, as well as the valuation of the arcs of a fuzzy graph. While each of these concepts has a rote one-point coverage random set representation, the challenge is to determine whether a nontrivial correspondence exists between the theorems and procedures associated with these fuzzy concepts and natural crisp counterparts applied to random sets, as in the case of the extension principle. If such correspondences could be found for either the currently used or suitably modified definitions, then an even stronger case can be made for justifying the very direction of fuzzification of the crisp concepts involved. Much work remain in this essentially unexplored area.

I.R. Goodman
Space & Naval Warfare Systems Center
San Diego, CA 92152

Preface

In 1965, L.A. Zadeh introduced the concepts of a fuzzy subset of a set as a way for representing uncertainty. Zadeh's ideas stirred researchers worldwide. His ideas have been applied to a wide range of scientific areas. Here we consider the area of mathematics which is normally known as discrete structures. Since texts on discrete structures cover a wide variety of subjects, the topics appearing in such books do not appear in the depth they would in books specializing in these topics. This book deals with fuzzy logic, fuzzy switching functions, fuzzy decision trees, fuzzy networks, fuzzy Petri nets, fuzzy path problems, fuzzy automata, and fuzzy languages. As in crisp books on discrete structures, the topics in this book do not appear in the depth they would in books specializing in these topics. The book should be of interest to research mathematicians, engineers, and computer scientists interested in applications.

In Chapter 1, we present just enough material on fuzzy logic to study fuzzy switching functions and their decomposition. We concentrate on the minimization of completely specified fuzzy functions. We touch briefly on the minimization of incompletely specified fuzzy functions, leaving the interested reader to pursue the issue elsewhere, say in Kandel and Lee, [4]. The chapter concludes with a section on the solution of fuzzy logical inequalities.

Chapter 2 is concerned with fuzzy decision trees. We begin the chapter with a review of crisp decision trees. We introduce the idea of a fuzzy decision tree and present a branch-bound-backtrack algorithm which has an effective backtracking mechanism leading to the optimal solution while requiring usually only $O(\log n)$ time, where n is the number of decision classes. In the Appendix, this algorithm is described in C++.

In Chapter 3, we examine network flow problems with various fuzzy arc capacities, e.g., integer flows with two-sided fuzzy capacity constraints and real-valued flows with fuzzy flows with fuzzy arc capacities. In this chapter, we also begin with a review of the standard material on network flow problems. We conclude the chapter with the introduction of the notion of a fuzzy Petri net for rule based decision making.

In Chapter 4, we concentrate on the most basic network problem, the shortest path problem. A fuzzy shortest path can be found, but it may not correspond to an actual path in the network. This problem is overcome with models based on fuzzy shortest paths and multiple objective functions. Algorithms for these models based on classical dynamic programming procedures and multi-criteria optimization are also presented.

The literature is quite extensive on the fuzzification of automata and languages. There has been a wide variety of approaches applied to this fuzzification. It is our purpose in Chapter 5 to present a sample of these approaches. In this chapter, all fuzzy machines have some form of output. We begin by introducing the notion of a max-product machine. The machine here is defined so that the next state and output are handled by one transition function. Irreducibility and minimality are studied for max-product machines. We then present the notion of a maximin machine. Here too, the next state and output are handled by one transition function. The reduction of maximin machines is examined. We also present the notion of a context-free max-product grammar and a context-free fuzzy language. We close the chapter with the presentation of a fuzzy language on a free monoid.

In Chapter 6, we concentrate on algebraic fuzzy automata. In this chapter, the fuzzy automata are without an output. Some topics we consider are semigroups of fuzzy finite state machines, homomorphisms, admissible relations, fuzzy transformation semigroups, submachines, fuzzy regular languages, and fuzzy recognizers.

The authors are grateful to the editorial and production staffs of Physica-Verlag, especially Janusz Kacprzyk, Martina Bihn, and Gabriele Keidel. We are indebted to Paul Wang and Hu Cheng-ming for their support of fuzzy mathematics. We are also appreciative of the support of Fr. Albert Agresti, Dean Creighton College of Arts and Sciences, Dr. and Mrs. George Haddix, benefactors of our research center, and Lynn Schneiderman of the Creighton Reinert Alumni Library. The first author dedicates the book to his parents and his daughter Shelly. The second author dedicates the book to his mother Margaret, his sister Rosemary, and his nephew John.

D. S. Malik
John N. Mordeson

Contents

1

Fuzzy Logic Functions

1.1 Sets

In this section, we present certain definitions that will be used through out the book. Sets will always be denoted by capital letters. For example, we use the notation $\mathbb{N}$ for the set of positive integers, $\mathbb{Z}$ for the set of integers, $\mathbb{Q}$ for the set of rational numbers, $\mathbb{R}$ for the set of real numbers, and $\mathbb{R}^+$ for the set of positive real numbers. For $n \in \mathbb{N}$, we let $\mathbb{N}_n = \{1, 2, \ldots, n\}$.

We assume that the reader is familiar with the basics of set theory. Nevertheless, we give a brief review of some basics of set theory. We think of a **set** as a collection of objects. A set S with only a finite number of elements is called a **finite** set; otherwise S is called an **infinite** set. We let $|S|$ denote the number of elements of S.

Given a set S, we use the notation $x \in S$ and $x \notin S$ to mean x is a member of S and x is not a member of S, respectively.

A set A is said to be a **subset** of a set S if every element of A is an element of S. In this case, we write $A \subseteq S$ and say that A is contained in S. If $A \subseteq S$, but $A \neq S$, then we write $A \subset S$ and say that A is properly contained in S or that A is a **proper subset** of S.

The **null set** or **empty set** is the set with no elements. We denote the empty set by $\emptyset$.

We also describe sets in the following manner. Given a set S, the notation

$$A = \{x \mid x \in S, P(x)\}$$

or

$$A = \{x \in S \mid P(x)\}$$

means that A is the set of all elements x of S such that x satisfies the property P.

We can combine sets in several ways.

Definition 1.1.1 *The* ***union*** *of two sets A and B, written $A \cup B$, is defined to be the set*

$$A \cup B = \{x \mid x \in A \text{ or } x \in B\}.$$

Definition 1.1.2 *The* ***intersection*** *of two sets A and B, written $A \cap B$, is defined to be the set*

$$A \cap B = \{x \mid x \in A \text{ and } x \in B\}.$$

The union and intersection for any finite number of sets can be similarly defined. That is, suppose that $A_1, A_2, \ldots, A_n$ are n sets. The union of $A_1, A_2, \ldots, A_n$, denoted by $\cup_{i=1}^n A_i$ or $A_1 \cup A_2 \cup \ldots \cup A_n$, is the set of all elements x such that x is an element of some A_i, where $1 \leq i \leq n$. The intersection of $A_1, A_2, \ldots, A_n$, denoted by $\cap_{i=1}^n A_i$ or $A_1 \cap A_2 \cap \ldots \cap A_n$, is the set of all elements x such that $x \in A_i$ for all i, $1 \leq i \leq n$.

We say that a set I is an **index set** for a collection of sets $\mathcal{A}$ if for any $i \in I$, there exists a set $A_i \in \mathcal{A}$ and $\mathcal{A} = \{A_i \mid i \in I\}$.

The **union** of the sets A_i, $i \in I$, is defined to be the set $\{x \mid x \in A_i$ for at least one $i \in I\}$ and is denoted by $\cup_{i \in I} A_i$. The **intersection** of the sets A_i, $i \in I$, is defined to be the set $\{x \mid x \in A_i$ for all $i \in I\}$ and is denoted by $\cap_{i \in I} A_i$.

Definition 1.1.3 *Given two sets A and B, the* ***relative complement*** *of B in A, denoted by the set difference $A \backslash B$, is the set*

$$A \backslash B = \{x \mid x \in A, \text{ but } x \notin B\}.$$

We sometimes write $\overline{A}$ for the relative complement of a set A in the universal set.

If A and B are sets, we let $A \times B$ denote the **Cartesian cross-product** of A and B, i.e., $A \times B$ is the set of all ordered pairs (a, b), where $a \in A$ and $b \in B$. We let $\mathcal{P}(x)$ denote the **power set** of a set X, i.e., $\mathcal{P}(x)$ is the set of all subsets of X.

1.2 Relations

Definition 1.2.1 *A* ***binary relation*** *or simply a* ***relation*** *R from a set A into a set B is a subset of $A \times B$.*

Let R be a relation from a set A into a set B. If $(x, y) \in R$, we sometimes write xRy or $R(x) = y$. If $A = B$, then we call R a **binary relation** on A.

Definition 1.2.2 *Let R be a relation from a set A into a set B. Then the* ***domain*** *of R, denoted by $Dom(R)$, is defined to be the set*

$$\{x \mid x \in A \text{ and there exists } y \in B \text{ such that } (x, y) \in R\}.$$

The ***image*** *of R, denoted by $Im(R)$, is defined to be the set*

$$\{y \mid y \in B \text{ and there exists } x \in A \text{ such that } (x, y) \in R\}.$$

Definition 1.2.3 *Let R be a binary relation on a set A. Then R is called*
(1) ***reflexive*** *if for all $x \in A$, xRx,*
(2) ***symmetric*** *if for all $x, y \in A$, xRy implies yRx,*
(3) ***transitive*** *if for all $x, y, z \in A$, xRy and yRz imply xRz.*

Definition 1.2.4 *A binary relation E on a set A is called an* ***equivalence relation*** *on A if E is reflexive, symmetric, and transitive.*

Definition 1.2.5 *Let E be an equivalence relation on a set A. For all $x \in A$, let $[x]$ denote the set*

$$[x] = \{y \in A \mid yEx\}.$$

The set $[x]$ is called the ***equivalence class*** *(with respect to E) determined by x.*

In the following theorem, we prove some basic properties of equivalence classes.

Theorem 1.2.6 *Let E be an equivalence relation on the set A. Then*
(1) for all $x \in A$, $[x] \neq \emptyset$,
(2) if $y \in [x]$, then $[x] = [y]$, where $x, y \in A$,
(3) for all $x, y \in A$, either $[x] = [y]$ or $[x] \cap [y] = \emptyset$,
(4) $A = \cup_{x \in A}[x]$, i.e., A is the union of all equivalence classes with respect to E.

Proof. (1) Let $x \in A$. Since E is reflexive, xEx. Hence, $x \in [x]$ and so $[x] \neq \emptyset$.

(2) Let $y \in [x]$. Then yEx and by the symmetric property of E, xEy. Let $u \in [y]$. Then uEy. Since uEy and yEx, the transitivity of E implies that uEx. Hence, $u \in [x]$. Thus, $[y] \subseteq [x]$. Now let $u \in [x]$. Then uEx. Since uEx and xEy, uEy by transitivity and so $u \in [y]$. Hence, $[x] \subseteq [y]$. Consequently, $[x] = [y]$.

(3) Let $x, y \in A$. Suppose $[x] \cap [y] \neq \emptyset$. Then there exists $u \in [x] \cap [y]$. Thus, $u \in [x]$ and $u \in [y]$, i.e., uEx and uEy. Since E is symmetric and

uEy, we have yEu. Now yEu and uEx and so by the transitivity of E, yEx. This implies that $y \in [x]$. Hence, by (2), $[y] = [x]$.

(4) Let $x \in A$. Then $x \in [x] \subseteq \cup_{x \in A}[x]$. Thus, $A \subseteq \cup_{x \in A}[x]$. Also, $\cup_{x \in A}[x] \subseteq A$. Hence, $A = \cup_{x \in A}[x]$. ■

One of the objectives of this section is to study the relationship between an equivalence relation and a partition of a set. We now focus our attention on partitions.

Definition 1.2.7 *Let A be a set and $\mathcal{P}$ be a collection of nonempty subsets of A. Then $\mathcal{P}$ is called a* **partition** *of A if the following properties are satisfied:*

(1) for all B, $C \in \mathcal{P}$, either $B = C$ or $B \cap C = \emptyset$.

(2) $A = \cup_{B \in \mathcal{P}} B$. ■

The following theorem is immediate from Theorem 1.2.6.

Theorem 1.2.8 *Let E be an equivalence relation on the set A. Then*

$$\mathcal{P} = \{[x] \mid x \in A\}$$

is a partition of A. ■

By Theorem 1.2.8, given an equivalence relation E on a set A, the set of all equivalence classes forms a partition of A. We now prove that corresponding to any partition, we can associate an equivalence relation.

Theorem 1.2.9 *Let $\mathcal{P}$ be a partition of the set A. Define a relation E on A by for all $x, y \in A$, xEy if there exists $B \in \mathcal{P}$ such that x, $y \in B$. Then E is an equivalence relation on A and the equivalence classes are precisely the elements of $\mathcal{P}$.*

Proof. Since $\mathcal{P}$ is a partition of A, $A = \cup_{B \in \mathcal{P}} B$. First we show that E is reflexive. Let x be any element of A. Then there exists $B \in \mathcal{P}$ such that $x \in B$. Since $x, x \in B$, we have xEx. Hence, E is reflexive. We now show that E is symmetric. Let xEy. Then $x, y \in B$ for some $B \in \mathcal{P}$. Thus, y, $x \in B$ and so yEx. Hence, E is symmetric. We now establish the transitivity of E. Let $x, y, z \in A$. Suppose xEy and yEz. Then x, $y \in B$ and $y, z \in C$ for some B, $C \in \mathcal{P}$. Since $y \in B \cap C$, $B \cap C \neq \emptyset$. Also, since $\mathcal{P}$ is a partition and $B \cap C \neq \emptyset$, we have $B = C$ so that x, $z \in B$. Hence, xEz. This shows that E is transitive. Consequently, E is an equivalence relation.

We now show that the equivalence classes determined by E are precisely the elements of $\mathcal{P}$. Let $x \in A$. Consider the equivalence class $[x]$. Since $A = \cup_{B \in P} B$, there exists $B \in \mathcal{P}$ such that $x \in B$. We claim that $[x] = B$. Let $u \in [x]$. Then uEx and so $u \in B$ since $x \in B$. Thus, $[x] \subseteq B$. Also, since $x \in B$, we have yEx for all $y \in B$ and so $y \in [x]$ for all $y \in B$. This implies that $B \subseteq [x]$. Hence, $[x] = B$. Finally, note that if $C \in \mathcal{P}$, then $C = [u]$ for all $u \in C$. Thus, the equivalence classes are precisely the elements of $\mathcal{P}$. ■

The relation E in Theorem 1.2.9 is called the **equivalence relation** on A **induced by the partition** $\mathcal{P}$.

New relations can be constructed from existing relations. For example, given relations R and S from a set A into a set B, we can form relations $R \cap S$, $R \cup S$, $R \backslash S$, $(A \times B) \backslash R$ in a natural way. In all these relations, the domain and image of the relations under consideration are subsets of A and B, respectively. Now given a relation R from a set A into a set B and a relation S from B into a set C, there is a relation T from A into C that arises in a natural way as follows: $(a, c) \in T$ for $a \in A$ and $c \in C$ if and only if there exists $b \in B$ such that $(a, b) \in R$ and $(b, c) \in S$. This relation T is called the **composition** of R and S and is denoted by $S \circ R$. Note that to form the composition of R and S, we must have the domain of S and the image of R to be subsets of the same set. More formally we have the following definition.

Definition 1.2.10 *Let R be a relation from a set A into a set B and S be a relation from B into a set C. The **composition** of R and S, denoted by $S \circ R$, is the relation from A into C defined by*

$$x(S \circ R)z \text{ if there exists } y \in B \text{ such that } xRy \text{ and } ySz$$

for all $x \in A$, $z \in C$.

Let R be a relation on a set A. Recursively, we define a relation R^n, $n \in \mathbb{N}$, as follows:

$$\begin{aligned} R^1 &= R \\ R^n &= R \circ R^{n-1} \text{ if } n > 1. \end{aligned}$$

Definition 1.2.11 *Let R be a relation from a set A into a set B. The **inverse** of R, denoted by R^{-1}, is the relation from B into A defined by*

$$xR^{-1}y \text{ if } yRx$$

for all $x \in B$, $y \in A$.

1.3 Functions

In this section, we review some of the basic properties of functions.

Definition 1.3.1 *Let A and B be nonempty sets. A relation f from A into B is called a **function** (or **mapping**) from A into B if*

(1) $Dom(f) = A$ *and*
(2) for all $(x, y), (x', y') \in f$, $x = x'$ *implies* $y = y'$.
When (2) is satisfied by a relation f, *we say that* f *is* ***well defined*** *or* ***single-valued.***

We use the notation $f : A \to B$ to denote a function f from a set A into a set B. For $(x, y) \in f$, we usually write $f(x) = y$ and say that y is the **image** of x under f and x is a **preimage** of y under f.

Definition 1.3.2 *Let* f *be a function from a set* A *into a set* B. *Then*
(1) f *is called* **one-one** *if for all* $x, x' \in A$, $f(x) = f(x')$ *implies* $x = x'$.
(2) f *is called* **onto** B *(or* f **maps** A **onto** B*) if* $Im(f) = B$.

Let A be a nonempty set. The function $i_A : A \to A$ defined by $i_A(x) = x$ for all $x \in A$ is a one-one function of A onto A. i_A is called the **identity map** on A.

Let A, B, and C be nonempty sets and $f : A \to B$ and $g : B \to C$. By Definition 1.2.10, the **composition** $\circ$ of f and g, written $g \circ f$, is the relation from A into C defined as follows:

$$g \circ f = \{(x, z) \mid x \in A,\ z \in C, \text{ there exists } y \in B \text{ such that } f(x) = y \text{ and } g(y) = z\}.$$

Let $f : A \to B$ and $g : B \to C$ and $(x, z) \in g \circ f$, i.e., $(g \circ f)(x) = z$. Then by the definition of composition of functions, there exists $y \in B$ such that $f(x) = y$ and $g(y) = z$. Now

$$z = g(y) = g(f(x)).$$

Hence, $(g \circ f)(x) = g(f(x))$.

Theorem 1.3.3 *Suppose that* $f : A \to B$ *and* $g : B \to C$. *Then*
(1) $g \circ f : A \to C$, *i.e.,* $g \circ f$ *is a function from* A *into* C.
(2) If f *and* g *are one-one, then* $g \circ f$ *is one-one.*
(3) If f *is onto* B *and* g *is onto* C, *then* $g \circ f$ *is onto* C.

Proof. (1) Let $x \in A$. Since f is a function and $x \in A$, there exists $y \in B$ such that $f(x) = y$. Now since g is a function and $y \in B$, there exists $z \in C$ such that $g(y) = z$. Thus, $(g \circ f)(x) = g(f(x)) = g(y) = z$, i.e., $(x, z) \in g \circ f$. Hence, $x \in \text{Dom}(g \circ f)$. This shows that $A \subseteq \text{Dom}(g \circ f)$. But $\text{Dom}(g \circ f) \subseteq A$ and so $\text{Dom}(g \circ f) = A$. Next, we show that $g \circ f$ is well defined.

Suppose that $(x, z) \in g \circ f$, $(x_1, z_1) \in g \circ f$ and $x = x_1$, where $x, x_1 \in A$ and $z, z_1 \in C$. By the definition of composition of functions, there exist $y, y_1 \in B$ such that $f(x) = y$, $g(y) = z$, $f(x_1) = y_1$ and $g(y_1) = z_1$. Since f is a function and $x = x_1$, we have $y = y_1$. Similarly, since g is a function and $y = y_1$, we have $z = z_1$. Thus, $g \circ f$ is well defined. Hence, $g \circ f$ is a function from A into C.

(2) Let $x, x' \in A$. Suppose $(g \circ f)(x) = (g \circ f)(x')$. Then $g(f(x)) = g(f(x'))$. Since g is one-one, $f(x) = f(x')$. Since f is one-one, $x = x'$. Thus, $g \circ f$ is one-one.

(3) Let $z \in C$. Then there exists $y \in B$ such that $g(y) = z$ since g is onto C. Since f is onto B, there exists $x \in A$ such that $f(x) = y$. Thus, $(g \circ f)(x) = g(f(x)) = g(y) = z$. Hence, $g \circ f$ is onto C. ■

Theorem 1.3.4 *Let $f : A \to B$, $g : B \to C$, and $h : C \to D$. Then*

$$h \circ (g \circ f) = (h \circ g) \circ f.$$

That is, composition of functions is associative.

Proof. First note that $h \circ (g \circ f) : A \to D$ and $(h \circ g) \circ f : A \to D$. Let $x \in A$. Then $[h \circ (g \circ f)](x) = h((g \circ f)(x)) = h(g(f(x))) = (h \circ g)(f(x)) = [(h \circ g) \circ f](x)$. Thus, $h \circ (g \circ f) = (h \circ g) \circ f$. ■

1.4 Fuzzy Sets

Let X be a set and A be a subset of X. The **characteristic function** of A is the function χ of X into $\{0, 1\}$ defined by $\chi(x) = 1$ if $x \in A$ and $\chi(x) = 0$ if $x \notin A$. The characteristic function can be used to indicate either members or nonmembers of a subset of X. This notion can be generalized in a way that introduces the notion of a fuzzy subset of X.

Definition 1.4.1 *A **fuzzy subset** μ of X is a function of X into the closed interval* $[0, 1]$, [10].

Let μ be a fuzzy subset of a set X. For all $x \in X$, $\mu(x)$ can be thought of as the degree of membership of x in μ.We sometimes use the notion μ_A for a fuzzy subset of X, where A is thought of as a fuzzy set and μ_A gives the grade of membership of elements of X in A. At times A, may be merely a description of a fuzzy subset μ of X. If X is finite, say $X = \{x_1, x_2, \ldots, x_n\}$, we sometimes write $A = \mu_A(x_1)/x_1 + \mu_A(x_2)/x_2 + \ldots + \mu_A(x_n)/x_n$, where $\mu_A(x_i)/x_i$ is the pair 'grade of membership/element' and '+' is meant in a set theoretic sense.

Definition 1.4.2 *Let μ be a fuzzy subset of X.*

*(i) Let $\alpha \in [0, 1]$. Define $\mu_\alpha = \{x \in X \mid \mu(x) \geq \alpha\}$. We call μ_α an **α-cut**.*

*(ii) The **support** of μ is defined to be the set*

$$Supp(\mu) = \{x \in X \mid \mu(x) > 0\}.$$

We let $\mathcal{FP}(X)$ denote the **fuzzy power set** of X, i.e., the set of all fuzzy subsets of X.

Definition 1.4.3 *Let μ and ν be fuzzy subsets of a set X. Define $\overline{\mu}$, $\mu \cap \nu$, $\mu \cup \nu$ as follows:*

$$\begin{aligned} \overline{\mu}(x) &= 1 - \mu(x), \\ (\mu \cap \nu)(x) &= \min\{\mu(x), \nu(x)\}, \\ (\mu \cup \nu)(x) &= \max\{\mu(x), \nu(x)\}, \end{aligned}$$

for all $x \in X$. $\overline{\mu}$ is called the **complement** *of μ.*

We some times use $\wedge$ to denote min and infimum and $\vee$ to denote max and supremum. Using these symbols, $(\mu \cap \nu)(x) = \mu(x) \wedge \nu(x)$ and $(\mu \cup \nu)(x) = \mu(x) \vee \nu(x)$.

We can extend the notion of union and intersection to a family of fuzzy subsets of X. Let $\{\mu_i\}_{i \in I}$ be a family of fuzzy subsets of X. Define $\cap_{i \in I} \mu_i$ and $\cup_{i \in I} \mu_i$ as follows:

$$\begin{aligned} (\cap_{i \in I} \mu_i)(x) &= \wedge_{i \in I} \mu_i(x), \\ (\cup_{i \in I} \mu_i)(x) &= \vee_{i \in I} \mu_i(x). \end{aligned}$$

Thus if I is a finite set, say $|I| = \mathbb{N}_n$, then $\cap_{i \in I} \mu_i = \mu_1 \cap \mu_2 \cap \ldots \cap \mu_n$ and $\cup_{i \in I} \mu_i = \mu_1 \cup \mu_2 \cup \ldots \cup \mu_n$. In this case, we sometimes write,

$$(\cap_{i \in I} \mu_i)(x) = \mu_1(x) \wedge \mu_2(x) \wedge \ldots \wedge \mu_n(x)$$

and

$$(\cup_{i \in I} \mu_i)(x) = \mu_1(x) \vee \mu_2(x) \vee \ldots \vee \mu_n(x).$$

For a detailed study of fuzzy sets, the reader is referred to [3, 5-11].

1.5 Semigroups

In this section, we provide some fundamental results on semigroups which are needed in our presentation of fuzzy automata and fuzzy languages.

Let X be a nonempty set. Then a function from $X \times X$ into X is called a **binary operation** on X. If $*$ is a binary operation on X, then the pair $(X, *)$ is called a **mathematical system**. If $(X, *)$ is a mathematical system such that $\forall a, b, c \in X$, $(a*b)*c = a*(b*c)$, then $*$ is called associative and $(X, *)$ is called a **semigroup**. Let $(X, *)$ be a mathematical system. If there exists $e \in X$ such that $\forall a \in X$, $a * e = a = e * a$, then e is called an **identity** of $(X, *)$ and $(X, *)$ is said to have an identity. If $(X, *)$ has an identity, then it is easily seen that the identity is unique. A semigroup $(X, *)$ is said to be a **monoid** if it has an identity.

Let $(X, *)$ be a semigroup and let S be a nonempty subset of X. Then $(S, *)$ is called a **subsemigroup** of $(X, *)$ if $(S, *)$ is a mathematical system.

(Here for $(S, *)$, we mean $*$ restricted to $S \times S$.) Let $(X, *)$ be a monoid and let $(S, *)$ be a subsemigroup of $(X, *)$. If e is the identity of $(X, *)$ and $e \in S$, then $(S, *)$ is called a **submonoid** of $(X, *)$. We often write X for $(X, *)$ when the operation $*$ is understood.

We next state some well-known theorems (without proofs) and definitions.

Theorem 1.5.1 *The intersection of any collection of subsemigroups (submonoids) of a semigroup (monoid) X is a subsemigroup (submonoid) of X.*

Definition 1.5.2 *Let X be a semigroup (moniod) and let S be a subset of X. Let $\langle S \rangle$ denote the intersection of all subsemigroups (submonoids) of X which contain S. Then $\langle S \rangle$ is called the subsemigroup (submonoid) of X* ***generated*** *by S. If $\langle S \rangle = X$, then S is called a set of* ***generators*** *for X.*

Let $(X, *)$ be a semigroup and let e be an element not in X. Denote $X \cup \{e\}$ by X'. Extend $*$ from $X \times X$ into X to $*'$ from $X' \times X'$ into X' as follows: $a *' b = a * b$ if $a, b \in X$; $a *' e = a = e *' a$ if $a \in X$; $e *' e = e$. Then $(X', *')$ is a monoid and X is a subsemigroup of X'. Even if X were a monoid, it is not a submoniod of X'.

Definition 1.5.3 *Let $(X, *)$ and $(Y, \cdot)$ be semigroups. A function f from X into Y is called a* ***homomorphism*** *if $\forall a, b \in X$, $f(a * b) = f(a) \cdot f(b)$. Let f be a homomorphism of X into Y. If f is one-one, then f is called a* ***monomorphism****. If f is onto Y, then f is called an* ***epimorphism****. If f is both a monomorphsim and epimorphsim, then f is called an* ***isomorphim*** *and X and Y are said to be* ***isomorphic****.*

Theorem 1.5.4 *Let X, Y, and Z be semigroups. If f is a homomorphism of X into Y and g is a homomorphism of Y into Z, then $g \circ f$ is a homomorphism of X into Z.*

Let S be a set. By a **free semigroup** on the set S, we mean a semigroup F together with a function $f : S \to F$ such that for any semigroup X and every function $g : S \to X$, there exists a unique homomorphism $h : F \to X$ such that $h \circ f = g$.

Theorem 1.5.5 *Let F be a free semigroup on a set S together with a function $f : S \to F$. Then f is one-one and $\langle f(S) \rangle = F$.*

Proof. Let $a, b \in S$ be such that $a \neq b$. Let X be a semigroup containing more that one element and choose a function $g : S \to X$ such that $g(a) \neq g(b)$. Now $h \circ f = g$. Since $h(f(a)) = g(a) \neq g(b) = h(f(b))$, it follows that $f(a) \neq f(b)$. Hence f is one-one.

We now show that $\langle f(S) \rangle = F$. Let $\langle f(S) \rangle = A$. Then the function f defines a function $g : S \to A$ such that $i \circ g = f$, where i is the identity homomorphism $i : A \to F$. By the definition of a free semigroup, there exists a homomorphism $h : F \to A$ such that $h \circ f = g$. Consider the diagram:

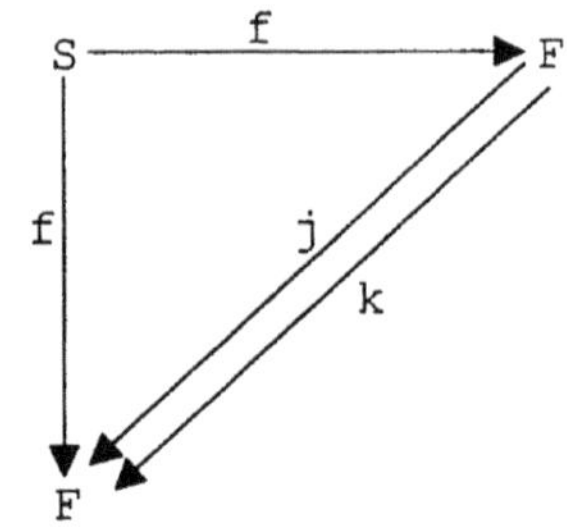

where j is the identity isomorphism and $k = i \circ h$. Since $j \circ f = f$, $k \circ f = i \circ h \circ f = i \circ g = f$, from the uniqueness, it follows that $i \circ h = k = j$. The inclusion homomorphism i must be an epimorphism. Hence $A = X$ and so $f(S)$ generates F. ■

Theorem 1.5.6 (Uniqueness) *Let (F, f) and (F', f') be free semigroups on the same set S. Then there exists a unique isomorphism $j : F \to F'$ such that $j \circ f = f'$.*

Proof. Since (F, f) is a free semigroup on the set S, there exists a homomorphism $j : F \to F'$ such that $j \circ f = f'$. Similarly, there exists a homomorphism $k : F' \to F$ such that $k \circ f' = f$. Consider the composition $h = k \circ j$ and the identity isomorphism i of F. In the diagram,

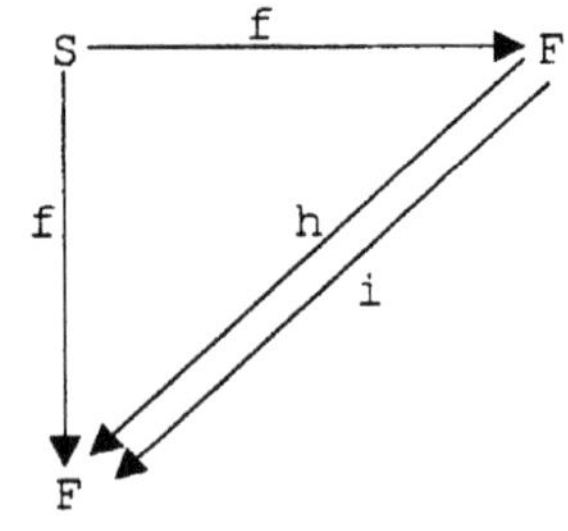

we have $h \circ f = k \circ j \circ f = k \circ f' = f$, $i \circ f = f$. From the uniqueness, it follows that $k \circ j = h = i$. Since i is an isomorphism, it follows that j is a monomorphism. Similarly, we can show that $j \circ k$ is the identity endomorphism on F'. Hence j is also an epimorphism. Hence j is an isomorphism. ■

Theorem 1.5.7 (Existence) *Let S be a set. Then there exists a free semigroup on S.*

Proof. Let F denote the set of all finite sequences of elements (repetition allowed) of S. Let $\alpha = (a_1, \ldots, a_m)$, $\beta = (b_1, \ldots, b_n) \in F$. Define

$$\alpha\beta = (a_1, \ldots, a_m, b_1, \ldots, b_n).$$

Clearly, for any $\alpha, \beta, \gamma \in F$, $\alpha(\beta\gamma) = (\alpha\beta)\gamma$. Under this operation, F is a semigroup.

Define $f : S \to F$ as follows: For all $a \in S$, define $f(a) = (a)$, the sequence consisting of the element a. We now show that (F, f) is a free semigroup on S.

Let $g : S \to X$ be an arbitrary function from S into a semigroup X. Define $h : F \to X$ by $h(a_1, \ldots, a_m) = g(a_1) \ldots g(a_m)$ for all $(a_1, \ldots, a_m) \in F$. Clearly, h is a homomorphism. Let $a \in S$. Then $(h \circ f)(a) = h(f(a)) = h((a)) = g(a)$. Hence $h \circ f = g$.

Now let $k : F \to X$ be a homomorphism such that $k \circ f = g$. We show that $h = k$. Let $(a_1, \ldots, a_m) \in F$. Then $k(a_1, \ldots, a_m) = k((a_1) \ldots (a_m)) = k((a_1)) \ldots k((a_m)) = k(f(a_1)) \ldots k(f(a_m)) = g(a_1) \ldots g(a_m) = h(a_1, \ldots, a_m)$. Hence $h = k$. Consequently, (F, f) is free semigroup on S. ■

Theorem 1.5.8 *Let X be a semigroup and let S be a set of generators of X. Then every element of X can be written as the product of a finite sequence of element in S.*

Proof. Let (F, f) be the free semigroup on S as constructed in the proof of Theorem 1.5.7. Then by the definition of a free semigroup, there exists a homomorphism $h : F \to X$ such that $h \circ f = g$, where g is the inclusion map $g : S \to X$.

Next, we prove that h is an epimorphism. Consider the image $h(F)$. Now $h(F)$ is a subsemigroup of X. Since $S = g(S) = (h \circ f)(S) = h(f(S)) \subseteq h(F)$ and S generates X, it follows that $h(F) = X$. Thus h is an epimorphism.

Let $x \in X$. Then there exists $(a_1, \ldots, a_m) \in F$ such that $h(a_1, \ldots, a_m) = x$. Now $x = h(a_1, \ldots, a_m) = h((a_1) \ldots (a_m)) = h((a_1)) \ldots h((a_m)) = h(f(a_1)) \ldots h(f(a_m)) = g(a_1) \ldots g(a_m) = a_1 \ldots a_m$, where $a_1, \ldots, a_m \in S$. ■

Let S be a set. Then S determines a unique free semigroup (F, f). Since $f : S \to F$ is injective, we may identify S with its image $f(S)$ in F. We can then regard S to be a subset of F such that S generates F. Also, any function $g : S \to X$, where X is a semigroup extends to a unique homomorphism $h : F \to X$. We will refer to F as the free semigroup generated by S.

Consider the monoid $F^* = F \cup \{e\}$, where F is the free semigroup generated by S. Then every function $g : S \to X$, where X is monoid extends to a unique proper homomorphism $h : F^* \to X$. This monoid F^* is called the **free monoid generated** by the set S.

1.6 Fuzzy Logic

In the remainder of this chapter, we study fuzzy switching functions in combinational switching systems. We consider the minimization and decomposition of them. Our approach is patterned after that of [4].

We begin with the notion of a well-formed formula in two-valued logic. In propositional logic, also known as sentential logic, a proposition is a declarative sentence that is either true of false, but not both. For convenience, we shall use upper case symbols to denote propositions. For example,

$A \equiv$ Sun is hot,

$B \equiv$ Snow is white.

The symbols, such as A and B, that we use to denote propositions are called **(ground) atomic formulas** or **atoms.**

Definition 1.6.1 *In the propositional logic,* ***well-formed formulas (WFF)*** *or* ***formulas*** *(for short) are defined recursively as follows:*

(i) An atom is a well-formed formula.

(ii) If U is a well-formed formula, then $\overline{U}$ (not U) is a well-formed formula.

(iii) If U and V are well-formed formulas, then

(a) $(U \cdot V)$, (U and V), is a well-formed formula.

(b) $(U + V)$, (U or V), is a well-formed formula.

(c) $(U \rightarrow V)$, (if U, then V), is a well-formed formula.

(iv) All well-formed formulas are generated by applying the above rules.

Since an atom in a propositional logic is a single unit, its structure and composition are suppressed ([1, p. 26]) and so there are many ideas that cannot be treated in this simple way. To represent more complex situations the notion of first order logic has been defined. For a review of propositional and first order logic, see [1].

The binary approach to the treatment of switching theory is sometimes lacking in its description of systems in the real world. This approach is due in part to the relative simplicity of designing binary switching systems, and also to the fact that basic switching modules in common use are two positional. Thus the variables in Boolean logic are assumed to be two-valued. However, the attributes of system variables are often ambiguously defined due to real-word constraints. Artificial intelligence and related subjects are cases of which such attributes occur. In this section, we study fuzzy logic as a means for such investigations. Fuzzy logic is based on the concepts of fuzzy sets and symbolic logic.

Let S be a well-formed formula in fuzzy logic. The truth value of S can assume any value in the interval $[0, 1]$. The truth value of S indicates the degree of truth represented by the formula.

We assume that well-formed formulas (formulas for short) are defined to be exactly the same as those in two-valued logic.

Let $T(S)$ denote the truth value of a formula S, i.e., $T(S) \in [0,1]$. The following rules can be used to evaluate a formula in fuzzy logic.

(1) $T(S) = T(A)$ if $S = A$ and A is a ground atomic formula.

(2) $T(S) = 1 - T(R)$ if $S = \overline{R}$.

(3) $T(S) = \min\{T(S_1), T(S_2)\}$ if $S = S_1 \cdot S_2$

(4) $T(S) = \max\{T(S_1), T(S_2)\}$ if $S = S_1 + S_2$.

(5) $T(S) = \inf\{T(B(x)) \mid x \in D\}$ if $S = (\forall x)B$ and D is the domain of x.

(6) $T(S) = \sup\{T(B(x)) \mid x \in D\}$ if $S = (\exists x)B$ and D is the domain of x.

Remark 1.6.2 *If D is a finite set, then (5) and (6) become, respectively,*

(5)′ $T(S) = T(B(a_1) \cdot \ldots \cdot B(a_n))$ if $S = (\forall x)B$ and x assumes values $a_1, \ldots, a_n$.

(6)′ $T(S) = T(B(a_1) + \ldots + B(a_n))$ if $S = (\exists x)B$ and x assumes values $a_1, \ldots, a_n$.

Since two-value logic is a special case of fuzzy logic, all the rules stated above also apply to two-valued logic.

Example 1.6.3 *(i)Consider $S = (P \cdot Q) + R$. Assume $T(P) = 0.3$, $T(Q) = 0.7$, and $T(R) = 0.6$. Then*

$$\begin{aligned} T(S) &= \max\{\min\{T(P), T(Q)\}, T(R)\} \\ &= \max\{\min\{0.3, 0.7\}, 0.6\} \\ &= \max\{0.3, 0.6\} \\ &= 0.6. \end{aligned}$$

(ii)Consider $S = (P + \bar{Q}) \cdot (R)$. Assume $T(P) = 0.4$, $T(Q) = 0.2$, and $T(R) = 0.7$. Then

$$\begin{aligned} T(S) &= \min\{\max\{T(P), 1 - T(Q)\}, T(R)\} \\ &= \min\{\max\{0.4, 1 - 0.2\}, 0.7\} \\ &= \min\{0.8, 0.7\} \\ &= 0.7. \end{aligned}$$

Example 1.6.4 *(i) Consider $S = (\forall x)P(x)$. Assume the domain of x is $\{a_1, a_2, a_3, a_4\}$ and $T(P(a_1)) = 0.8$, $T(P(a_2)) = 0.4$, $T(P(a_3)) = 0.5$ and $T(P(a_4)) = 0.3$. Then*

$$\begin{aligned} T(S) &= T(P(a_1) \cdot P(a_2) \cdot P(a_3) \cdot P(a_4)) \\ &= \min\{P(a_1), P(a_2), P(a_3), P(a_4)\} \\ &= \min\{0.8, 0.4, 0.5, 0.3\} \\ &= 0.3. \end{aligned}$$

(ii) Consider $S = (\exists x)P(x)$. Assume the domain of x is $\{a_1, a_2, a_3, a_4\}$ and $T(P(a_1)) = 0.4$, $T(P(a_2)) = 0.2, T(P(a_3)) = 0.9$ and $T(P(a_4)) = 0.6$.

Then

$$
\begin{aligned}
T(S) &= T(P(a_1) + P(a_2) + P(a_3) + P(a_4)) \\
&= \max\{P(a_1), P(a_2), P(a_3), P(a_4)\} \\
&= \max\{0.4, 0.2, 0.9, 0.6\} \\
&= 0.9.
\end{aligned}
$$

Example 1.6.5 *Consider* $S = (\forall x)(\exists y)P(x,y)$. *Assume* $D = \{a_1, a_2\}$ *and* $T(P(a_1,a_1)) = 0.2$, $T(P(a_1,a_2)) = 0.6$, $T(P(a_2,a_1)) = 0.5$, $T(P(a_2,a_2)) = 0.8$. *Note that* $T((\exists y)P(x,y)) = T(P(x,a_1) + P(x,a_2))$. *Therefore*

$$
\begin{aligned}
T((\forall x)(\exists y)P(x,y)) &= T(P(a_1,a_1) + P(a_1,a_2))\cdot \\
&\quad (P(a_2,a_1) + P(a_2,a_2))) \\
&= \min\{\max\{P(a_1,a_1), P(a_1,a_2)\}, \\
&\quad \max\{P(a_2,a_1), P(a_2,a_2)\}\} \\
&= \min\{\max\{0.2, 0.6\}, \max\{0.5, 0.8\}\} \\
&= \min\{0.6, 0.8\} \\
&= 0.6.
\end{aligned}
$$

Suppose A is a statement in a problem solving system. If two-valued logic is used in the system, then we store the statement A, if $T(A) = 1$, otherwise we store $\overline{A}$. In fuzzy logic, the statement A is stored if $T(A) \geq T(\bar{A})$. That is, in fuzzy logic, we store A if $T(A) \geq 1 - T(A)$, or equivalently if $T(A) \geq 0.5$.

We can use this concept to define "**satisfiability**" in fuzzy logic.

Definition 1.6.6 *Let S be a formula in fuzzy logic and let I be an interpretation of S.*

*(i) I is said to **satisfy** S if $T(S) \geq 0.5$ under I.*

*(ii) I is said to **falsify** S if $T(S) \leq 0.5$ under I.*

*(iii) S is said to be **unsatisfiable** if it is falsified by every interpretation of it.*

In the above definition, if $T(S) = 0.5$ under I, then I both satisfies and falsifies S.

Note that these definitions are compatible with those in two-valued logic. However, in fuzzy logic "not satisfying" is different from "falsifying", and "not falsifying" is different from "satisfying".

Definition 1.6.7 *Let S be a formula in two-valued logic.*

*(i) S is called **valid** if $T(S) = 1$ under all of its possible two-valued assignments.*

*(ii) S is called **invalid** if it is not valid.*

*(iii) S is called **inconsistent** if $T(S) = 0$ under all of its possible two-valued assignments.*

*(iv) S is called **consistent** if it is not inconsistent.*

Definition 1.6.8 *Let S be a formula in fuzzy logic.*

(i) S is called ***fuzzily valid*** *if $T(S) \geq 0.5$ under all possible assignments.*

(ii) S is called ***fuzzily inconsistent*** *if $T(S) \leq 0.5$ under all possible assignments.*

One of the problems in any logic is to determine the consistency of formulas. In two-valued logic, since there are a finite number of distinct assignments for a formula, an exhaustive search can be utilized to determine the consistency of the formula. Suppose S is a formula in two-valued logic. If S is 1 under all possible two-valued assignments, then S is valid. If S is 0 under all possible two-valued assignments, then S is inconsistent.

However, since in fuzzy logic there are infinite number of possible assignments for a formula, an exhaustive examination of all possible assignments may be neither possible nor desirable. Therefore, the determination of non-exhaustive methods to determine consistency of formulas in fuzzy logic is very important.

In order to determine whether a formula A is valid or inconsistent in fuzzy logic, the simplest approach involves expanding A into conjunctive and disjunctive normal forms. Before defining these normal forms, we need the following definitions.

Definition 1.6.9 *(i) A* ***literal*** *is a variable x_i or $\overline{x_i}$, the complement of x_i.*

(ii) A ***clause*** *is a disjunction of one or more literals.*

(iii) A ***phrase*** *is a conjunction of one or more literals.*

Definition 1.6.10 *Let S be a formula in fuzzy logic.*

(i) S is said to be in ***conjunctive normal form*** *if*

$$S = C_1 \cdot C_2 \cdot \ldots \cdot C_m, \; m \geq 1$$

and C_i is a clause for all i, $1 \leq i \leq m$.

(ii) S is said to be in ***disjunctive normal form*** *if*

$$S = P_1 + P_2 + \ldots + P_m, \; m \geq 1$$

and P_i is a phrase for all i, $1 \leq i \leq m$.

By using distributive laws and De Morgan's laws, a formula in two-valued logic can be expressed in conjunctive and disjunctive normal forms. Since these laws also hold in fuzzy logic, formulas in fuzzy logic can also be expressed in conjunctive and disjunctive normal form.

Theorem 1.6.11 *Let C be a clause. Then C is fuzzily valid if and only if C contains a complementary pair of literals.*

Proof. Suppose that C contains a complimentary pair of literals. Let $C = L_1 + L_2 + \ldots + L_n$. Assume L_i and L_j form such a complementary pair.

Then $T(L_i) = 1- T(L_j)$. For every possible assignment, either $T(L_i)$ or $T(L_j)$ will be greater than or equal to 0.5. Therefore, $\max\{T(L_i), T(L_j)\} \geq 0.5$ for all possible assignments. Now

$$T(C) = \max\{T(L_1), T(L_2), \ldots, T(L_n)\} \geq \max\{T(L_i), T(L_j)\} \geq 0.5.$$

Thus C is fuzzily valid.

Conversely, suppose that C is fuzzily valid. Suppose C does not contain any pair of complimentary literals. Consider the assignment in which every literal of C is assigned a truth value smaller than 0.5. $T(C)$ will be smaller than 0.5 under this assignment, so C is not fuzzily valid. This is contradictory to the assumption that C is fuzzily valid. ∎

The following theorem, which is the dual of the previous theorem can be proved in a similar manner.

Theorem 1.6.12 *Let P be a phrase. P is fuzzily inconsistent if and only if P contains a complementary pair of literals.* ∎

We can effectively use Theorems 1.6.11 and 1.6.12 to determine the consistency of formulas in fuzzy logic. For, let S be a formula in fuzzy logic. To determine whether S is fuzzily valid, we first expand S into a conjunctive normal form:

$$S = C_1 \cdot C_2 \cdot \ldots \cdot C_n.$$

Then S is fuzzily valid if and only if every C_i is fuzzily valid. Now we can use Theorems 1.6.11 to determine the validity of each C_i. Similarly, to determine whether S is fuzzily inconsistent, we first expand S into disjunctive normal form:

$$S = P_1 + P_2 + \ldots + P_n.$$

Then S is fuzzily inconsistent if and only if every phrase P_i is fuzzily inconsistent and the fuzzy inconsistency of a phrase can be established through Theorem 1.6.12.

1.7 Fuzzy Functions and Decomposition

This section deals with the study of fuzzy switching functions in combinational switching systems. We undertake this study by means of a suitable fuzzy algebra and techniques for their minimization and decomposition. A fuzzy function can usually be represented by several expressions. We establish criteria for the minimality of a fuzzy function and then develop techniques for obtaining a minimal expression for any such function. It is possible to simplify fuzzy expressions by means of algebraic manipulations. However, this approach does not provide an algorithm and is ineffective for expressions of even a small number of variables. The methods presented in this section are systematic.

We first consider fuzzy functions and normal forms. By using the basic fuzzy logic AND, OR, and inverter circuits, by associating the multiplication factors to the variables at different stages, we can realize any fuzzy logic function.

The electronic implementation of fuzzy functions is easily carried out by utilizing, for instance, diode-transistor logic. The operations $x \cdot y$ and $x+y$ have exactly the same electronic implementation on the AND and OR logic operations of two-valued logic. The logic operation complementation is easily implemented by making use of a transistor inverter.

However, the practical usefulness of the fuzzy function concept and its electronic implementation are questionable. A decision method is not provided by a fuzzy function in the same way that a two-valued (or n-valued) logic function does with its "one-zero", "yes-no", or true-false" states. In order for knowing that a fuzzy function has a particular value at a point to be significant, it must be related to some outcome which is dependent upon that fact. A possible way for introducing this quality into a fuzzy system is by partitioning the interval $[0,1]$ into a finite number of classes in the following manner:

$$\begin{array}{ll} \text{Class 1:} & \alpha_1 \le x \le 1 \\ \text{Class 2:} & \alpha_2 \le x < \alpha_1 \\ & \vdots \\ \text{Class } n\text{:} & 0 \le x < \alpha_{n-1} \end{array} \tag{1.1}$$

where $1 > \alpha_1 > \alpha_2 > \ldots > \alpha_{n-1} > 0$. A fuzzy variable or a fuzzy function can now be identified with one of these n classes according to the value they take in $[0,1]$. This partitioning of $[0,1]$ into a finite number of classes allows one to use the properties of n-valued logic in the treatment of fuzzy logic systems.

We may now give certain meanings to the different classes. For example, in the case where $n = 3$, we may assume that an object (a) belongs to a set if it is in "Class 1", (b) does not belong to a set if it is in "Class 3", and (c) its status remains undecided if it is in "Class 2".

We consider two functional representations in the analysis of fuzzy functions, namely, the "sum-of-products" and "product-of-sums" forms. Any other functional form is a combination of these two forms.

We illustrate the analysis of fuzzy functions of sum-of-products form by use of examples.

Example 1.7.1 *(i) Consider the fuzzy function*

$$f(x,y) = x + \overline{y}. \tag{1.2}$$

Let $f(x,y)$ *belong to class* M. *Then*

$$a_M \leq f(x,y) < a_{M-1} \tag{1.3}$$

That is,

$$x + \overline{y} \geq a_M \tag{1.4}$$

and

$$x + \overline{y} < a_{M-1} \tag{1.5}$$

From equation (1.4), $x \geq a_M$ *or* $\overline{y} \geq a_M$. *This is equivalent to* $x \geq a_M$ *or* $y \leq 1 - a_M$. *Also, from equation (1.5), i.e.,* $x + \overline{y} < a_{M-1}$, *it follows that* $x < a_{M-1}$ *and* $\overline{y} < a_{M-1}$, *which is equivalent to* $x < a_{M-1}$ *and* $y > 1 - a_{M-1}$. *Thus, if the fuzzy function* f *given by equation (1.2) is to belong to the class* M, *(i.e.,* $a_M \leq f(x,y) < a_{M-1}$*), then* x *and* y *must satisfy the following conditions:*

$$\text{Group 1}: [x \geq a_M \quad \text{or} \quad y \leq 1 - a_M] \tag{1.6}$$

and

$$\text{Group 2}: [x < a_{M-1} \quad \text{and} \quad y > 1 - a_{M-1}]. \tag{1.7}$$

(ii) Consider the fuzzy function

$$f(x,y,z) = x + \overline{y} + z.$$

Then f *belongs to class* M *if and only if* $a_M \leq f(x,y) < a_{M-1}$, *i.e.,* $a_M \leq x + \overline{y} + z < a_{M-1}$. *Solving the inequalities* $a_M \leq x + \overline{y} + z$ *and* $x + \overline{y} + z < a_{M-1}$ *as in (i), we obtain*

$$\text{Group 1}: [x \geq a_M \quad \text{or} \quad y \leq 1 - a_M \quad \text{or} \quad x \geq a_M]$$

and

$$\text{Group 2}: [x < a_{M-1} \quad \text{and} \quad y > 1 - a_{M-1} \quad \text{and} \quad z < a_{M-1}].$$

Thus the fuzzy function f *belongs to class* M *if and only if* x, y, *and* z *satisfy the conditions of Group1 and Group 2.*

Example 1.7.2 *Consider the fuzzy function*

$$f(x,y,z) = x\overline{y} + \bar{x}y\overline{z}. \tag{1.8}$$

Then f belongs to class M if and only if $a_M \leq f(x,y,z) < a_{M-1}$, i.e., $a_M \leq x\overline{y} + \overline{x}y\overline{z} < a_{M-1}$. Solving the inequality $a_M \leq x\overline{y} + \overline{x}y\overline{z}$, we obtain

$$\text{Group 1} = \left[\left\{ \begin{array}{c} x \geq a_M \\ \text{and } y \leq 1 - a_M \end{array} \right\} \text{ or } \left\{ \begin{array}{c} x \leq 1 - a_M \\ \text{and } y \geq a_M \\ \text{and } z \leq 1 - a_M \end{array} \right\} \right]. \tag{1.9}$$

Similarly, solving the inequality $x\overline{y} + \overline{x}y\overline{z} < a_{M-1}$ we obtain

$$\text{Group 2} = \left[\left\{ \begin{array}{c} x < a_{M-1} \\ \text{or } y > 1 - a_{M-1} \end{array} \right\} \text{ and } \left\{ \begin{array}{c} x > 1 - a_{M-1} \\ \text{or } y < a_{M-1} \\ \text{or } z > 1 - a_{M-1} \end{array} \right\} \right]. \tag{1.10}$$

Thus, the fuzzy function f belongs to class M, if and only if x, y, and z satisfy the conditions of Group 1 and Group 2.

Next, we illustrate the product-of-sums form.

Example 1.7.3 *Consider the fuzzy function*

$$f(x,y,z) = (\overline{x} + \overline{y})(x + y + \overline{z}). \tag{1.11}$$

Now, $f(x,y,z)$ belongs to class M if and only if $a_M \leq f(x,y,z) < a_{M-1}$, i.e., $a_M \leq (\overline{x} + \overline{y})(x + y + \overline{z}) < a_{M-1}$. This is equivalent to $a_M \leq (\overline{x} + \overline{y})(x + y + \overline{z})$ and $(\overline{x} + \overline{y})(x + y + \overline{z}) < a_{M-1}$.

Let us first solve $a_M \leq (\overline{x} + \overline{y})(x + y + \overline{z})$. Now, $a_M \leq (\overline{x} + \overline{y})(x + y + \overline{z})$ if and only if $a_M \leq (\overline{x} + \overline{y})$ and $a_M \leq (x + y + \overline{z})$. From this it follows that the fuzzy variables x, y, and z must satisfy the conditions given as

$$\text{Group 1} = \left[\left\{ \begin{array}{c} x \leq 1 - a_M \\ \text{or } y \leq 1 - a_M \end{array} \right\} \text{ and } \left\{ \begin{array}{c} x \geq a_M \\ \text{or } y \geq a_M \\ \text{or } z \leq 1 - a_M \end{array} \right\} \right]. \tag{1.12}$$

Similarly, solving the inequality $(\overline{x} + \overline{y})(x + y + \overline{z}) < a_{M-1}$, we obtain

$$\text{Group 2} = \left[\left\{ \begin{array}{c} x > 1 - a_{M-1} \\ \text{and } y > 1 - a_{M-1} \end{array} \right\} \text{ or } \left\{ \begin{array}{c} x < a_{M-1} \\ \text{and } y < a_{M-1} \\ \text{and } z > 1 - a_{M-1} \end{array} \right\} \right]. \tag{1.13}$$

Thus the fuzzy function f belongs to class M if and only if the fuzzy variables x, y, and z satisfy the conditions given in Groups 1 and 2.

Next, we consider a fuzzy function which is neither sum-of-products form nor product-of-sums form.

Example 1.7.4 *Consider the fuzzy function*

$$f(x,y,z,w) = x\overline{y}(\overline{z}+w) + \overline{x}y + \overline{z}w. \tag{1.14}$$

Now f belongs to class M if and only if $a_M \leq f(x,y,z,w) < a_{M-1}$. That is, $a_M \leq x\overline{y}(\overline{z}+w) + \overline{x}y + \overline{z}w \ < a_{M-1}$. Even though f is neither sum-of-products form nor product-of-sums form, we can still solve the two inequalities $a_M \leq x\overline{y}(\overline{z}+w) + \overline{x}y + \overline{z}w$ and $x\overline{y}(\overline{z}+w) + \overline{x}y + \overline{z}w \ < a_{M-1}$ to obtain the following two groups whose conditions must be satisfied by the fuzzy variables x, y, z, w.

$$\begin{array}{rcl} \textit{Group 1} & = & \left[\left\{ \begin{array}{l} x \geq a_M \\ \textit{and } \ y \leq 1 - a_M \\ \textit{and } \ (z \leq 1 - a_M \ \textit{or } w \geq a_M \end{array} \right\}\right] \\ & & \textit{or} \left[\left\{ \begin{array}{l} x \leq 1 - a_M \\ \textit{and } \ y \geq a_M \end{array} \right\} \ \textit{or} \ \left\{ \begin{array}{l} z \leq 1 - a_M \\ \textit{and } \ w \geq a_M \end{array} \right\}\right] \end{array} \tag{1.15}$$

and

$$\begin{array}{rcl} \textit{Group 2} & = & \left[\left\{ \begin{array}{l} x < a_{M-1} \\ \textit{or } \ y > 1 - a_{M-1} \\ \textit{or } \ (z < a_{M-1} \ \textit{ and } \ w > 1 - a_{M-1} \end{array} \right\}\right] \\ & & \textit{and} \left[\left\{ \begin{array}{l} x > 1 - a_{M-1} \\ \textit{or } \ y < a_{M-1} \end{array} \right\} \ \textit{and} \ \left\{ \begin{array}{l} z > 1 - a_{M-1} \\ \textit{or } \ w < a_{M-1} \end{array} \right\}\right]. \end{array} \tag{1.16}$$

When a phrase is in a disjunctive normal form, it corresponds to a logic gate and each literal corresponds to an input line. The relation between the cost of a logic gate and the cost of an input line depends on the type of gates used. However, in practice the cost of an additional input line on an already existing gate is much less than the cost of an additional logic gate. For this reason, the elimination of gates will be the primary objective of the minimization process. This leads to the following definition of a minimal expression.

Definition 1.7.5 *A disjunctive normal form is called a* ***minimal complexity form*** *if there exists*

(i) no other equivalent form involving fewer phrases, and

(ii) no other equivalent form involving the same number of phrases, but a fewer total number of literals.

Definition 1.7.6 *A phrase f* ***subsumes*** *another phrase g if f contains all the literals of g.*

Definition 1.7.7 *Let F, F_1, and F_2 be fuzzy functions.*

(i) F_1 ***covers*** *F_2, denoted $F_1 \supseteq F_2$, if the value that F_1 assumes is greater than or equal to the value that F_2 assumes for all values of the domain.*

(ii) Let G be a product of literals. G is said to be an ***implicant*** *of F if F covers G.*

(iii) Let P be an implicant of F. P is a ***prime implicant*** *of F if the deletion of any literal from P results in a new product which is not covered by F.*

Clearly, the definitions of implicant and prime implicant are without any restrictions on the form of representation of a fuzzy function, e.g., sum-of-products form or product-of-sums form. In addition, a phrase subsuming another phrase becomes a special case of Definition 1.7.7.

Let f be an implicant of a fuzzy function F. f is said to be a **fuzzy prime implicant (FPI)** of F if it subsumes no other fuzzy implicant of F.

It can be easily verified that the following identities hold in Boolean algebra.

$$x + x = x \qquad x \cdot x = x \tag{1.17}$$

$$x + 0 = x \qquad x \cdot 1 = x \tag{1.18}$$

$$x + 1 = 1 \qquad x \cdot 0 = 0 \tag{1.19}$$

$$x + \overline{x} = 1 \qquad x \cdot \overline{x} = 0 \tag{1.20}$$

To minimize a Boolean function, these identities together with the distributive and absorption laws form the set of rules. It can be easily shown that identities (1.17), (1.18), and (1.19), and the distributive and absorption laws are also true in fuzzy algebra. However, the identities in (1.20) do not hold generally in fuzzy algebra and so cannot be applied to minimize a fuzzy function. This is illustrated by the following example.

Example 1.7.8 *Let $F = x + \overline{x}y$. Since in Boolean algebra $x + \overline{x} = 1$,*

$$F = (x + \overline{x})(x + y) = x + y.$$

However, this is not so in fuzzy algebra. For example, suppose $x = 0.3$ and $y = 0.8$. Then $x + y = 0.8$, but

$$x + \overline{x}y = \max\{0.3, \min\{0.7, 0.8\}\} = 0.7.$$

In certain cases the fuzzy version of identities (1.20) can be applied. We discuss these cases in the following results.

Lemma 1.7.9 *Let P be a phrase of fuzzy literals from the set $\{x_1, x_2, \ldots, x_n\}$. Let $\widehat{\alpha} = x_k + \overline{x}_k$, $1 \le k \le n$. Then $\widehat{\alpha}$ can be appended to P without affecting the general value of the phrase if there exists a variable x_i and its complement $\overline{x}_i$ in P for some i, $1 \le i \le n$.*

Proof. Let $P = \alpha x_i \beta \overline{x}_i \gamma$ for some i, where α, β, and γ are conjunctions of literals from the set $\{x_1, x_2, \ldots, x_n\}$. Now $x_i \overline{x}_i \le 0.5$ implies that $P \le 0.5$. Since $\widehat{\alpha} \ge 0.5$, it follows that $\widehat{\alpha}$ can be appended to P without changing the value of P. ■

Theorem 1.7.10 *Let P be a phrase of fuzzy literals from the set $\{x_1, x_2, \ldots, x_n\}$. Let $\widehat{\alpha} = x_k + \overline{x_k}$, $1 \le k \le n$. Suppose neither x_k nor $\overline{x_k}$ is in P. Then $\widehat{\alpha}$ can be appended to P without affecting the general value of the phrase if and only if there exists a variable x_i and its complement $\overline{x}_i$ in P for some i, $1 \le i \le n$.*

Proof. The if part follows from Lemma 1.7.9. Conversely, suppose for all i, $i \ne k$, either x_i or $\overline{x_i}$ is not in P. Suppose $\widehat{\alpha} = x_k + \overline{x_k} = t < 1$. Let s be a real number such that $t < s \le 1$. Assign the value s to the literals of P. Then the value of P for this assignment is s. However, the value of $P\widehat{\alpha}$ for this assignment is t, which is a contradiction. Hence there must exists a variable x_i and its complement $\overline{x_i}$ in P for some i, $1 \le i \le n$. ■

The following example shows that the converse of Lemma 1.7.9 is not true without the assumption in Theorem 1.7.10 that neither x_k nor $\overline{x_k}$ is in P.

Example 1.7.11 *Let $P = xy$ and the set of literals be $\{x, y\}$. Clearly, $x = x(x + \overline{x})$ and $y = y(y + \overline{y})$. Now $P = P(x + \overline{x}) = P(y + \overline{y})$. However, P neither contains $x\overline{x}$ nor $y\overline{y}$.*

Lemma 1.7.12 *Let C be a clause of fuzzy literals from the set $\{x_1, x_2, \ldots, x_n\}$. Let $\widehat{\beta} = x_k\overline{x_k}$, $1 \le k \le n$. Then $\widehat{\beta}$ can be appended to C without affecting the general value of the clause if there exists a variable x_i and its complement $\overline{x}_i$ in C for some i, $1 \le i \le n$.*

Proof. Let $C = \alpha + x_i + \beta + \overline{x}_i + \gamma$ for some i, where α, β, and γ are disjunctions of literals from the set $\{x_1, x_2, \ldots, x_n\}$. Now $x_i + \overline{x}_i \ge 0.5$ implies that $P \le 0.5$. Since $\widehat{\beta} \le 0.5$, it follows that $\widehat{\beta}$ can be appended to C without changing the value of C. ■

The following theorem, which is dual to Theorem 1.7.10, can be proved in a similar manner.

Theorem 1.7.13 *Let C be a clause of fuzzy literals from the set $\{x_1, x_2, \ldots, x_n\}$. Let $\widehat{\beta} = x_k\overline{x_k}$, $1 \le k \le n$. Suppose neither x_k nor $\overline{x_k}$ is in C. Then $\widehat{\beta}$ can be appended to C without affecting the general value of the* *clause if and only if there exists a variable x_i and its complement $\overline{x}_i$ in C for some i, $1 \leq i \leq n$.* ■

If F is of the form $F = (\sum_j x_j\overline{x}_j\gamma_j)\beta$, $1 \leq j \leq n$, where β and γ_j are formulas in $\{x_1, x_2, \ldots, x_n\}$, $j = 1, 2, \ldots, n$, then $F \leq 0.5$. This is a trivial case, where one formula in the conjunction is ≤ 0.5 and so F is ≤ 0.5. The following theorem gives a more general case.

Theorem 1.7.14 *Let $\{F_1, F_2, \ldots, F_m\}$ be a set of fuzzy formulas over $x_1, x_2, \ldots, x_n$, and let F be a conjunction of formulas from this set. A disjunction F_d, of any formula F_k and its complement $\overline{F}_k$, can be appended to or deleted from the conjunction representing F without affecting the value of F, if there exist functions F_s and F_t in the conjunction representing F such that $\overline{F}_s$ is subsumed by F_t.*

Proof. Let F_s and F_t be in the conjunction representing F such that $\overline{F}_s$ is subsumed by F_t. Then $F_sF_t \leq 0.5$ since

$$\overline{F_sF_t} = \overline{F_s} + \overline{F_t} = \overline{F_s} + F_t + \overline{F_t} \geq 0.5. \quad ■$$

Theorem 1.7.15 *Let $\{F_1, F_2, \ldots, F_m\}$ be a set of fuzzy formulas over $x_1, x_2, \ldots, x_n$ and let F be a disjunction of formulas from this set. A conjunction F_d of any formula F_r and its complement $\overline{F_r}$ can be appended to or deleted from the disjunction representing F without affecting the value of F, if there exist functions F_i and F_j in the disjunction representing F such that F_i is subsumed by $\overline{F_j}$.*

Proof. $F_i + F_j \geq 0.5$ since $F_i + F_j = F_i + \overline{F_j} + F_j$. ■

It is possible to specify a fuzzy function by means of an arbitrary sum-of-products expression which is not a minimal sum and which can contain phrases which are not FPIs. Next, we present a method to derive the set of all FPIs of a fuzzy function by means of the fuzzy iterative consensus.

Definition 1.7.16 *Let R and Q be two phrases over the fuzzy variables $x_1, x_2, \ldots, x_n$. The* ***fuzzy consensus*** *of R and Q, written $R\psi Q$, is defined to be the set of phrases $\{R_iQ_i\}$, where $R = x_iR_i$ and $Q = \overline{x_i}Q_i$ (or $R = \overline{x_i}R_i$ and $Q = x_iQ_i$) and $x_i \in \{x_1, x_2, \ldots, x_n\}$, if the phrase R_iQ_i includes the conjunction $x_j\overline{x}_j$ for at least one $j \in \{1, 2, \ldots, n\}$. If the phrase R_iQ_i does not include $x_j\overline{x}_j$ for any $j \in \{1, 2, \ldots, n\}$, then $R\psi Q = \{R_iQ_ix_j\overline{x}_j \mid j = 1, 2, \ldots, n\}$, where $x_i \in \{x_1, x_2, \ldots, x_n\}$.*

If none of the above occurs then we say that $R\psi Q = \emptyset$. (Any repeated literals or phrases are removed from the fuzzy consensus of R and Q.)

Example 1.7.17 *(i) If $R = x_1\overline{x}_1x_2x_3$ and $Q = x_1\overline{x}_1\overline{x}_2x_3$, then*

$$R\psi Q = \{x_1\overline{x}_1x_3, x_1\overline{x}_1x_2\overline{x}_2x_3\}.$$

(ii) If $R = x_1x_2$ *and* $Q = \overline{x}_1x_2$, *then*

$$R\psi Q = \{x_2\overline{x}_2, x_1\overline{x}_1x_2\}.$$

(iii) If $R = \overline{x}_1$ *and* $Q = x_1x_2\overline{x}_2$, *then*

$$R\psi Q = \{x_2\overline{x}_2\}.$$

(iv) If $R = x_1\overline{x}_2$ *and* $Q = \overline{x}_2x_3$, *then*

$$R\psi Q = \emptyset.$$

Theorem 1.7.18 *Let* R, Q, *and* W *denote phrases. If* $W \in R\psi Q$, *then* $R + Q \supseteq W$.

Proof. Since $W \in R\psi Q$ by hypothesis, W includes the conjunction $x_j\overline{x}_j$ for at least one j, $j \in \{1, 2, \dots, n\}$ else $W = 0$. If $W = 0$, then clearly $R + Q \supseteq W$. Suppose $W \neq 0$. If either

$$\begin{array}{llll} R \supseteq W & \text{and} & Q \supseteq W & \text{or} \\ R \supseteq W & \text{and} & Q \not\supseteq W & \text{or} \\ R \not\supseteq W & \text{and} & Q \supseteq W, & \end{array} \tag{1.21}$$

then clearly $R + Q \supseteq W$.

Suppose that $R \not\supseteq W$ and $Q \not\supseteq W$. Since $W \neq 0$, $R\psi Q \neq \emptyset$. Now $R = x_iR_i$ and $Q = \overline{x_i}Q_i$ or $R = \overline{x_i}R_i$ and $Q = x_iQ_i$ for some i. Since $W \in R\psi Q$, either $W = R_iQ_i$, if R_iQ_i contains a conjunction $x_j\overline{x_j}$ for some j or $W = R_iQ_ix_j\overline{x_j}$ for some j. Let R/W be the product of those literals of R which are not present in W and Q/W be the product of those literals of Q which are not present in W. It follows that $R/W = (\overline{Q/W})$. Also since R/W and Q/W are both phrases, R/W must be a single literal, say x_i, and Q/W must be $\overline{x}_i$. Since $W \in R\psi Q$, by Theorem 1.7.10,

$$W = W(x_i + \overline{x}_i) = Wx_i + W\overline{x}_i. \tag{1.22}$$

Since

$$R + Q \supseteq Wx_i \tag{1.23}$$

and

$$R + Q \supseteq W\overline{x}_i, \tag{1.24}$$

it follows that $R + Q \supseteq W$. ∎

The phrases added whenever $x_j\overline{x}_j$ is not included in R_iQ_i are fuzzy prime implicants which are not needed. This can be seen from the discussion which follows.

Theorem 1.7.19 *Let $R = x_iR_i$ and $Q = \overline{x}_iQ_i$. Suppose R_iQ_i does not include $x_j\overline{x}_j$ for any j, $j \in \{1, 2, \ldots, n\}$. Then*

$$R + Q + (R\psi Q) = R + Q.$$

Proof. $R+Q+(R\psi Q) = x_iR_i + \overline{x}_iQ_i + \{R_iQ_ix_j\overline{x}_j \mid j = 1, 2, \ldots, n\} = x_iR_i + \overline{x}_iQ_i + \{R_iQ_ix_j\overline{x}_j(x_i + \overline{x}_i) \mid j = 1, 2, ..., n\} = x_iR_i + \overline{x}_iQ_i + \{x_iR_iQ_ix_j\overline{x}_j + \overline{x}_iQ_iR_ix_j\overline{x}_j \mid j = 1, 2, \ldots, n\} = x_iR_i + \overline{x}_iQ_i = R + Q$. ■

Definition 1.7.20 *Let f be a phrase over the literals $\{x_1, x_2, \ldots, x_n\}$. Then f is called a* ***type-1 phrase*** *if f contains a conjunction of the form $x_j\overline{x}_j$ for at least one $j \in \{1, 2, \ldots, n\}$. Otherwise f is called a* ***type-2 phrase****.*

Clearly a type-1 phrase cannot be subsumed by type-2 phrases. However, a type-1 phrase can subsume a type-2 phrase.

If members of the set $\{R_iQ_i\}$ do not include a conjunction of the form $x_j\overline{x}_j$ for at least one $j \in \{1, 2, \ldots, n\}$, two situations must be considered.

(1) Suppose that R and Q are both type-2 phrases. Then $\{R_iQ_ix_j\overline{x}_j \mid j = 1, 2, ..., n\}$ is a set of type-1 phrases covered by $R + Q$ and thus this set is not needed.

(2) Suppose that R is a type-1 phrase and Q is a type-2 phrase. In order for members of $\{R_iQ_i\}$ not to include a conjunction $x_j\overline{x}_j$ for any $j = 1, 2, \ldots, n$, R must be of the form $\alpha x_i\overline{x}_i\beta$ and Q must be of the form $\gamma x_i\delta$ (or $\gamma'\overline{x}_i\delta'$), where the phrase $\alpha x_i\beta\gamma\delta$ (or $\alpha\overline{x}_i\beta\gamma'\delta'$) is a type-2 phrase. Hence

$$\begin{aligned} R_i &= \alpha x_i\beta \text{ (or } \alpha\overline{x}_i\beta) \\ Q_i &= \gamma\delta \qquad \text{(or } \gamma'\delta') \end{aligned}$$

and clearly the set $\{R_iQ_ix_j\overline{x}_j \mid j = 1, 2, \ldots, n\}$ is covered by Q and is not needed.

Example 1.7.21 *Let $f(x_1, x_2, x_3) = x_1x_2 + \overline{x}_2x_3$. If the variables are Boolean, then the Boolean consensus will add the phrase x_1x_3. However, in the fuzzy case $x_1\overline{x}_1$ and $x_3\overline{x}_3$ do not appear in $x_1x_2\overline{x}_2x_3$ and thus*

$$\begin{aligned} x_1x_2\psi\overline{x}_2x_3 &= \{x_1x_3x_1\overline{x}_1,\ x_1x_3x_2\overline{x}_2,\ x_1x_3x_3\overline{x}_3\} \\ &= \{x_1\overline{x}_1x_3,\ x_1x_2\overline{x}_2x_3,\ x_1x_3\overline{x}_3\}. \end{aligned}$$

Since the phrase $x_1x_2\overline{x}_2x_3$ subsumes both x_1x_2 and $\overline{x}_2x_3$, it is not a fuzzy prime implicant. However, the phrases $x_1\overline{x}_1x_3$ and $x_1x_3\overline{x}_3$ are fuzzy prime implicants, even though not essential.

Theorem 1.7.22 *Let $f(x_1, x_2, \ldots, x_n) = \prod_k (x_k + \overline{x}_k + \sigma_k)$, where σ_k is any function in $x_1, x_2, ..., x_n$ and $k \in \{1, 2, \ldots, n\}$. Then*

$$f(x_1, x_2, \ldots, x_n) = \prod_k (x_k + \overline{x}_k + \sigma_k) + \sum_j x_j\overline{x}_j\gamma_j,$$

where γ_j is some arbitrary function in $x_1, x_2, \ldots, x_n$ and $j \in \{1, 2, \ldots, n\}$.

Proof.

$$\prod_k (x_k + \overline{x}_k + \sigma_k) = \min\{\max\{x_k, \overline{x}_k, \sigma_k\} \mid k = 1, 2, \ldots, n\} \geq 0.5$$

$$\sum_j x_j \overline{x}_j \gamma_j = \max\{\min\{x_j, \overline{x}_j, \gamma_j\} \mid j = 1, 2, \ldots, n\} \leq 0.5. \blacksquare$$

It is possible for the fuzzy consensus to add fuzzy prime implicants which include variables not even presented in the original fuzzy switching function. For example, let $f(x_1, x_2) = x_1 + \overline{x}_1$. Then

$$f(x_1, x_2) = x_1 + \overline{x}_1 + x_2\overline{x}_2,$$

where all three phrases are fuzzy prime implicants. However,

$$x_2\overline{x}_2 = x_2\overline{x}_2(x_1 + \overline{x}_1) = x_2\overline{x}_2x_1 + x_2\overline{x}_2\overline{x}_1$$

and these two latter phrases are included in x_1 and $\overline{x}_1$, respectively. Thus $f_{\min}(x_1, x_2) = x_1 + \overline{x}_1$and $x_2\overline{x}_2$ is not an essential fuzzy prime implicant of the function.

We now consider the minimization of completely specified fuzzy functions. The basis for the method of fuzzy integrated consensus and minimization of fuzzy switching functions is formed from the axioms of fuzzy algebra and Theorem 1.7.10. We show below that the successive addition of fuzzy consensus phrases to a sum-of-products expression and the removal of phrases which are included in other phrases ($x + xy = x$) will result in an expression which represents the function as the sum of all its FPIs.

Example 1.7.23 *(i) Let* $f(x_1, x_2) = \overline{x}_1x_2 + x_1x_2\overline{x}_2$. *Now*

$$\overline{x}_1x_2 \psi x_1x_2\overline{x}_2 = \{x_2\overline{x}_2, x_1\overline{x}_1x_2\}.$$

Adding $x_2\overline{x}_2$ *and* $x_1\overline{x}_1x_2$ *to* $f(x_1, x_2)$, *we get*

$$f(x_1, x_2) = \overline{x}_1x_2 + x_1x_2\overline{x}_2 + x_2\overline{x}_2 + x_1\overline{x}_1x_2.$$

Since $x_1x_2\overline{x}_2$ *subsumes* $x_2\overline{x}_2$ *and* $x_1\overline{x}_1x_2$ *subsumes* $\overline{x}_1x_2$, *we get* $f(x_1, x_2) = \overline{x}_1x_2 + x_2\overline{x}_2$.

(ii) Let $f(x_1, x_2, x_3) = x_1x_2x_3 + \overline{x}_1x_2x_3$. *Now* $x_1x_2x_3 \psi \overline{x}_1x_2x_3 = \{x_2x_3\}$. *Adding* x_2x_3 *to* $f(x_1, x_2, x_3)$, *we get*

$$f(x_1, x_2, x_3) = x_1x_2x_3 + \overline{x}_1x_2x_3 + x_2x_3 = x_2x_3.$$

(iii) Let

$$f(x_1, x_2, x_3) = x_1x_2\overline{x}_2x_3 + \overline{x}_1x_2\overline{x}_2x_3.$$

Now

$$x_1x_2\overline{x}_2x_3\psi\overline{x}_1x_2\overline{x}_2x_3 = \{x_2\overline{x}_2x_3, x_1\overline{x}_1x_2\overline{x}_2x_3\}.$$

Adding $x_2\overline{x}_2x_3$ *and* $x_1\overline{x}_1x_2\overline{x}_2x_3$ *to* f, *we get*

$$f(x_1, x_2, x_3) = x_1x_2\overline{x}_2x_3 + \overline{x}_1x_2\overline{x}_2x_3 + x_2\overline{x}_2x_3 + x_1\overline{x}_1x_2\overline{x}_2x_3.$$

Thus $f(x_1, x_2, x_3) = x_2\overline{x}_2x_3$ *since* $x_2\overline{x}_2x_3$ *includes all other phrases.*

Theorem 1.7.24 *A sum-of-products expression* $F = P_1 + P_2 + \ldots + P_r$ *for the function* $f(x_1, x_2, \ldots, x_n)$ *is the sum of all the FPIs of* $f(x_1, x_2, \ldots, x_n)$ *if and only if*

(i) no phrase includes any other phrase, $P_j \nsubseteq P_i$ *for any* i *and* $j, i \neq j$, *where* $i, j \in \{1, 2, \ldots, r\}$, *and*

(ii) the fuzzy consensus of any two phrases, $P_i\psi P_j$, *either does not exist* ($P_i\psi P_j = \emptyset$) *or every phrase that belongs to the set describing* $P_i\psi P_j$ *is included in some other phrase from* $\{P_k \mid k = 1, 2, \ldots, r\}$.

Proof. Suppose that $F = P_1 + P_2 + \ldots + P_r$ is a sum-of-products expression, but not the sum of all FPIs of $f(x_1, x_2, ..., x_n)$.

We show that there exists P_i and P_j such that either $P_j \subseteq P_i$ or that some phrase from the set $P_i\psi P_j$ is not included in any P_k. Since F is not the sum of all FPIs, either some of the P_js are not FPIs or some FPIs of $f(x_1, x_2, \ldots, x_n)$ are missing from F. Suppose P_j is not an FPI for some j. Then there exists an FPI, say T, of $f(x_1, x_2, \ldots, x_n)$ such that $P_j \subseteq T$. If $T = P_i$ for some i, then $P_j \subseteq P_i$. If $T \neq P_i$ for all i, then at least one of the FPIs is missing from F.

Suppose that there is a FPI, say T, of $f(x_1, x_2, ..., x_n)$ such that T is not included in F. Since T is an FPI and $T \neq P_i$ for all i, $T \nsubseteq P_i, \forall i$, $i = 1, 2, \ldots, r$. Now T must be covered by the representation of the function. Thus either $T \subseteq P_i$ for some i or T is covered by a set of phrases generated with respect to Theorem 1.7.14. Since $T \nsubseteq P_i$, $\forall i$, T is covered by a set of phrases generated with respect to Theorem 1.7.14. Then T must include at least one conjunction of the form $x_j\overline{x}_j$ for some j, $j \in \{1, 2, \ldots, n\}$.

It may be possible to add some literals to T, forming a phrase T' which still has the property that $T' \nsubseteq P_i$ for all i. Now there may be more phrases satisfying the requirements placed on T'. If so, T' is defined as one of those phrases which satisfies the requirements given and contains as many literals as any other phrase satisfying these requirements. On the other hand, if it is not possible to join an additional variable to T without having the resulting phrase included in one of the P_i's, then T' is defined to be T.

Thus T' satisfies the following.

(1) T' contains no variables other than $x_1, x_2, \ldots, x_n$.

(2) $T' \subseteq T$.

(3) $T' \nsubseteq P_i$, for any i, $i = 1, 2, \ldots, r$.

(4) No phrase exists having more literals than T' and satisfying (1), (2), and (3) above.

From the definition of T', it follows that one of the variables will be missing from T'. If T' had all the set of variables appearing in it, condition (3) cannot be satisfied in the definition of T'.

Suppose that the variable x_k, for some k, is missing from T'. Consider the phrase $T'x_k$. Clearly, $T'x_k$ satisfies (1) and (2). If $T'x_k$ satisfies (3), then T' does not satisfy (4), which is a contradiction of the definition of T'. Thus $T'x_k \subseteq P_l$ for some l. Similarly, $T'\overline{x}_k \subseteq P_s$ for some s. Since $T'x_k \subseteq P_l$ and $T' \not\subseteq P_l$, the literal x_k must appear in P_l. Thus $P_l = x_k Q_l$ for some phrase Q_l. Similarly, $P_s = \overline{x}_k Q_s$ for some phrase Q_s. Since $T'x_k \subseteq x_k Q_l = P_l$ and neither Q_l nor T' contains x_k, $T' \subseteq Q_l$. Similarly, $T' \subseteq Q_s$. Hence $T' \subseteq Q_l Q_s$ and so the phrase $Q_l Q_s \neq 0$. Since $T' \subseteq Q_l Q_s$, $Q_l Q_s$ includes $x_j \overline{x}_j$ for some j. Further since $T'Q_s$ belongs to the set describing $P_l \psi P_s$, $P_l \psi P_s \neq \emptyset$.

Next, we show that $Q_l Q_s \not\subseteq P_i$ for all i. Suppose $Q_l Q_s \subseteq P_i$ for some i,. Then $T' \subseteq Q_l Q_s$ implies that $T' \subseteq P_i$, which contradicts (3). This argument can be repeated for every phrase in the set describing $P_l \psi P_s$.

Conversely, suppose that $F = P_1 + P_2 + \ldots + P_r$ is the sum of all the FPIs of $f(x_1, x_2, \ldots, x_n)$. Since for all i, P_i is an FPI, P_i is not included in P_j for all j, $j \neq i$. Suppose that for some i, j, $i \neq j$, $P_i \psi P_j$ exists. Then the phrases in the set $P_i \psi P_j$ are included in $f(x_1, x_2, \ldots, x_n)$. By the definition of an FPI, any phrase which is included in $f(x_1, x_2, \ldots, x_n)$ must be included in some FPI of $f(x_1, x_2, \ldots, x_n)$. Thus every phrase belonging to the set describing $P_i \psi P_j$ must be included in some other phrase from $\{P_1, P_2, \ldots, P_r\}$. ∎

The identity

$$xy + \overline{x}z + yz = xy + \overline{x}z \tag{1.25}$$

which is the basic Boolean consensus, does not hold in fuzzy logic. For example, let $x = 0.6$, $y = 0.7$, and $z = 0.8$. Then $xy + \overline{x}z + yz = 0.7$ and $xy + \overline{x}z = 0.6$. Thus $xy + \overline{x}z + yz \neq xy + \overline{x}z$. For fuzzy logic, we have the following theorem.

Theorem 1.7.25 *Let x, y and z be fuzzy variables. Then*

$$x\overline{x}y + x\overline{y}z = x\overline{x}y + x\overline{y}z + x\overline{x}z. \tag{1.26}$$

Proof. By Theorem 1.7.10, $x\overline{x}z = x\overline{x}z(y + \overline{y}) = x\overline{x}yz + x\overline{x}\overline{y}z$. The phrase $x\overline{x}yz$ is included in $x\overline{x}y$ and the phrase $x\overline{x}\overline{y}z$ is included in $x\overline{y}z$. Hence $x\overline{x}y + x\overline{y}z + x\overline{x}z = x\overline{x}y + x\overline{y}z + x\overline{x}yz + x\overline{x}\overline{y}z = x\overline{x}y + x\overline{y}z$. ∎

In order to use Theorem 1.7.24 to form the sum of all FPIs from a sum-of-products expression, the original expression must be converted into one in which every phrase generated by the fuzzy consensus of any two phrases is included in some other phrase, and in which no phrase includes any other. The following algorithm accomplishes this.

To use the algorithm, express $f(x_1, x_2, \ldots, x_n)$ in disjunctive normal form.

Algorithm Compute the FPIs of a function [4].

Input: The set of phrases representing $f(x_1, x_2, \ldots, x_n)$.
Output:The set of FPIs of $f(x_1, x_2, \ldots, x_n)$.
Step 1: Compare each phrase with every other phrase in the expression and remove any phrase which subsumes any other phrase.
Step 2: Add the fuzzy consensus of any two phrases to the expression, provided that the fuzzy consensus phrases are not included in some other phrases.
Step 3: Remove any phrases that subsume phrases added in Step 2.

Step 2 and 3 are iteratively repeated. The process terminates when all possible consensus operations have been exhausted. The remaining phrases are all of the FPIs. A phrase of the form $x_j\overline{x}_jAB$ is always included in the set describing $x_j\overline{x}_jA\psi\, x_j\overline{x}_jB$. This phrase, called a **secondary consensus**, need not be formed since it is always included in both of the initiating phrases.

Example 1.7.26 *Let*

$$f(x_1, x_2, x_3, x_4) = x_1x_2\overline{x}_2\overline{x}_3 + x_1x_2\overline{x}_2\overline{x}_4 + x_1x_2\overline{x}_2x_3x_4.$$

Let $R_1 = x_1x_2\overline{x}_2\overline{x}_3$, $R_2 = x_1x_2\overline{x}_2\overline{x}_4$, *and* $R_3 = x_1x_2\overline{x}_2x_3x_4$. *Then* $R_1\psi R_3 = \{a_1, a_2\}$ *and* $R_2\psi R_3 = \{b_1, b_2\}$, *where* $a_1 = x_1x_2\overline{x}_2x_3\overline{x}_3x_4$, $a_2 = x_1x_2\overline{x}_2x_4$, $b_1 = x_1x_2\overline{x}_2x_3x_4\overline{x}_4$, *and* $b_2 = x_1x_2\overline{x}_2x_3$.

Now phrases a_2 *and* b_2 *include phrases* a_1 *and* b_1, *respectively. Adding* $R_1\psi R_3$ *and* $R_2\psi R_3$ *to* $f(x_1, x_2, x_3, x_4)$ *yields, after removal of subsuming phrases,*

$$f(x_1, x_2, x_3, x_4) = x_1x_2\overline{x}_2\overline{x}_3 + x_1x_2\overline{x}_2\overline{x}_4 + x_1x_2\overline{x}_2x_4 + x_1x_2\overline{x}_2x_3.$$

Let $R_4 = x_1x_2\overline{x}_2\overline{x}_3$, $R_5 = x_1x_2\overline{x}_2\overline{x}_4$, $R_6 = x_1x_2\overline{x}_2x_4$, *and* $R_7 = x_1x_2\overline{x}_2x_3$. *Then* $R_4\psi R_7 = \{x_1x_2\overline{x}_2x_3\overline{x}_3, x_1x_2\overline{x}_2\}$. *The phrase* $x_1x_2\overline{x}_2x_3\overline{x}_3$ *is included in the phrase* $x_1x_2\overline{x}_2$. *We add* $R_4\psi R_7$ *to the function and delete the subsuming phrases to obtain* $f(x_1, x_2, x_3, x_4) = x_1x_2\overline{x}_2$.

Usually it is a minimal sum rather than the sum of all FPIs of a fuzzy function that is desired.

In a sum-of-products form, each phrase corresponds to a gate and each literal to an input line. The type of gate determines the ratio between the cost of a gate and the cost of an input line. However, in practice an additional gate is more expensive than an additional input line on a gate which already exists. Consequently, the elimination of gates will be the main objective of the minimization process. Recall the definition of a minimal expression given in, Definition 1.7.5.

Theorem 1.7.27 *A minimal sum-of-products expression consists of a sum of phrases representing FPIs.*

Proof. Consider the sum-of-products expression representing a function $f(x_1, x_2, \dots, x_n)$. Any phrase which is not an FPI is included in some FPI by definition. Thus we can replace this phrase in the set by any FPI including it. Since the FPI is of higher dimension, its representation will have fewer literals than that of the phrase it replaces. ∎

Once all FPIs have been found using the fuzzy iterated consensus algorithm, the best set of FPIs must be found. We next determine a method to find certain FPIs that can be eliminated from the representation of the function without changing the function. Note that the minimum canonical sum-of-products form of a fuzzy function is not necessarily the union of all its FPIs

Definition 1.7.28 *A phrase is called a* ***fundamental phrase*** *if it does not include a conjunction of the form $x_j\overline{x}_j$ for at least one j, $j = 1, 2, \dots, n$, or if it includes such a conjunction, the phrase contains as many variables (complemented, uncomplemented, or both) as the function.*

Definition 1.7.29 *An FPI that includes a fundamental phrase which is not included in any other FPI is called an* ***essential FPI(EFPI)*** *and must be included in the corresponding minimal sum. (Actually, an EFPI must be included in all irredundant sums.)*

In order to find a minimal sum of FPIs, a fuzzy prime implicant table can be constructed in manner similar to the construction of a prime implicant table in Boolean algebra. Each column in the table is headed by a fundamental phrase of $f(x_1, x_2, \dots, x_n)$. Each row is headed by the FPIs of $f(x_1, x_2, \dots, x_n)$.

An asterisk is placed at the intersection of a row and column if the corresponding FPI includes the corresponding fundamental phrase. With respect to the table, the basic requirement on the minimal-sum phrases is that each column must have an asterisk in at least one of the rows which corresponds to minimal-sum phrases.

If a column contains exactly one asterik, the column corresponds to a distinguished fundamental phrase (it is included in only one FPI) and the row in which the asterisk occurs corresponds to an EFPI. Rows and columns corresponding to EFPIs and distinguished fundamental phrases are called essential rows and distinguished columns, respectively. The essential rows are checked and the labels of the distinguished columns are checked. If all the columns have been covered by the checked rows, all the fundamental phrases are included in the EFPIs and the sum of the EFPIs is the minimal sum-of-products expression. If not all of the fundamental phrases are covered by EFPIs, a reduced table is produced. In this table, all essential rows and all distinguished columns are eliminated.

After the table has been reduced, the least expensive set of FPIs that covers the remainder of the function must be found. This is done by applying to the reduced table the same methods of reduction as used in the minimization of Boolean functions.

Example 1.7.30 *Let*

$$f(x_1, x_2, x_3, x_4) = x_1x_2\overline{x}_2 + \overline{x}_1x_2x_3x_4 + x_2\overline{x}_2x_3x_4. \tag{1.27}$$

Clearly, all three phrases are FPIs. Expanding into fundamental phrases, we obtain

$$\begin{aligned} &x_1x_2\overline{x}_2 \to \{x_1x_2\overline{x}_2x_3x_4, x_1x_2\overline{x}_2\overline{x}_3x_4, x_1x_2\overline{x}_2x_3\overline{x}_4, x_1x_2\overline{x}_2\overline{x}_3\overline{x}_4\},\\ &\overline{x}_1x_2x_3x_4 \to \{\overline{x}_1x_2x_3x_4\}\\ &x_2\overline{x}_2x_3x_4 \to \{x_1x_2\overline{x}_2x_3x_4, \overline{x}_1x_2\overline{x}_2x_3x_4\}. \end{aligned} \tag{1.28}$$

	a	b	c	d	e	f
$x_1x_2\overline{x}_2$	*	*	*	*		
$\overline{x}_1x_2x_3x_4$					*	*
$x_2\overline{x}_2x_3x_4$	*				*	

In this table, $a = x_1x_2\overline{x}_2x_3x_4$, $b = x_1x_2\overline{x}_2\overline{x}_3x_4$, $c = x_1x_2\overline{x}_2x_3\overline{x}_4$, $d = x_1x_2\overline{x}_2\overline{x}_3\overline{x}_4$, $e = \overline{x}_1x_2\overline{x}_2x_3x_4$, *and* $f = \overline{x}_1x_2x_3x_4$. *(Recall that each column in the table is headed by a fundamental phrase and each row is headed by the FPIs.)*

From the above table, it can be seen that the phrases $x_1x_2\overline{x}_2$ *and* $\overline{x}_1x_2x_3x_4$ *are EFPIs. In this example, all the fundamental phrases are included in the EFPIs and* $x_1x_2\overline{x}_2 + \overline{x}_1x_2x_3x_4$ *is the minimal sum-of-products expression for the function.*

Minimization of Incompletely Specified Fuzzy Functions

Up to this point the fuzzy switching functions considered have been completely specified for every combination of the variables. There are situations where a function assumes a grade of membership for some combinations, but it may assume unspecified grade memberships for other combinations. Combinations for which the grade membership of the function is not specified are called **don't care combinations**.

We next formulate the incompletely specified single-output minimization problem.

Given a fuzzy function F^* in the variables $x_1, x_2, \ldots, x_n$, constructed as a union of a completely specified subfunction and a don't care subfunction, find the function F in the same variables such that F is the minimal complexity form of F^*.

The next theorem shows how to obtain the FPIs of an incompletely specified fuzzy function.

Theorem 1.7.31 *Let F^* be an incompletely specified fuzzy function in the variables $x_1, x_2, \ldots, x_n$. Suppose that F^* is expressed in disjunctive normal form. A phrase P is an FPI of F^* if and only if P is an FPI of F_s+F_d, where F^* is represented by a pair of fuzzy functions F_s and F_d such that F_s and F_d are the completely specified and the don't-care subfunctions, respectively.*

Proof. (1) Assume P_1 is an FPI of F^* and P_1 is not an FPI of F_s+F_d. Then there exists P_2 such that P_2 is an FPI of F_s+F_d and P_1 subsumes P_2. Since P_2 is an implicant of F^*, P_1 cannot be an FPI of F^*. This is impossible. Hence P_1 is an FPI of F_s+F_d.

(2) Similarly, assume P_1 is an FPI of F_s+F_d and P_1 is not an FPI of F^*. Then there exists P_2 such that P_2 is an FPI of F^* and P_2 subsumes P_1. Since P_1 is an implicant of F^*, part (1) implies that, if P_2 is an FPI of F^*, then P_2 is an FPI of F_s+F_d. Hence, P_1, P_2 are FPIs of F_s+F_d and P_2 subsumes P_1. This is impossible. Hence P_1 is an FPI of F^*. ∎

From this theorem, we see that in order to obtain the FPIs of an incompletely specified fuzzy function, it is sufficient to consider the FPIs of the fuzzy function described by F_s+F_d. This is an extended result with respect to Boolean incompletely specified functions.

In order to determine a minimal sum, the FPIs of the incompletely specified fuzzy function are found. To accomplish this, one must consider both the phrases of F_s and F_d. The phrases of F_d are regarded as phrases of F_s during the formation of the FPIs.

We see that it is not necessary to include all the fundamental phrases of F_d in the minimal representation of the function. Hence, we can find all the FPIs of F_s+F_d using the Algorithm. However, in constructing the fuzzy prime implicant table, each column in the table corresponds only to a fundamental phrase of F_s. Thus columns are used only for the disjuncts of F_s and not of F_d, namely, each column in the table corresponds to a fundamental phrase of F_s. At the left of each row are listed the FPIs of F_s+F_d.

Example 1.7.32 *Let*

$$F_s = \overline{x}_1 x_2 + x_3 x_4 + x_1 x_2 \overline{x}_2 + x_1 x_2 x_4 + x_1 \overline{x}_3 x_4 \tag{1.29}$$

and

$$F_\Phi = x_2 + \overline{x}_1\overline{x}_2 + x_3 + \overline{x}_3 + x_4 + \overline{x}_4 + x_3\overline{x}_4. \tag{1.30}$$

By the previous procedure, we have that the FPIs of $F_s + F_\Phi$ *are*

$$x_2, x_3, \overline{x}_3, x_4, \overline{x}_4, \overline{x}_1\overline{x}_2. \tag{1.31}$$

		$\surd$						$\surd$	
		a	b	c	d	e	f	g	h
$\surd$	x_2	*		*	*	*	*	*	
	x_3		*	*		*			
	$\overline{x}_3$				*		*		*
$\surd$	x_4	*	*	*				*	*
	$\overline{x}_4$					*	*		
	$\overline{x}_1\overline{x}_2$								

In this table, $a = \overline{x}_1x_2$, $b = x_3x_4$, $c = x_1x_2\overline{x}_2x_3x_4$, $d = x_1x_2\overline{x}_2\overline{x}_3x_4$, $e = x_1x_2\overline{x}_2x_3\overline{x}_4$, $f = x_1x_2\overline{x}_2\overline{x}_3\overline{x}_4$, $g = x_1x_2x_4$, *and* $h = x_1\overline{x}_3x_4$.

From the above table.

$$F = F^*_{\min} = x_2 + x_4. \tag{1.32}$$

From the construction of the algorithm, we see that the algorithm is complete.

We have touched briefly on the minimization of incompletely specified fuzzy functions. A more detailed account can be found in [4].

1.8 Solution of Fuzzy Logic Inequalities

Let $f : [0,1]^n \to [0,1]$ be a fuzzy logic function. In this section, we determine a method of solving fuzzy logic inequalities involving fuzzy logic functions. The results are essentially those of [2]. Let n be a positive integer and let $a_1, a_2, \ldots, a_{n-1} \in [0,1]$ be such that $1 > a_1 > a_2 > \ldots > a_{n-1} \geq 0$, where $a_0 = 0$, if $n = 1$. Define the sets of n-tuples $R_1, R_2, \ldots, R_n$ as follows:

$$\begin{aligned} R_1 &= \{(x_1, x_2, \ldots, x_n) \mid a_1 < f(x_1, x_2, \ldots, x_n) \leq 1\}, \\ &\vdots \\ R_i &= \{(x_1, x_2, \ldots, x_n) \mid a_i < f(x_1, x_2, \ldots, x_n) \leq a_{i-1}\}, \\ &\vdots \\ R_n &= \{(x_1, x_2, \ldots, x_n) \mid 0 \leq f(x_1, x_2, \ldots, x_n) \leq a_n\}. \end{aligned}$$

We develop a method to determine R_i, $i = 1, 2, \ldots, n$.

Define the sets S_i, $i = 1, 2, \ldots, n+1$, as follows:

$$\begin{aligned}
S_{n+1} &= \emptyset, \\
S_n &= \{(x_1, x_2, \ldots, x_n) \mid f(x_1, x_2, \ldots, x_n) \leq a_{n-1}\}, \\
&\vdots \\
S_i &= \{(x_1, x_2, \ldots, x_n) \mid f(x_1, x_2, \ldots, x_n) \leq a_{i-1}\}, \\
&\vdots \\
S_2 &= \{(x_1, x_2, \ldots, x_n) \mid f(x_1, x_2, \ldots, x_n) \leq a_1\}, \\
S_1 &= [0, 1].
\end{aligned}$$

Clearly $R_i = S_i \backslash S_{i+1}$, $i = 1, 2, \ldots, n$. Hence it suffices to determine the sets S_i, $i = 1, 2, \ldots, n+1$.

Recall that a fuzzy logic function f of n-variables is a function from $[0, 1]^n$ into $[0, 1]$ defined recursively in the following manner:
(1) Let $a \in [0, 1]$. Define $f(x_1, x_2, \ldots, x_n) = a \ \forall\ x_1, x_2, \ldots, x_n \in [0, 1]$. Then f is a fuzzy logic function called a **constant function.**
(2) For $i = 1, 2, \ldots, n$, let $f_i(x_1, x_2, \ldots, x_n) = x_i$. Then f is a fuzzy logic function called a **projection function.**
(3) If g and h are fuzzy logic functions, then $g + h$, gh, and $\bar{g}$ are fuzzy logic functions, where

$$\begin{aligned}
(g + h)(x_1, x_2, \ldots, x_n) &= g(x_1, x_2, \ldots, x_n) + h(x_1, x_2, \ldots, x_n) \\
(gh)(x_1, x_2, \ldots, x_n) &= g(x_1, x_2, \ldots, x_n) h(x_1, x_2, \ldots, x_n) \\
\bar{g}(x_1, x_2, \ldots, x_n) &= 1 - g(x_1, x_2, \ldots, x_n).
\end{aligned}$$

Theorem 1.8.1 *Let f be a function from $[0, 1]^n$ into $[0, 1]$. Then the following conditions are equivalent:*
(i) f is a fuzzy logic function;
(ii) $f(x_1, x_2, \ldots, x_n) = \sum_{S \in \mathcal{F}} (c_S \Pi S)$;
(iii) $f(x_1, x_2, \ldots, x_n) = \Pi_{S \in \mathcal{F}} (c_S + \sum S)$;
*where $\mathcal{F}$ is the power set of $\{x_1, x_2, \ldots, x_n, \overline{x_1}, \overline{x_2}, \ldots, \overline{x_n}\}$ and $c_S \in [0, 1]$. The expression (ii) is called the **disjunctive normal form** of f and $c_S \Pi S$ is called an **elementary conjunctive normal form.** The expression in (iii) is called the **conjunctive normal form** of f and $c_S + \sum S$ is called an **elementary disjunctive normal form.***

Proof. (i)$\Rightarrow$(ii) The proof is by induction of n, the number of variables. If $n = 1$, then the implication is clear. Suppose the implication is true for $n = k - 1$, $k \geq 2$. Let $n = k$. Then there exist fuzzy logic functions $g_i : [0, 1]^{k-1} \to [0, 1]$, $i = 1, 2, 3, 4$ such that

$$\begin{aligned}
f(x_1, x_2, \ldots, x_k) &= g_1(x_2, \ldots, x_k) x_1 + g_2(x_2, \ldots, x_k) x_1 \overline{x_1} \\
&\quad + g_3(x_2, \ldots, x_k) \overline{x_1} + g_4(x_2, \ldots, x_k).
\end{aligned} \tag{1.33}$$

By the induction hypothesis,

$$g_i(x_2, \ldots, x_k) = \sum_{S_i \in \mathcal{F}'} (c^{(i)} \Pi S_i), \qquad i = 1, 2, 3, 4, \tag{1.34}$$

where $\mathcal{F}'$ is the power set of $\{x_2, \ldots, x_n, \overline{x_2}, \ldots, \overline{x_n}\}$. By making the appropriate substitution for g_i in (1.33) using (1.34), we obtain

$$f(x_1, x_2, \ldots, x_k) = \sum_{S \in \mathcal{F}} (c_S \Pi S).$$

Hence by induction, the desired implication holds.

(ii)⇒(i): Immediate.

The equivalence of (i) and (iii) follows in a similar manner. ■

Definition 1.8.2 *Let α be the function from $[0,1]^2$ into $[0,1]$ defined as follows:* $\forall a, b \in [0,1]$

$$a\alpha b = \vee\{x \in [0,1] \mid ax \leq b\}.$$

In Definition 1.8.2, we have that

$$a\alpha b = \begin{cases} 1 & \text{if } a \leq b \\ b & \text{if } a > b. \end{cases}$$

Theorem 1.8.3 *The following properties hold* $\forall a, b, c \in [0,1]$.

(i) $a(a\alpha b) \leq b$.

(ii) If $ac \leq b$, *then* $c \leq a\alpha b$.

(iii) If $a \leq b$, *then* $c\alpha a \leq c\alpha b$ *and* $a\alpha c \geq b\alpha c$.

Proof. (i) If $a > b$, then $a(a\alpha b) = ab \leq b$. If $a \leq b$, then $a(a\alpha b) = a1 = a \leq b$.

(ii) $a\alpha b = \vee\{x \mid ax \leq b\} \geq c$.

(iii) That $c\alpha a \leq c\alpha b$ is immediate. Since $\{x \mid ax \leq c\} \supseteq \{x \mid bx \leq c\}$, it follows that $a\alpha c \geq b\alpha c$. ■

Theorem 1.8.4 *The following properties hold* $\forall a, b, c \in [0,1]$.

(i) $(a\alpha b)c \leq a\alpha(bc)$.

(ii) $a\alpha b \leq (ca)\alpha(cb)$.

(iii) $(a\alpha b)(b\alpha c) \leq a\alpha c$.

Proof. (i) It is immediate that $a(a\alpha b) \leq b$ and so $a(a\alpha b)c \leq bc$. Thus by Theorem 1.8.3(ii), $(a\alpha b)c \leq a\alpha(bc)$.

(ii) If $ax \leq b$, then $(ca)x \leq cb$. Thus $a\alpha b \leq (ca)\alpha(cb)$.

(iii) We have that $(a\alpha b)(b\alpha c) \leq a\alpha(b(b\alpha c)) \leq a\alpha c$, where the first inequality holds by (i) and the second inequality holds by use of Theorem 1.8.3(i) to see that $b(b\alpha c) \leq c$. ■

Theorem 1.8.5 *The following statements are equivalent* $\forall a, b, c \in [0,1]$.

(i) $\frac{1}{2}ac \leq b$.

(ii) $\overline{a\alpha b} \leq c\alpha b$.

(iii) $\overline{ac\alpha b} \leq \overline{bc\alpha b}$.

Proof. (i)⇒(ii): Suppose that $c \le b$. Then $\overline{a\alpha b} \le 1 = c\alpha b$. Suppose that $c > b$. Then $\frac{1}{2}a \le b$. If $a \le b$, then $\overline{a\alpha b} = 0 \le c\alpha b$. If $a > b$, then $\frac{1}{2} \le b$ and so $\overline{a\alpha b} = \bar{b} \le \frac{1}{2} \le b = c\alpha b$.

(ii)⇒(iii): Suppose that $c \le b$. Then $\overline{c\alpha b} = \bar{1} = 0$. Hence $a\overline{c\alpha b} \le b(\overline{c\alpha b})$. Suppose that $c > b$. Then $\overline{c\alpha b} = \bar{b}$. If $a \le b$, then $a\overline{c\alpha b} \le b\overline{c\alpha b}$. Suppose that $a > b$. Then $a\alpha b = b$. Since $c > b$, $c\alpha b = b$. Now $\overline{a\alpha b} = \bar{b}$. Thus by hypothesis $\bar{b} \le b$. Hence $a\bar{b} \le b\bar{b}$ and so $a(\overline{c\alpha b}) \le b(\overline{c\alpha b})$.

(iii)⇒(i): If either $a \le b$ or $c \le b$, then the result is immediate. Suppose that $a > b$ and $c > b$. Then $c\alpha b = b$. Hence $a\bar{b} \le b\bar{b}$ by hypothesis. Since $a > b$, it follows that $\bar{b} \le b$. Thus $\frac{1}{2} \le b$. ∎

We now consider the fuzzy logic inequality,

$$f(x_1, x_2, \ldots, x_n) \le p,$$

where $p \in [0,1]$. Suppose that f is a fuzzy logic function in one variable. If we write f in disjunctive normal from, we have

$$f(x) = ax + c\bar{x} + d \le p,$$

where $a, c, d \in [0,1]$.

Theorem 1.8.6 *For all $a, c, d, p \in [0,1]$, $\{x \in [0,1] \mid ax + c\bar{x} + d \le p\} \ne \emptyset$ if and only if $\frac{1}{2}ac + d \le p$. If $\frac{1}{2}ac + d \le p$, then $\{x \in [0,1] \mid ax + c\bar{x} + d \le p\} = [\overline{c\alpha p}, a\alpha p]$.*

Proof. Suppose that $\{x \in [0,1] \mid ax + c\bar{x} + d \le p\} \ne \emptyset$. Then $\exists x_0 \in [0,1]$ such that $ax_0 + c\overline{x_0} + d \le p$. Clearly, $d \le p$, $ax_0 \le p$, and $c\overline{x_0} \le p$. Hence $x_0 \le a\alpha p$ and $x_0 \ge \overline{c\alpha p}$. Thus $\overline{c\alpha p} \le a\alpha p$. By Theorem 1.8.5, $\frac{1}{2}ac + d \le p$. We also have that $\{x \mid ax + c\bar{x} + d \le p\} = \{x \mid ax + c\bar{x} \le p\} = \{x \mid ax \le p\} \cap \{x \mid c\bar{x} \le p\} = [0, a\alpha p] \cap [\overline{c\alpha p}, 1] = [\overline{c\alpha p}, a\alpha p]$.

Conversely, suppose that $\frac{1}{2}ac + d \le p$. As in previous arguments, $\{x \mid ax + c\bar{x} + d \le p\} = [0, a\alpha p] \cap [\overline{c\alpha p}, 1]$. By Theorem 1.8.5 and the hypothesis, we have that $\overline{c\alpha p} \le a\alpha p$. Hence $\{x \in [0,1] \mid ax + c\bar{x} + d \le p\} = [\overline{c\alpha p}, a\alpha p] \ne \emptyset$. ∎

We now consider the fuzzy logic inequality

$$ax + bx\bar{x} + c\bar{x} + d \le p,$$

where $a, b, c, d, p \in [0,1]$.

Theorem 1.8.7 *Let $a, b, c, d, p \in [0,1]$. Let $S = \{x \in [0,1] \mid ax + bx\bar{x} + c\bar{x} + d \le p\}$, $S_1 = \{x \in [0,1] \mid (a+b)x + c\bar{x} + d \le p\}$, and $S_2 = \{x \in [0,1] \mid ax + (b+c)\bar{x} + d \le p\}$. Then $S \ne \emptyset$ if and only if $\frac{1}{2}ac + d \le p$ and $S = S_1 \cap S_2$.*

Proof. Since $bx\bar{x} \le p$ if and only if $bx \le p$ or $b\bar{x} \le p$, it is clear that

$$S = \{x \mid (a+b)x + c\bar{x} + d \le p\} \cup \{x \mid ax + (b+c)\bar{x} + d \le p\}.$$

Since $\{x \mid (a+b)x + c\bar{x} + d \le p\} \neq \emptyset$ if and only if $\frac{1}{2}(a+b)c + d \le p$ and $\{x \mid ax + (b+c)\bar{x} + d \le p\} \neq \emptyset$ if and only if $\frac{1}{2}a(b+c) + d \le p$ by Theorem 1.8.6, $S \neq \emptyset$ if and only if $(\frac{1}{2}(a+b)c + d)(\frac{1}{2}a(b+c) + d) \le p$ if and only if $\frac{1}{2}(a+b)c \cdot \frac{1}{2}a(b+c) + d \le p$ if and only if $\frac{1}{2}ac + d \le p$, where the final equivalence holds since $(a+b)ca(b+c) = ac$. ■

Let f be a fuzzy logic function of n variables. Assume that f is in disjunctive normal form. We now consider the inequality

$$f_1(x_1, x_2, \ldots, x_n) \le p, \tag{1.35}$$

where $f_1 = f$ and $p \in [0,1]$. Then we can express f_1 in the form

$$\begin{aligned} f_1(x_1, x_2, \ldots, x_n) &= g_1(x_2, \ldots, x_n)x_1 + g_2(x_2, \ldots, x_n)x_1\overline{x_1} \\ &\quad + g_3(x_2, \ldots, x_n)\overline{x_1} + g_4(x_2, \ldots, x_n), \end{aligned}$$

where g_j is a fuzzy logic function of $n-1$ variables, $j = 1, 2, 3, 4$. Then by Theorem 1.8.7, inequality (1.35) has a solution if and only if

$$\frac{1}{2}g_1(x_2, \ldots, x_n)g_3(x_2, \ldots, x_n) + g_4(x_2, \ldots, x_n) \le p.$$

Let f_2 be the fuzzy logic function defined as follows:

$$f_2(x_2, \ldots, x_n) = \frac{1}{2}g_1(x_2, \ldots, x_n)g_3(x_2, \ldots, x_n) + g_4(x_2, \ldots, x_n).$$

Suppose f_2 is in disjunctive normal form. Then we consider the inequality $f_2(x_2, \ldots, x_n) \le p$. Hence we see that we can recursively define fuzzy logic functions f_i of $n - i + 1$ variables to obtain the following system of inequalities

$$\begin{aligned} f_1(x_1, x_2, \ldots, x_n) &\le p \\ f_2(x_2, \ldots, x_n) &\le p \\ &\vdots \\ f_i(x_i, \ldots, x_n) &\le p \\ &\vdots \\ f_n(x_n) &\le p \\ f_{n+1} &\le p. \end{aligned} \tag{1.36}$$

Theorem 1.8.8 *The fuzzy logic inequality (1.35) has a solution if and only if the inequality $f_i(x_i, \ldots, x_n) \le p$ of (1.36) has a solution for $i = 1, 2, \ldots, n, n+1$.*

Proof. Suppose that (1.35) has a solution. Then $\exists x_1^*, x_2^*, \ldots, x_n^* \in [0,1]$ such that $f_1(x_1^*, x_2^*, \ldots, x_n^*) \leq p$, i.e., $g_1(x_2^*, \ldots, x_n^*)x_1^* + g_2(x_2^*, \ldots, x_n^*)x_1^*\overline{x_1^*} + g_3(x_2^*, \ldots, x_n^*)\overline{x_1^*} + g_4(x_2^*, \ldots, x_n^*) \leq p$. If we consider the left-hand of this latter inequality as a fuzzy logic function of one variable x_1, then by Theorem 1.8.7,

$$\frac{1}{2}g_1(x_2^*, \ldots, x_n^*)g_3(x_2^*, \ldots, x_n^*) + g_4(x_2^*, \ldots, x_n^*) \leq p,$$

i.e.,

$$f_2(x_2^*, \ldots, x_n^*) \leq p.$$

Continuing in this manner, we reach the desired conclusion. Conversely, suppose $f_i(x_i, \ldots, x_n) \leq p$ has a solution for $i = 1, 2, \ldots, n, n+1$. Then $\exists x_n^* \in [0,1]$ such that $f_n(x_n^*) \leq p$ since $f_{n+1} \leq p$. Since $f_n(x_n^*) \leq p$, $\exists x_{n-1}^* \in [0,1]$ such that $f_n(x_{n-1}^*, x_n^*) \leq p$. By continuing this procedure, we reach the conclusion that $\exists x_1^*, x_2^*, \ldots, x_n^* \in [0,1]$ such that

$$f_1(x_1^*, x_2^*, \ldots, x_n^*) \leq p. \blacksquare$$

Example 1.8.9 *Consider the fuzzy logic inequality*

$$.6x_1x_2 + .8\overline{x_2}\overline{x_1} + .7\overline{x_2} \leq .6. \tag{1.37}$$

Then we can rewrite (1.37) *in the form*

$$(.6x_2)x_1 + (.8\overline{x_2})\overline{x_1} + .7\overline{x_2} \leq .6 \tag{1.38}$$

By Theorem 1.8.7, (1.38) *has a solution if and only if*

$$\frac{1}{2}(.6x_2 \cdot .8\overline{x_2}) + .7\overline{x_2} \leq .6$$

or equivalently if and only if

$$\frac{1}{2}x_2\overline{x_2} + .7\overline{x_2} \leq .6$$

i.e., if and only if

$$0x_2 + .5x_2\overline{x_2} + .7\overline{x_2} + 0 \leq .6. \tag{1.39}$$

By Theorem 1.8.7, (1.39) *has a solution if and only if*

$$\frac{1}{2}(0)(.7) \leq .6. \tag{1.40}$$

Since (1.40) *holds,* (1.39) *has a solution. Here* $a = 0$, $b = .5$, $c = .7$, $d = 0$, *and* $p = .6$. *Since* $a < p$, $a\alpha p = 1$. *Also, since* $c = .7 > p$, $c\alpha p = p = .6$

and so $\overline{c\alpha p} = .4$. *Hence the solution set for* (1.39) *is* $[\overline{c\alpha p}, a\alpha p] = [.4, 1]$. *We can state this as* $x_2 = .4 + p_2$, *where* $p_2 \in [0,1]$. *Hence* (1.38) *becomes*

$$.6(.4+p_2)x_1 + .8\overline{(.4+p_2)}\overline{x_1} + .7(\overline{.4+p_2}) \leq .6$$

or

$$(.4+.6p_2)x_1 + .8(.6\cdot\overline{p_2})\overline{x_1} + .7(\overline{.4+p_2}) \leq .6$$

or

$$(.4+.6p_2)x_1 + .6\overline{p_2}\overline{x_1} + .7(\overline{.4+p_2}) \leq .6.$$

Thus by Theorem 1.8.6 *with* $a = .4 + .6p_2$ *and* $c = .6\overline{p_2}$, *we arrive at the solution set* $[\overline{(.6)\overline{p_2}\alpha(.6)}, (.4+.6p_2)\alpha(.6)]$.

Example 1.8.10 *Consider the fuzzy logic inequality*

$$.6x_1x_2 + .8x_2\overline{x_1} + .7\overline{x_2} \leq .6. \tag{1.41}$$

Then we can rewrite (1.41) *in the form*

$$(.6x_2)x_1 + (.8x_2)\overline{x_1} + .7\overline{x_2} \leq .6. \tag{1.42}$$

By Theorem 1.8.7, (1.42) *has a solution if and only if*

$$\frac{1}{2}(.6x_2 \cdot .8x_2) + .7\overline{x_2} \leq .6$$

or equivalently if and only if

$$.5x_2 + .7\overline{x_2} \leq .6. \tag{1.43}$$

By Theorem 1.8.7, (1.43) *has a solution if and only if*

$$\frac{1}{2}(.5)(.7) \leq .6. \tag{1.44}$$

Since (1.44) *holds,* (1.43) *has a solution and in fact the solution set for* (1.43) *is* $[.4, 1]$. *We can state this as* $x_2 = .4 + p_2$, *where* $p_2 \in [0,1]$. *Hence* (1.42) *becomes*

$$.6(.4+p_2)x_1 + .8(.4+p_2)\overline{x_1} + .7(\overline{.4+p_2}) \leq .6$$

or

$$(.4+.6p_2)x_1 + (.4+.8p_2)\overline{x_1} + .7(\overline{.4+p_2}) \leq .6.$$

Thus by Theorem 1.8.7 *with* $a = .4 + .6p_2$ *and* $c = .4 + .8p_2$, *we arrive at the solution set* $[\overline{(.4+.8p_2)\alpha(.6)}, (.4+.6p_2)\alpha(.6)]$.

1.9 References

1. Chang, C.L., *Symbolic Logic and Mechanical Theorem Proving*, Academic Press, New York, 1973.

2. Cheng Yangjin, Fuzzy logical inequalities, preprint.

3. Dubios, D. and Prade, H., *Fuzzy Sets and Systems: Theory and Applications*, Academic Press Inc., Mathematics in Science and Engineering, Vol. 144, Orlando, Florida, 1980.

4. Kandel, A. and Lee, S.C., *Fuzzy Switching and Automata: Theory and Applications*, Crane Russak and Company, Inc., New York, Edward Arnold (publishers) Ltd., London, 1979.

5. Kandel, A., *Fuzzy Mathematical Techniques with Applications*, Addison-Wesley, 1986.

6. Kaufmann, A., *Introduction to the Theory of Fuzzy Subsets*, Vol 1. Academic Press, Inc. Orlando, Florida, 1973.

7. Klir, G.J., U. St. Clair, U.H., and Yuan, B., *Fuzzy Set Theory, Foundations and Applications,* Prentice Hall, Upper Saddle River, N.J., 1997.

8. Klir, G.J. and Folger, T.A., *Fuzzy Sets, Uncertanity and Information*, Prentice Hall, Englewood Cliffs, N.J., 1988.

9. Klir, G.J. and Yuan, B., *Fuzzy Sets and Fuzzy Logic*, Theory and Applications, Prentice Hall, Upper Saddle River, N.J., 1995

10. Zadeh, L.A., Fuzzy sets. *Information and Control,* 8 (1965) 338-353.

11. Zimmermann, H.-J., *Fuzzy Set Theory and its Applications*, Kluwer Acsdemic Publishers, Boston, 1991.

2

Decision Trees

2.1 Decision Trees

Consider the binary tree of Figure 2.1. This tree gives an algorithm for choosing a bad coin. Beginning at the root and choosing the appropriate edge, we arrive at a terminal vertex that chooses the bad coin. Such a tree is called a **decision tree**. In this section, we use decision trees to specify algorithms and to obtain lower bounds on the worst case time for sorting as well as solving certain coin problems.

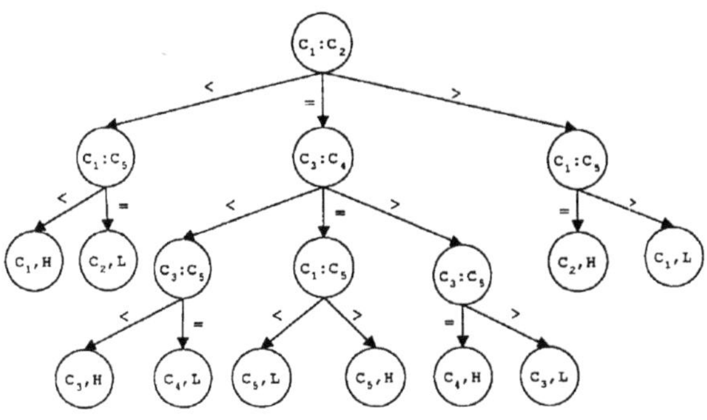

Figure 2.1

Let us review the five coins problem. There are five coins that appear identical. One of the coins is either heavier or lighter than the others. All others weigh the same and a specified coin is given to be a good coin. The problem is to identify the bad coin and determine whether it is heavier or lighter than the others using a pan balance. The pan balance compares the weights of two sets of coins.

Suppose the coins are labeled C_1, C_2, C_3, C_4, C_5 and the coin C_5 is a good coin. An algorithm to solve the five coin problem is given in Figure 2.1 as a decision tree. For all i and j, the labeling $C_i : C_j$ of a node in the tree means that the coin C_i is placed in the left pan and the coin C_j is placed in the right pan. The labeling C_i, L of a node means that the coin C_i is light and the labeling C_i, H means that the coin C_i is heavy. An edge labeled $<$ means that the left side of the pan balance is heavier than the right side, an edge labeled $>$ means that the right side of the pan balance is heavier that the left side and an edge labeled $=$ means that the two sides balance.

We begin at the root with $C_1 : C_2$. If the left side is heavy than the right side, then either C_1 is the heavier coin or C_2 is the lighter coin. Next we compare C_1 with C_5 and immediately determine whether the bad coin is C_1 or C_2 and whether it is heavy or light since we know that C_5 is the good coin. The terminal vertices give the solution. For example, when we compare C_1 with C_5, we follow the edge to the terminal vertex labeled C_2, L, which tells us that the bad coin is C_2 and that it is lighter than the others.

Suppose the worst-case time to solve a coin-weighting problem is defined to be the number of weighings required in the worst case. Then the worst-case time can be determined from the decision tree and the worst-case time is equal to the height of the tree. Since the height of the decision tree of Figure 2.1 is 3, the worst-case time for this algorithm is equal to 3.

We can use decision trees to show that the algorithm given in Figure 2.1 to solve the five-coin puzzle has worst-case time less than 3.

Suppose that there is an algorithm that solves the five-coins puzzle in the worst case in two or fewer weighings. The algorithm can be described by a decision tree. Since the worst-case time is 2 or less, the height of the decision tree is two or less. Now each internal vertex has at most three children and so such a tree can have at most nine terminal vertices. This can be seen from Figure 2.2.

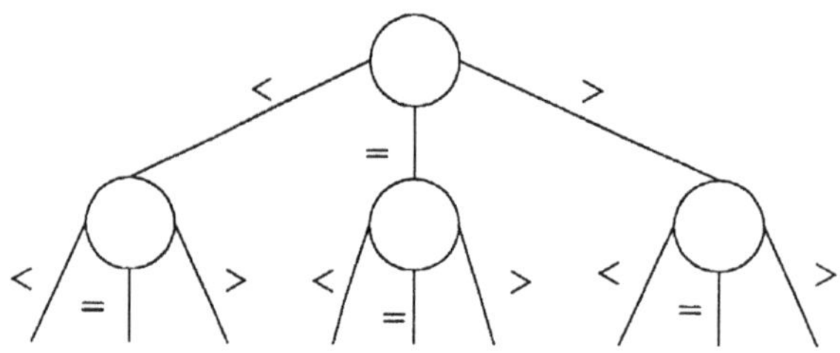

Figure 2.2

Since terminal vertices correspond to possible outcomes and the decision tree of height 2 or less, the tree represents at most nine outcomes. However the five-coins puzzle has 10 outcomes.

$$\begin{array}{lllll} C_1, L, & C_1, H, & C_2, L, & C_2, H, & C_3, L, \\ C_3, H, & C_4, L, & C_4, H, & C_5, L, & C_5, H. \end{array}$$

This is impossible. Hence no algorithm that solves the five-coins puzzle has worst-case time less than 3, and the algorithm of Figure 2.1 is optimal.

We now see how a decision tree can be used to give a lower bound for the worst-case time to solve a problem. Sometimes, the lower bound cannot be attained.

Consider the four-coins puzzle. The rules are the same as for the five -coins puzzle except that the number of coins is reduced by one. Since there are now eight outcomes rather than 10, it follows that any algorithm to solve the four-coins puzzle requires at least two weighings in the worst case. For this puzzle, we cannot conclude that at least three weighings are required in the worst case. However, it can be seen that three weighings are required.

The first weighing either compares two coins against two coins or one coin against one coin. If we begin by comparing two coins against two coins, the decision tree can account for at most six outcomes as can be seen by Figure 2.3.

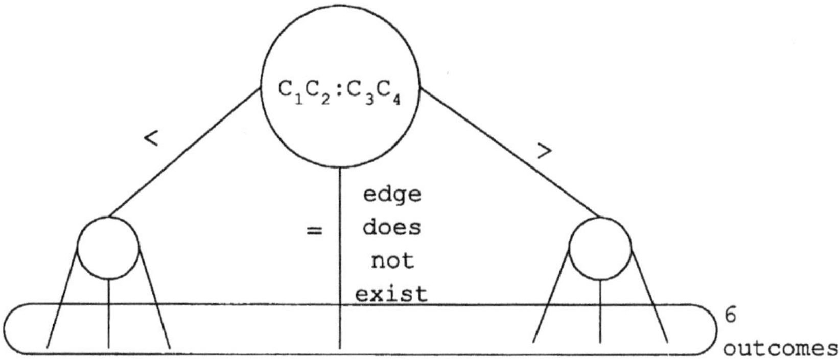

Figure 2.3

Since there are eight outcomes, no algorithm that begins by comparing two coins against two coins can solve the problem in two weighings or less in the worst case. If we begin by comparing one coin against one coin and the coins balance, the decision tree can account for only three outcomes. Four outcomes are possible after identifying two good coins and so no algorithm that begins by comparing one coin against one coin can solve the problem in two weighings or less in the worst case. Therefore, any algorithm that solves the four-coin puzzle requires at least three weighings in the worst case.

2.2 Fuzzy Decision Tree Algorithms

We discuss certain theoretical aspects of fuzzy decision trees and their applications. The results of this section are from [5] and [7]. We present a branch-bound-backtrack algorithm which has an effective backtracking mechanism leading to the optimal solution while still requiring usually only $O(\log n)$ time, where n is the number of decision classes. This is obtained via pruning subtrees unlikely to be traversed and by installing tree-transversal pointers.

The decision tree is a well known technique used in pattern recognition for making classification decisions. Its main virtue lies in the fact that a large number of classes can be maintained while at the same time minimizing the time for making the final decision by a series of small local decisions. In fact, very often it takes only $O(\log n)$ time to reach one of the n possible classes, since a tree of n leaves usually has depth $\log n$.

We assume that the reader is familiar with the definition of a tree, node, branch, son, father, descendant, root, leaf, subtree, path, and path length [9, Section 2.3]. The fuzzy algebra used is similar to those given in [3, 15]. A systematic treatment can be found in [4].

Definition 2.2.1 *A **fuzzy decision function** f_x at node x is a function $f_x : X \to [0,1]^k$, where k is an integer ≥ 2, X is the input and the k- tuple is the labels (decision $l_i = l(x_i)$ of the outgoing branches $(x, x_i), i = 1, \ldots, k$, where x_i is the ith son of the node x. A $0-1$ **decision function** is a fuzzy decision function that can assume only integer values $0, 1$: $f_x : X \to \{0,1\}^k$, with exactly one element of the k-tuple equal to 1.*

The input in Definition 2.2.1 could be a digitized picture or a voice spectograph.

Definition 2.2.2 *A **decision tree** T_r is a tree with root r such that each nonleaf node i has a corresponding k-tuple decision function f_i and k ordered sons $i_1, \ldots, i_k$. A **fuzzy decision tree** is one with fuzzy decision functions, and a binary decision tree is one with $0-1$ decision functions. A complete (or, completely balanced) k-ary decision tree is a decision tree in which the decision function at each nonleaf node is k-ary and each path from the root to a leaf has the same length.*

Definition 2.2.3 *The **decision path** (x, y) is the path of the decision tree from node x to node y. The decision path $\bar{x}$ is the decision path $(root, x)$. The decision path $(x_0, x_1, \ldots, x_n)$ is the path from node x_0 to node x_n via nodes x_i, x_{i+1}, $i = 0, 1, \ldots, n-1$.*

A complicated decision represented by a decision tree is always made up of several simple decisions, each represented by a node in the tree.

It is realistic to assume that the time required for making a k-tuple decision (i.e., for evaluating a k-tuple decision function) is of $O(k)$.

Definition 2.2.4 *The (**decision**) **value** $L(x)$ of a decision path x is*

$$L(x) = \prod_{y \in path\ x} l(y),$$

where $\prod$ denotes a suitable 'product' of the decision values (labels) of the branches composing it.

In Definition 2.2.4, $\prod$ is product in the prob model or is the minimum function defined as in the max-min model.

It can be shown [7] that given the same decision tree T with the same decision values, the $0-1$, fuzzy max-min, and fuzzy prob criteria may all lead to different decisions.

Definition 2.2.5 *Two decision trees are **equivalent** if they give the same decisions.*

It can be shown that the max-min decision tree and the prob decision tree are mutually reducible to one another in $O(n)$ time, where n is the number of decision classes.

Theorem 2.2.6 (4) *There exists a max-min decision tree T_1 if and only if there exists an equivalent prob decision tree of the same structure (but different labels). Furthermore, they are mutually reducible in linear time with respect to the number of decision classes.* ∎

We refer the reader to [2] for the top-down decision algorithm for $0-1$ decision trees and to [10] for more insight on the subject. The top down algorithm looks very attractive since it has good time complexity $O(\log n)$. However, it is not, in general, applicable to any fuzzy decision tree.

Consider the example in Figure 2.4.

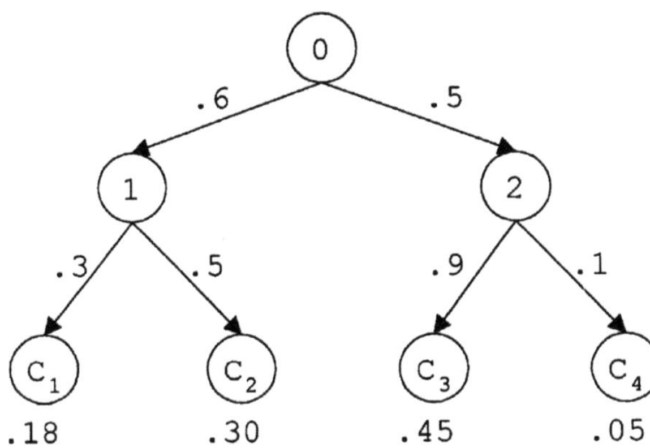

Figure 2.4

If at each decision node we pick the highest value branch, then we will end up with class C_2. Clearly, class C_3 is a better choice, since the path to class C_3 has decision value $0.45 > 0.3$ value of the path to class C_2. There is a guaranteed bottom-up algorithm for general fuzzy decision trees. The mechanism is essentially the same as the well-known tournament method for finding the maximum of n numbers, but it has a time complexity $O(n)$.

It follows that $O(n)$ is the theoretical least upper bound of the time complexity of a general fuzzy decision algorithm. Consider a complete tree whose branches are all labeled 0.1 except one which is labeled 0.9 at the bottom. Then the decision values of all paths are $(0.1)^d$ except for one which is $(0.1)^{d-1}(0.9) > (0.1)^d$. However, this 0.9 label may be at any of the n branches at the bottom. Hence all n branches must be examined before the best decision can be found, taking $O(n)$ time.

The top-down algorithm is efficient, but may cause errors when applied to fuzzy decision trees. The bottom-up algorithm is general enough, but not efficient. We will present now an algorithm for fuzzy decision trees whose complexity is usually $O(\log n)$, but still gives the optimal solution. We call it the branch-bound-backtrack algorithm (BBB), since it belongs to the family of branch-and-bound methods [11].

We link the sons $(x_1, \ldots, x_k)$ of each node x from left to right in descending order $(x_{i_1}, \ldots, x_{i_k})$, where $i_1, \ldots, i_k$ is a sorting of $1, \ldots, k$, of their decision function values $f_x(I)$, i.e., $l(x_{i_1}) \geq \ldots \geq l(x_{i_k})$. Descendants of each node are linked downwards.

Let $L(x)$ be the label value of x designating the decision value of path bottom up to this point.

Definition 2.2.7 *A node x is called a* ***younger brother*** *of w if x and w have the same father and $L(w) \geq L(x)$.*

The following theorem presents the idea behind the algorithm.

Theorem 2.2.8 *Let x be a node on the right side of the path from r to y in tree T_r. Then there exists some subtree T_x (by construction) such that there exists a path to the right of or below node x with decision value $> L(y)$ if and only if $L(x) > L(y)$.*

Proof. Since x is on the right of the path from r to y, there exists a node w which is an elder brother of x and an ancestor of y.

Suppose that there exists a subtree T_x such that there exists a path to the right of or below node x with decision value $> L(y)$. Let z be the node to which this path is defined. Then z is a descendent of x, or a younger brother of x, or a descendent of x's younger brother. First suppose that z is a descendent of x. Then $L(z) = L(x) \cdot (\text{value of the path}(x, z)) \leq L(x)$. Hence $L(x) > L(y)$. If z is younger brother of x, then by definition, $L(x) \geq L(z)$ and so $L(x) > L(y)$. Suppose that z is a descendent of x's younger brother, say x'. Then as before $L(z) \leq L(x')$. Also $L(x') \leq L(x)$ and so $L(x) > L(y)$.

Conversely, suppose that $L(x) > L(y)$. Let z be a child of x (by construction) such that (value of the path (x, z)) $l(z, x) = 1$. Then $L(z) =$

$L(x) \cdot 1 = L(x)$. Hence $L(z) = L(x) > L(y)$. ■

BBB Algorithm[5, 7]

Goal: To find the best decision path and its corresponding decision value for a given decision tree T_r (with unknown input).

Input: I, T_r

Output: A leaf x defining a best decision path with the best decision value $L(x)$.

Data-Structure: T_r is represented by an array of nodes, each node x consisting of four fields, including one attribute and three links:

(1) $l(x)$ is the label (decision value) of the branch connecting the father of x to x.

(2) $FT(x)$ points to the father x.

(3) $ES(x)$ points to the eldest son of x.

(4) $YB(x)$ points to the next younger brother of x.

A null pointer is indicated by a 0 and so $FT(r) = 0, YB(r) = 0, L(r) = 1$. The procedure is

$$x \leftarrow BBB(I, r, 0),$$

where $BBB(I, r, d)$ is a recursive function. Its goal is to give a leaf defining a best decision path for a subtree T_r with decision value $> d$.

The BBB algorithm is:

(1) if r is a leaf, then return $BBB \leftarrow r$; $x \leftarrow r$.

(2) $(l_1, \ldots, l_k) \leftarrow f_x(I)$ (assuming $l_1 \geq \ldots \geq l_k$).

(3) $l \leftarrow L(x)l_1$;
    if $l \leq d$ then do
        $L(x) \leftarrow 0$;
        $FT(x_1) \leftarrow x$;
        $x \leftarrow x_1$;
        $y \leftarrow 0$;
        go to (6);
    end;
    $ES(x) \leftarrow x_1$;
    $FT(x_1) \leftarrow x$;
    $L(x_1) \leftarrow l$;
    if x_1 is a leaf then go to (4).
    For $j = 2, \ldots, k$ do
        $l \leftarrow L(x)l_j$;
        if $l \leq d$ then do
            $YB(x_{j-1}) \leftarrow 0$;
            go to (4);
        end;
        $FT(x_j) \leftarrow x$;
        $YB(x_{j-1}) \leftarrow x_j$;

```
        L(x_j) ← l;
        if x_j is a leaf then do
          YB(x_j) ← 0;
          go to (4);
        end;
      end;  //end for
      YB(x_k) ← 0
  (4) x ← ES(x);
      if x is not a leaf then go to (2).
  (5) y ← x.
  (6) If FT(x) = r, then return BBB ← y; x ← FT(x).
  (7) If YB(x) = 0, then go to (6);
      x ← YB(x);
      if (y != 0)
        L(x) ≤ L(y), then go to (6);
      else
        L(x) <= d, then go to (6);   //d is the formal parameter
  (8)  if (y != 0)
        p ← BBB(I, x, L(y));
      else
        p ← BBB(I, x, d);   //d is the formal parameter
      if p ≠ 0, then y ← p;
      YB(x) = 0;
      go to (7).
```

Next we give some examples to show how the algorithm works.

Example 2.2.9 *Consider the decision tree with only the root node.*

Number of nodes = 1
BBBLeaf = 0
Path = 0
Decision Value = 1.000

Example 2.2.10 *Consider the decision tree of Figure 2.5.*

Number of nodes = 4
BBBLeaf = 1
Path = 0-1
Decision Value = 0.600

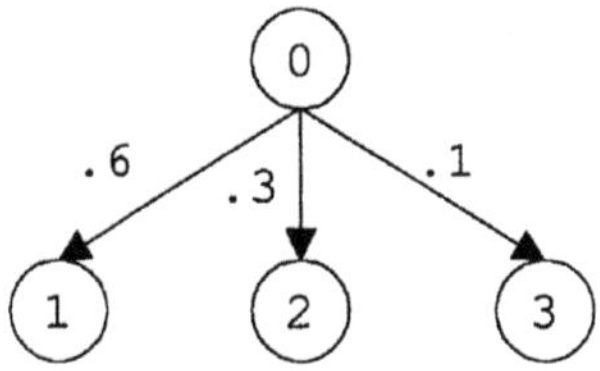

Figure 2.5

Example 2.2.11 *Consider the decision tree of Figure 2.6.*
Number of nodes = 12
Call No. 1: $x = 0$ (root), $dv = 0$.
Call No. 2: $x = 5$, $L(y) = 0.021$
Call No. 3: $x = 2$, $L(y) = 0.140$
Call No. 4: $x = 3$, $L(y) = 0.140$
BBBLeaf = 8
Path = 0-3-8
Decision Value = 0.200

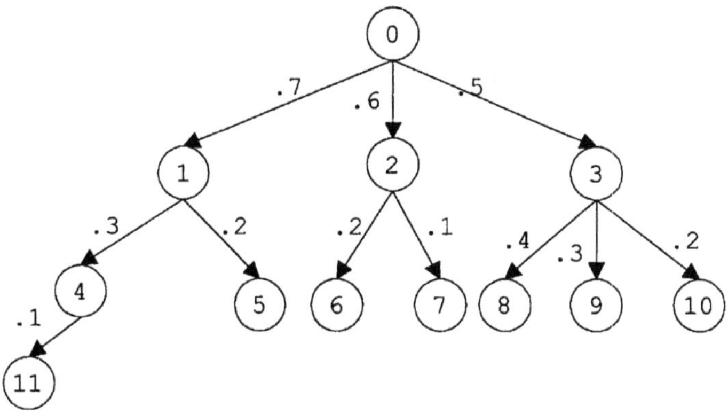

Figure 2.6

Example 2.2.12 *Consider the decision tree of Figure 2.7.*
Number of nodes = 30
Call No. 1: $x = 0$ (root), $dv = 0$.
Call No. 2: $x = 2$, $L(y) = 0.405$
Call No. 3: $x = 3$, $L(y) = 0.560$
Call No. 4: $x = 4$, $L(y) = 0.560$
BBBLeaf = 8
Path = 0-2-8
Decision Value = 0.560

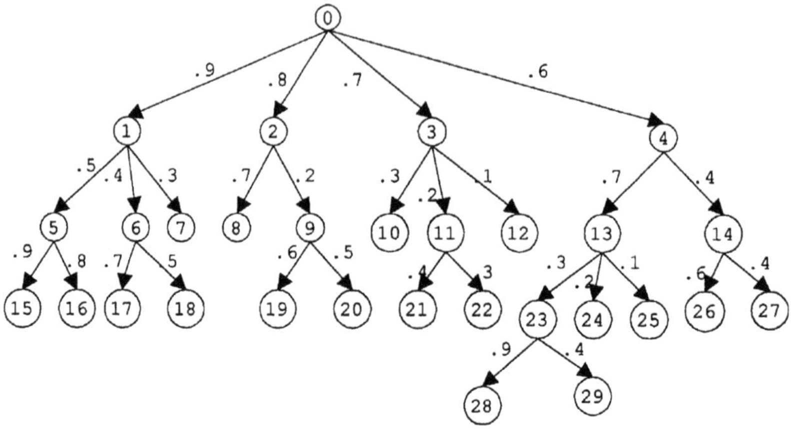

Figure 2.7

Example 2.2.13 *Consider the decision tree of Figure 2.8.*

Number of nodes = 31
Call No. 1: $x = 0$ (root), $dv = 0$.
Call No. 2: $x = 7$, $L(y) = 0.072$
Call No. 3: $x = 2$, $L(y) = 0.270$
Call No. 4: $x = 3$, $L(y) = 0.270$
Call No. 5: $x = 4$, $L(y) = 0.454$
BBBLeaf = 26
Path = 0-3-11-21-26
Decision Value = 0.454

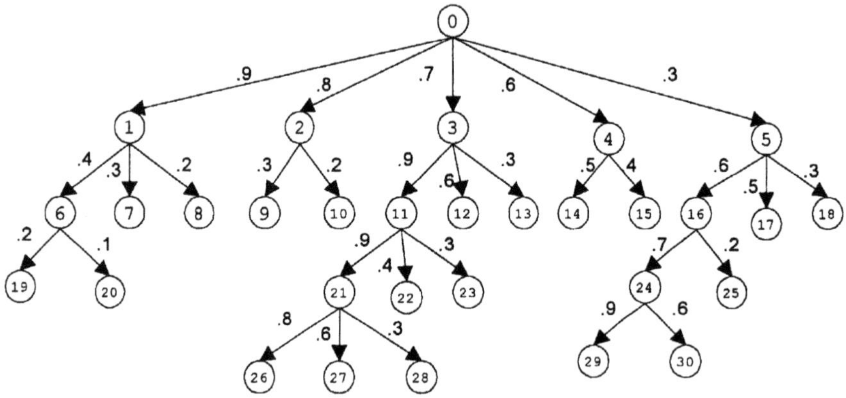

Figure 2.8

Example 2.2.14 *Consider the decision tree of Figure 2.9.*

Number of nodes = 31

Call No. 1: x = 0 (root), dv = 0.
Call No. 2: x = 2, L(y) = 0.729
BBBLeaf = 19
Path = 0-1-6-19
Decision Value = 0.729

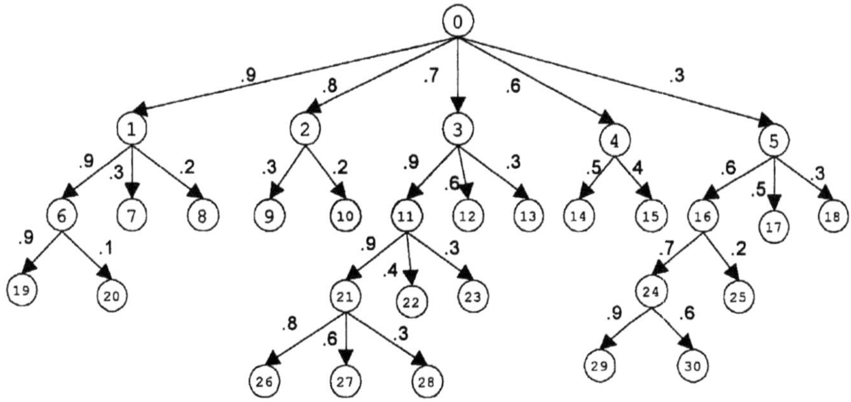

Figure 2.9

Example 2.2.15 *Consider the decision tree of Figure 2.10.*
Number of nodes = 15
Call No. 1: x = 0 (root), dv = 0.
Call No. 2: x = 2, L(y) = 0.270
Call No. 3: x = 7, L(y) = 0.270
BBBLeaf = 7
Path = 0-2-7
Decision Value = 0.640

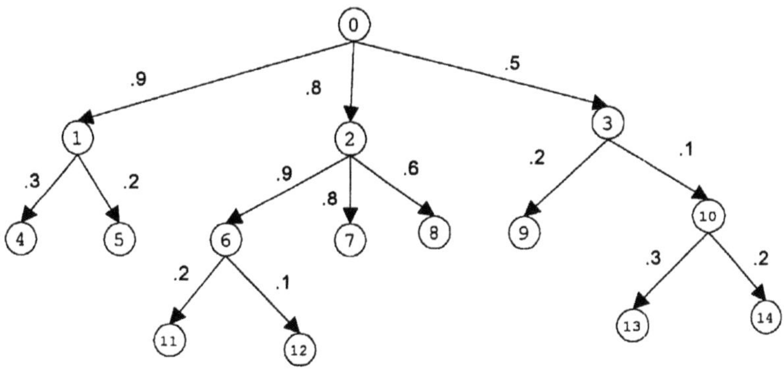

Figure 2.10

Example 2.2.16 *Consider the decision tree of Figure 2.11.*
Number of nodes = 27
Call No. 1: x = 0 (root), dv = 0.
Call No. 2: x = 2, L(y) = 0.180
Call No. 3: x = 3, L(y) = 0.320
Call No. 4: x = 16, L(y) = 0.320
Call No. 5: x = 10, L(y) = 0.340
Call No. 6: x = 4, L(y) = 0.340
BBBLeaf = 25
Path = 0-3-9-16-23-25
Decision Value = 0.340

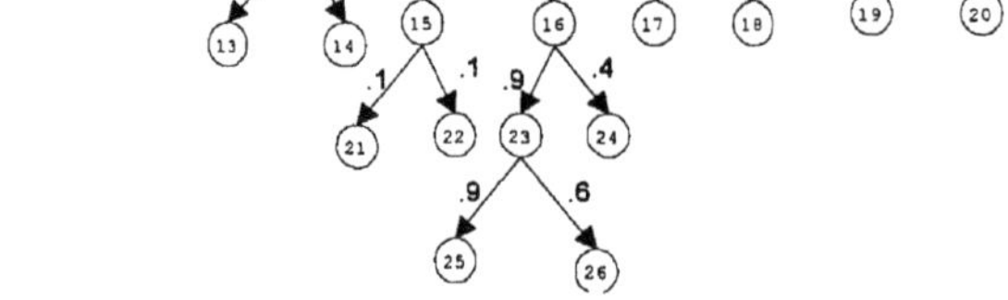

Figure 2.11

Remark 2.2.17 *A complete programming code of the above BBB algorithm in the programming language C++ is given in the appendix.*

Theorem 2.2.6 is the main idea behind the *BBB* algorithm for the prob model. It can be also applied for the max-min model as well: let x be a node with father t, descendent z_2, younger brother z_1 and a younger brother's descendent z_3 (see Figure 2.12 which displays pruning subtrees for the max-min model).

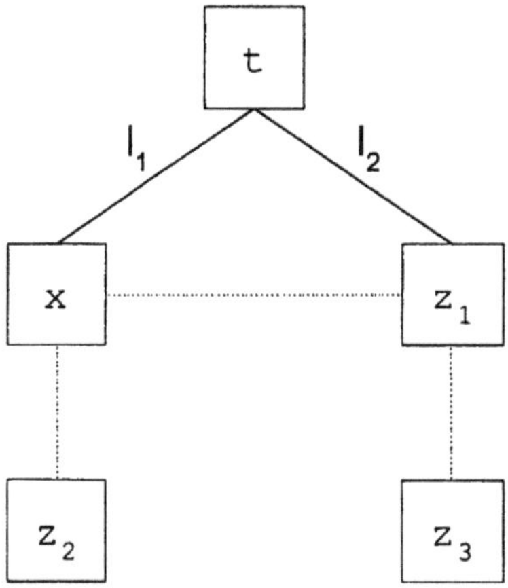

Figure 2.12

For the max-min model, the inequalities $L(z_2) \leq L(x)$ and $L(z_3) \leq L(z_1)$ must hold since $\min\{A, B\} \leq \min\{A\}$ for any two sets A and B. Let $L(t) = \min\{C\}$. Since by convention $l_1 \geq l_2$,

$$L(z_1) = \min\{C, l_2\} \leq \min\{C, l_1\} = L(x).$$

Thus $L(z_1)$, $L(z_2)$, $L(z_3) \leq L(x)$ and so the BBB algorithm can be applied to a fuzzy decision tree for the max-min model as well.

2.3 Analysis of the BBB Algorithm

We now give an analysis of the BBB algorithm. We assume that the decision tree T_r is finite.

We first consider the correctness of the algorithm.

Theorem 2.3.1 *The BBB algorithm given above is correct.*

1. Termination: The main procedure has a finite number of steps. Since every decision function is evaluated at most once, $BBB(I, r, d)$ terminates.
2. Completeness: By examining all the exits of BBB, some value is always returned when exiting.
3. Validity: (Every output is the output intended.) Note that BBB can exit only in steps 1 and 6. We prove validity by induction on the number of times BBB is recursively executed.

Basis: BBB exits in step 1. Then r is the only possible node.

Induction Step: The proof is broken into the following lemmas.

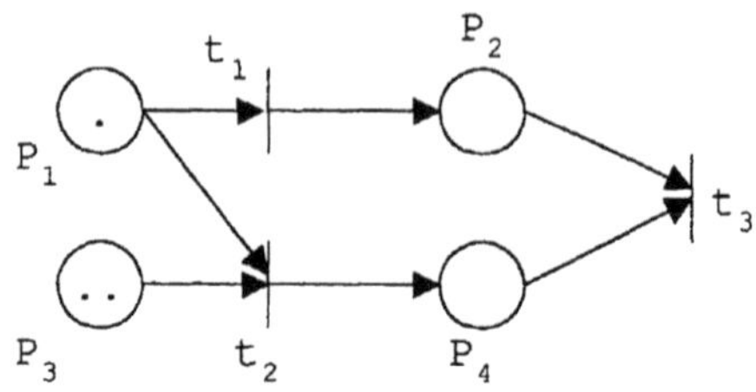

Figure 3.14

In modeling, the places represent **conditions**, the transitions represent **events**, and the presence of at least one token in a place (condition) indicates that condition is met.

Example 3.8.5 *Consider the following computer program:*
$A = 1$
$B = 2$
$A = A + 1$
$B = A + B$
It is modeled in Figure 3.15. Here the events (transitions) are the instructions, and the places represent the conditions under which an instruction can be executed.

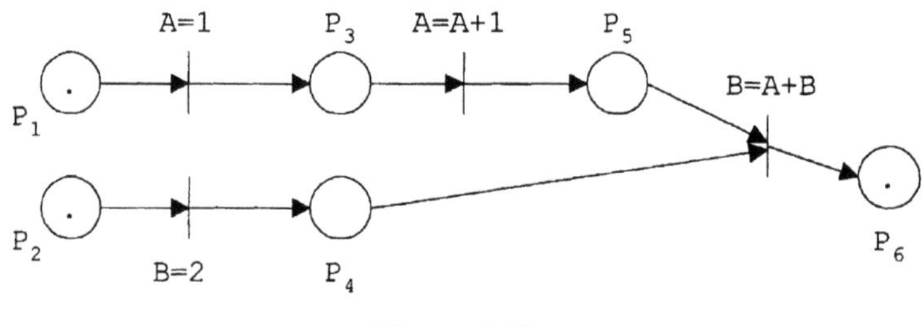

Figure 3.15

Definition 3.8.6 *If an edge in a Petri net is directed from place p to transition t, we say that p is an* ***input place*** *for transition t. An* ***output place*** *is defined similarly. If every input place for a transition t has at least one token, we say that t is* ***enabled.*** *A* ***firing*** *of an enabled transition removes one token from each input place and adds one token to each output place.*

Example 3.8.7 *In the Petri net of Figure 3.14, places p_1 and p_3 are input places for transition t_2. Transitions t_1 and t_2 are enabled, but transition t_3 is not enabled. If we fire transition t_2, we obtain the marked Petri net of*

Lemma 2.3.2 *For all x, if the decision value is $\leq d$, or undefined, or every son has a decision value $\leq d$, then $L(x) = 0$. If $L(x) \neq 0$, then $L(x) =$ the decision value.*

Proof. The proof is by induction on the length of the path to x, which is previously x_j for some j. Step 3 gives $L(x_j) \leftarrow L(x) \cdot l_j$ whenever a decision function is evaluated with $L(x_j) > d$. ■

Lemma 2.3.3 *$FT(x)$, $YB(x)$, $ES(x)$ are correct links.*

Proof. Consider step 2. Whenever a decision function is evaluated, all connecting links to and among sons are set up correctly. ■

Lemma 2.3.4 *Any path left of the path to y (the y in step 5) must have a decision value $\leq$ that of the path to y. (Note that initial portions of these paths may coincide.)*

Proof. Initially, y is the left most path. (See step 5.) Now y may be modified only in step 8, in which case we have that $p = BBB(I, x, L(y)) \neq 0$. By the induction hypothesis of the main correctness theorem, we have that $BBB(I, x, L(y))$ is executed correctly since $T_x \leq T_r$ x being a descendent of r. Thus $p \neq 0$ means $L(p) < L(y)$. ■

Lemma 2.3.5 *Any path to the right of the path to y must have a decision value $\leq d$ or $\leq$ that of y, else y will be modified subsequently.*

Proof. Let these two paths be defined by nodes z and y, respectively. Since the path to z is on the right of the path to y there must be some node w in both paths such that $YB(w)$ is in the path to z, but not in the path to y. Now $L(YB(w)) < L(w)$ by the sorting step 2. Furthermore, by the property of fuzzy decision functions, $0 \leq l_j \leq 1$ $\forall j$, and hence the label of any descendant must be equal or smaller by multiplication. Therefore, $L(z) \leq L(w)$. Hence nothing is lost by pruning the tree in step 3. ($YB(x_i) \leftarrow 0$ prunes the tree.) All other nodes are eventually traversed and labeled by the backtracking in steps 6-8 and recursion.

This completes the proof of the induction step of the main correctness theorem. ■

We next consider the complexity of the BBB algorithm.

Space required: Since each node require constant space, the total space require to store the tree is of $O(n)$.

Time Required (Worst Case): The whole tree is traversed. The main procedure takes constant time. We prove $BBB(I, r, d)$ takes $O(n)$ time by induction on the number of nodes n of T_r .

Basis: Root r is a leaf. Time required is constant.

Induction step: First consider steps 1-4 of the procedure. Now step 1 takes constant time. Step 2 is reachable only from step 1 or step 4 (recursion in step 8 is considered separately). It is reached only once from step 1. Thereafter step 2 is executed at most once for each new x (by $x \leftarrow ES(x)$, which makes x go downwards until x becomes a leaf). Clearly, each iteration takes constant time.

Step 3 takes at most constant time for each son x_{i_j}. Each node can be a son at most once. Each of step 4 and 5 is traversed at most once for each node.

Consider steps 6–8. Now x either goes up by $x \leftarrow FT(x)$, or goes right by $x \leftarrow YB(x)$, or both after each iteration. Each iteration takes constant time. Therefore total time is $O(n)$. By the induction hypothesis, $BBB(I, x, L(y))$ in step 8 takes O(number of nodes in T_x) time, where T_x is a smaller tree and all nodes in T_x have not been traversed yet. Consequently, the whole algorithm takes $O(n)$ time.

Time Required (Best Case) : T_r is a $0-1$ decision tree, for example. Here, BBB algorithm reduces to the top-down $0-1$ tree algorithm of time complexity $O(\log n)$ since only the depth-first evaluation part of the algorithm, and not the backtracking part, is executed. This is often the case in practice.

The most interesting aspect of the BBB algorithm is that it gives the best possible time complexity for both the worst and best cases considered above.

A comparison with other decision making techniques can be found in [5].

2.4 References

1. Adamo, J.H., Fuzzy decision trees, *Fuzzy Sets and Systems*, 4 (1980) 207-219.

2. Aho, A.V., Hopcroft, J.E, and Ullman, J.D., *The Design and Analysis of Computer Algorithms,* Addison-Wesley, Reading, MA, 1974.

3. Brown, J.G., A note on fuzzy sets, *Info. and Control,* 18 (1971) 32-39.

4. Chang, R.L.P., Application of fuzzy decision techniques to pattern recognition and curve fitting, Ph.D thesis, Dept. of EECS, Princeton Univ., 1976.

5. Chang, R.L.P. and Pavlidis, T., Fuzzy decision tree algorithms, *IEEE Trans. System Man Cybernet,* 7 (1977) 20-35.

6. Duda, R.O. and Hart, P.E., *Pattern Classification and Scene Analysis,* Wiley, New York, 1973.

7. Friedman, M., Bar-noy, T., Blau, M., and Kandel, A., Certain computational aspects of fuzzy decision trees, *Fuzzy Sets and Systems*, 28 (1988) 163-170.

8. Harary, F., *Graph Theory*, Addison Wesley, Reading, MA, 1971.

9. Knuth, D.E., *The Art of Computer Programming*, Vol.1, Addison Wesley, Reading, MA, 1969.

10. Knuth, D.E., Optimum binary search trees, *Acta Informatica*, 1 (1971) 14-25.

11. Lawler, E.L. and Wood, D.E., Branch- and-bound methods: A survey, *Operations Research*, 14 (1966) 699-719.

12. Pavlidis, T. and Ali, F., Computer recognition of hand written numerals by polygonal approximations, *IEEE Trans. Systems, Man, Cybern.*, 5 (1975) 610-614.

13. Siy, P. and Chen, C.S., Fuzzy logic approach to hand written character recognition problem, *Proc. IEEE Conf. Systems, Man, Cybern., Anaheim, Calif.*, 1971, pp. 113-117.

14. Zadeh, L.A., Outline of a new approach to the analysis of complex systems and decision processes, *IEEE Trans. Systems, Man, Cybern.*, 3 (1973) 28-44.

15. Zadeh, L.A., A fuzzy-algorithimic approach to the definition of complex or imprecise concepts, Memorandum ERL-M474, Electronic Research Lab., College of Eng., U.C. Berkeley, 1974.

3

Networks

3.1 Network Models

A **graph** (or **undirected graph**) G is a pair $G = (V, E)$, where V is a nonempty set and $E \subseteq V \times V$ with E symmetric. V is called the set of **vertices** or **nodes** and E is called the set of **edges** or **arcs**. Each edge $e \in E$ is associated with an unordered pair of vertices. Suppose e is a unique edge associated with the vertices i and j. Then we write $e = (i, j)$ or $e = (j, i)$.

A **directed graph** (or **digraph**) G is a pair $G = (V, E)$, where V is a nonempty set and $E \subseteq V \times V$. V is called the set of vertices or nodes and E is called the set of edges or arcs. Each edge $e \in E$ is associated with an ordered pair of vertices. Suppose e is a unique edge associated with the ordered pair (i, j). Then we write $e = (i, j)$.

Let $G = (V, E)$ be a graph directed or undirected. Let e be an edge associated with the pair of vertices i and j. Then we say that the edge e is **incident** on i and j, i and j are **incident** on e, and i and j are **adjacent vertices**.

Let $G = (V, E)$ be a graph. If there are distinct edges e_1 and e_2 associated with the vertex pair $\{i, j\}$, then e_1 and e_2 are called **parallel edges**. An edge incident on a single vertex is called a **loop**. A graph with neither loops nor parallel edges is called a **simple graph**. A **path** in a graph is a sequence of vertices $v_0, v_1, \ldots, v_n$ such that $v_i \neq v_{i+1}$ and $(v_i, v_{i+1}) \in E$, $i = 0, 1, \ldots, n-1$.

A weighted graph G is a 3-tuple, $G = (V, E, C)$, (V, E) is a graph and $C : E \to \mathbb{R}$. For each $e = (i, j) \in E$, $C(i, j) = C_{ij}$ is called the **weight of edge** (i, j).

Definition 3.1.1 *A* ***transport network*** *(or simply* ***network****) is a simple, weighted directed graph $G = (V, E, C)$ satisfying:*

(i) A designated vertex, called the **source**, *has no incoming edges.*

(ii) A designated vertex, called the **sink**, *has no outgoing edges.*

(iii) The weight C_{ij} of the directed edge (i, j), called the **capacity** *of (i, j), is a nonnegative number.*

Definition 3.1.2 *Let G be a transport network. Let C_{ij} denote the capacity of the directed edge (i, j). A* ***flow*** *F in G is a function $F : E \to \mathbb{R}$ such that $\forall (i, j) \in E$, $F(i, j) = F_{ij} \geq 0$,*

(i) $F_{ij} \leq C_{ij}$,

(ii) for each vertex j, which is neither the source nor the sink,

$$\sum_{i \in V} F_{ij} = \sum_{i \in V} F_{ji}. \tag{3.1}$$

F_{ij} is called the ***flow in edge*** *(i, j). If (i, j) is not an edge, we set $F_{ij} = 0$.*

In what follows, when we write $\sum_i F_{ij}$, the sum is assumed to be taken over all vertices.

For any vertex j,

$$\sum_i F_{ij}$$

is called the **flow into** j and

$$\sum_i F_{ji}$$

is called the **flow out of** j.

The property given by equation (3.1) is called **conservation of flow**.

Theorem 3.1.3 *Let $G = (V, E, C)$ be a network with source a and sink z. If F is a flow in G, then the flow out of the source a equals the flow into the sink z, i.e.,*

$$\sum_i F_{ai} = \sum_i F_{iz}.$$

Proof. We have

$$\sum_{j \in V} \sum_{i \in V} F_{ij} = \sum_{j \in V} \sum_{i \in V} F_{ji}$$

since each double sum is

$$\sum_{e \in E} F_e.$$

Now

$$\begin{aligned} 0 &= \textstyle\sum_{j\in V}(\sum_{i\in V} F_{ij} - \sum_{i\in V} F_{ji}) \\ &= (\textstyle\sum_{i\in V} F_{iz} - \sum_{i\in V} F_{zi}) + (\sum_{i\in V} F_{ia} - \sum_{i\in V} F_{ai}) + \\ &\quad \textstyle\sum_{j\in V, j\neq a,z}(\sum_{i\in V} F_{ij} - \sum_{i\in V} F_{ji}) \\ &= \textstyle\sum_{i\in V} F_{iz} - \sum_{i\in V} F_{ai} \end{aligned}$$

since $F_{zi} = 0 = F_{ia}$ $\forall i \in V$, and since by Definition 3.1.2(ii),

$$\sum_{i\in V} F_{ij} - \sum_{i\in V} F_{ji} = 0 \text{ if } j \in V\backslash\{a, z\}. \blacksquare$$

Theorem 3.1.3 motivates the following definition.

Definition 3.1.4 *Let F be a flow in a network G. The value*

$$\sum_i F_{ai} = \sum_i F_{iz}$$

is called the ***value of the flow*** *F.*

3.2 A Maximum Flow Algorithm

Definition 3.2.1 *Let G be a transport network. A* ***maximal flow*** *in G is a flow with maximum value.*

In general, there may be more than one flow having the same maximum value. In this section, we give an algorithm for finding a maximal flow. The basic idea of the algorithm is to start with some initial flow, e.g., the flow in each edge is equal to zero and iteratively increase the value of the flow until no more improvement is possible. The resulting flow will then be a maximal flow.

To increase the value of a given flow, a path from the source to the sink is found for which the flow along this path can be increased.

Throughout this section, G denotes a network with source a, sink z, and capacity C. Let us consider the edges of G to be undirected and let

$$P = (v_0, v_1, \ldots, v_n), \;\; v_0 = a, \;\; v_n = z$$

be a path from a to z in this undirected graph. (All paths in this section are with reference to the underlying undirected graph.) If an edge e in P is directed from v_{i-1} to v_i, we say that e is **properly oriented with respect**

to P; otherwise, we say that e is **improperly oriented with respect to** P.

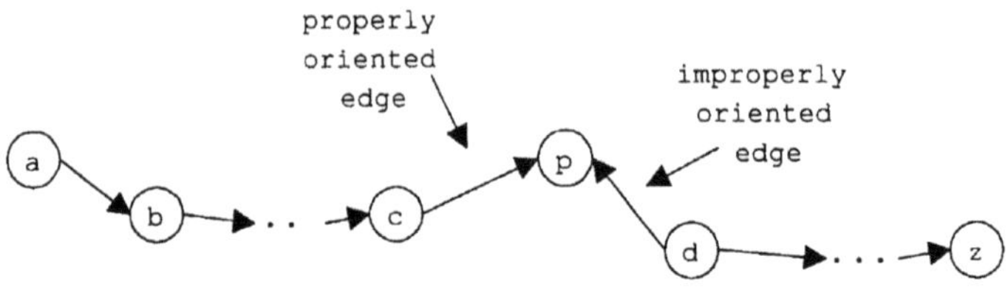

Figure 3.1

Let P be the path from a to z as shown in Figure 3.1. The edge (c, p) is properly oriented, while the edge (d, p) is improperly oriented on the path P.

If we can find a path P from the source to the sink in which every edge in P is properly oriented and the flow in each edge is less than the capacity of the edge, it is possible to increase the value of the flow.

Example 3.2.2 *Let P be the path from a to z in Figure 3.2. Since all edges in P are properly oriented, the value of the flow in this network can be increased by* 1. *The resulting figure is Figure 3.3.*

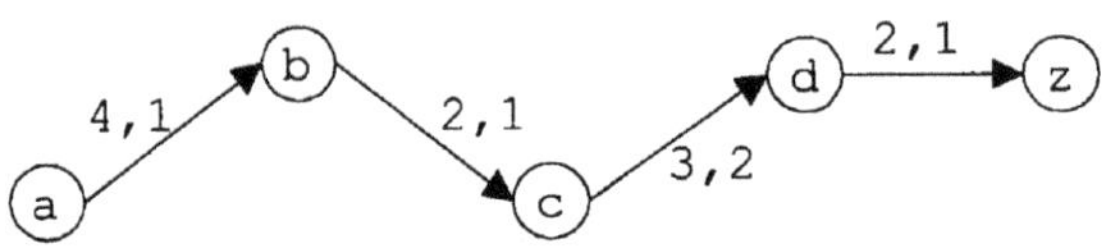

Figure 3.2

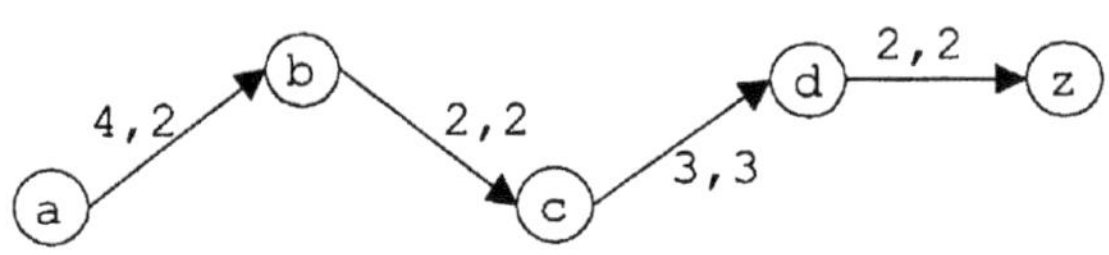

Figure 3.3

It is possible to increase the flow in certain paths from the source to the sink in which we have properly and improperly oriented edges. Let P be a path from a to z and let x be a vertex in P such that $x \neq a$ and $x \neq z$ as shown in Figure 3.4.

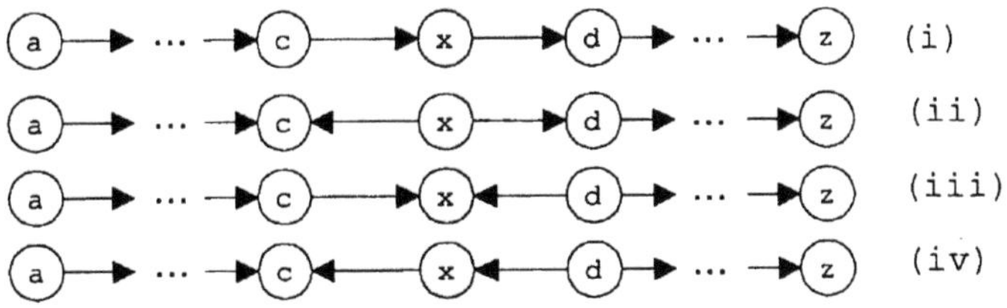

Figure 3.4

There are four possibilities for the orientations of the edges (c, x) and (x, d) incident on x.

Case (a): Both edges are properly oriented (Figure 3.4(i)). In this case, if we increase the flow in each edge by Δ, the flow into x will still equal the flow out of x.

Case (b): Edge (c, x) is improperly oriented and edge (x, d) is properly oriented (Figure 3.4(ii)). If we increase the flow in (x, d) by Δ, we must decrease the flow in (c, x) by Δ so that the flow into x will still equal the flow out of x.

Case (c): Edge (c, x) is properly oriented and edge (x, d) is improperly oriented (Figure 3.4(iii)). In this case, we increase the flow in (c, x) by Δ and decrease the flow in (x, d) by Δ.

Case (d): Both edges are improperly oriented (Figure 3.4(iv)). In this case, we decrease the flow in both edges by Δ.

In every case, the resulting edge assignments give a flow. The value of Δ must be chosen so that we have flow less than the capacity in a proper oriented edge and a nonzero flow in an improperly oriented edge.

Example 3.2.3 *Let P be the path from a to z as shown Figure 3.5. Now (a, b), (b, c), and (d, z) are properly oriented and edge (d, c) is improperly oriented. Thus, we decrease the flow by 1 in the edge (d, c) and increase the flow by 1 in the edges (a, b), (b, c), and (d, z) (see Figure 3.6). The value of the flow is 1 greater than that of original flow.*

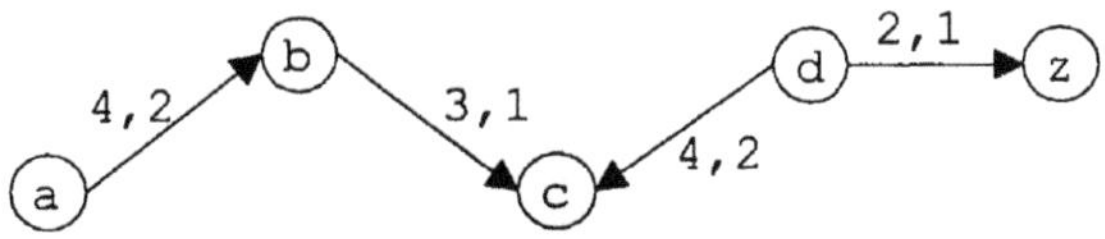

Figure 3.5

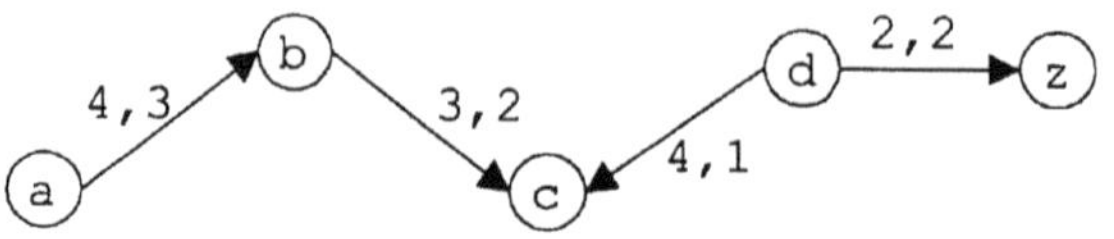

Figure 3.6

Examples 3.2.2 and 3.2.3 motivate the following theorem.

Theorem 3.2.4 *Let P be a path from a to z in a network G such that:*
(i) For each properly oriented edge (i, j) in P,

$$F_{ij} < C_{ij}.$$

(ii) For each improperly oriented edge (i, j) in P,

$$0 < F_{ij}.$$

Let

$$X_{ij} = \begin{cases} C_{ij} - F_{ij} & \text{for a properly oriented edge } (i,j) \in P \\ F_{ij} & \text{for an improperly oriented edge } (i,j) \in P. \end{cases}$$

Let $X = \{X_{ij} \mid (i, j) \in P\}$ and

$$\Delta = \min X.$$

Define

$$F^*_{ij} = \begin{cases} F_{ij} & \text{if } (i,j) \text{ is not in } P \\ F_{ij} + \Delta & \text{if } (i,j) \text{ is properly oriented in } P \\ F_{ij} - \Delta & \text{if } (i,j) \text{ is not properly oriented in } P. \end{cases}$$

Then F^* is a flow whose value is Δ greater than the value of F.

Proof. The argument preceding Example 3.2.3 shows that F^* is a flow. Since the edge (a, v) in P is increased by Δ, the value of F^* is Δ greater than the value of F. ■

We next show that if there are no paths satisfying the conditions of Theorem 3.2.4, the flow is maximal. Thus it is possible to construct an algorithm based on Theorem 3.2.4. The following steps form an outline for the algorithm.

1. Start with a flow, e.g., the flow in which the flow in each edge is 0.
2. Search for a path satisfying the conditions of Theorem 3.2.4. If no such path exists, stop; the flow is maximal.

3. Increase the flow through the path by Δ, where Δ is defined as in Theorem 3.2.4, and go to line 2.

In the formal algorithm, we search for a path satisfying the conditions of Theorem 3.2.4 while simultaneously keeping track of the quantities

$$C_{ij} - F_{ij},\ F_{ij}.$$

Algorithm Max-Flow Finding a Maximal Flow in a Network [25]

This algorithm finds a maximal flow in a network. The capacity of each edge is a nonnegative integer. $G = (V, E)$ is a network with source a and sink z and $V = \{a = v_0, \ldots, v_n = z\}$.

Input: Network G, capacity C, and n, the number of vertices in G.

Output: A maximal flow F

```
procedure maxFlow(G, C, v, n)
    // v's label is (predecessor(v), val(v))
    // Initially set flow in edge to 0.
    for each edge (i, j) do
        F_ij = 0
    while true do
    begin
        for i = 0 to n do
            begin
                predecessor(v_i) ← null
                val(v_i) ← null
            end
            // label a
        predecessor(a) ← –
        val(a) ← ∞
            // U is the set of unexamined, labeled vertices
        U ← {a}
            // continue until z is labeled
        while val(z) is null do
            begin
                if U is empty then // flow is maximal
                    return(F)
                choose v in U
```

```
            U ← U\{v}
            Δ ← val(v)
            for each edge (v, w) such that val(w) is null do
               if F_vw < C_vw then
                  begin
                  predecessor(w) ← v
                  val(w) ← min{Δ, C_vw − F_vw}
                  U ← U ∪ {w}
                  end
            for each edge (w, v) such that val(w) is null
               if F_wv > 0 then
               begin
                  predecessor(w) ← v
                  val(w) ← min{Δ, F_wv}
                  U ← U ∪ {w}
               end //if
      end
      // find path P from a to z on which to revise flow
      w_0 ← z
      k ← 0
      while w_k is not equal to a do
         begin
            w_{k+1} ← predecessor(w_k)
            k ← k + 1
         end //while
      P ← (w_k, w_{k+1}, ... , w_1, w_0)
      Δ ← val(z)
      for i = 1 to k do
      begin
         e ← (w_i, w_{i−1})
         if e is properly oriented in P then
            F_e ← F_e + Δ
         else
            F_e ← F_e − Δ
```

```
51.         end
52.      end
      end max_flow
```

If the capacities are allowed to be nonnegative rational numbers, then the algorithm also terminates. However, if arbitrary nonnegative real capacities are permitted and we allow the edge in line 20 to be examined in any order, then the algorithm may not terminate (see [20, p. 21-22]).

Algorithm max-flow is often referred to as the **labeling procedure.**

3.3 The Max Flow, Min Cut Theorem

In this section, we show that when Algorithm max-flow terminates, the flow in the network is maximal.

Let G be a network with source a and sink z. Let F be the flow in G when Algorithm max-flow terminates. Some vertices are labeled and some are unlabeled. Let P denote the set of labeled vertices and $\bar{P}$ denote the set of unlabeled vertices. Then $a \in P$ and $z \in \bar{P}$. Let $S = \{(v, w) \mid v \in P, w \in \bar{P}\}$. The set S is called a **cut** and the sum of the capacities of the edges in S is called the **capacity of the cut.** We denote the set S by $(P, \bar{P})$. We will show that this cut has a minimum capacity. In Theorem 3.3.9 below, we show that a minimal cut corresponds to a maximal flow. Hence the corresponding flow F is maximal.

Throughout this section, G is a network with source a and sink z. The capacity of edge (i, j) is C_{ij}.

Definition 3.3.1 *A* **cut** $(P, \bar{P})$ *in* G *consists of a set* P *of vertices and the complement* $\bar{P}$ *of* P, *with* $a \in P$ *and* $z \in \bar{P}$

Example 3.3.2 *Let* G *be the network of Figure* 3.7. *Let* $P = \{a, b, c\}$. *Then* $\bar{P} = \{d, e, z\}$ *and* $(P, \bar{P})$ *is a cut in* G.

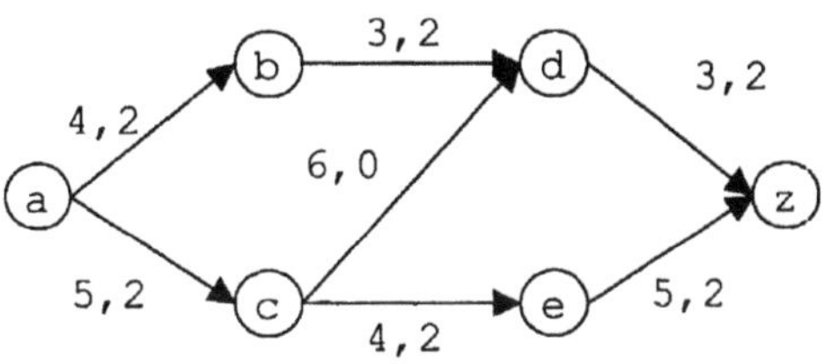

Figure 3.7

Example 3.3.3 *Let* G *be the network as shown in Figure* 3.8. *At the termination of Algorithm max-flow,* $P = \{a, b, d\}$ *and* $\bar{P} = \{c, e, z\}$.

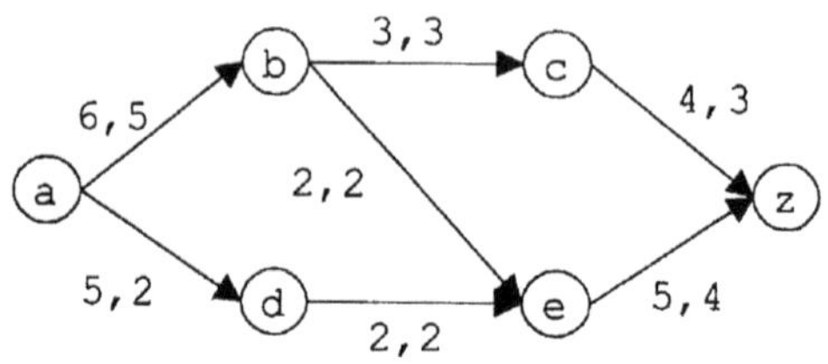

Figure 3.8

Definition 3.3.4 *The* ***capacity of the cut*** $(P, \bar{P})$ *is defined to be*

$$C(P, \bar{P}) = \sum_{i \in P} \sum_{j \in \bar{P}} C_{ij}.$$

Example 3.3.5 *(i) The capacity of the cut of Figure* 3.7 *is*

$$C_{bd} + C_{cd} + C_{ce} = 13.$$

(ii) The capacity of the cut of Figure 3.8 *is*

$$C_{bc} + C_{be} + C_{de} = 7.$$

Recall that the notation $\sum_i$ means the sum over all vertices i.

Theorem 3.3.6 *Let F be a flow in G and let $(P, \bar{P})$ be a cut in G. Then the capacity of $(P, \bar{P})$ is greater than or equal to the value of F; that is,*

$$\sum_{i \in P} \sum_{j \in \bar{P}} C_{ij} \geq \sum_i F_{ai}. \tag{3.2}$$

Proof. It follows that

$$\sum_{j \in P} \sum_{i \in P} F_{ji} = \sum_{i \in P} \sum_{j \in P} F_{ij},$$

since either side of the equation is the sum of F_{ij} over all $i, j \in P$. Now

$$\begin{aligned} \textstyle\sum_i F_{ai} &= \textstyle\sum_{j \in P} \sum_i F_{ji} - \sum_{j \in P} \sum_i F_{ij} \\ &= \textstyle\sum_{j \in P} \sum_{i \in P} F_{ji} + \sum_{j \in P} \sum_{i \in \bar{P}} F_{ji} \\ &\quad \textstyle- \sum_{j \in P} \sum_{i \in P} F_{ij} - \sum_{j \in P} \sum_{i \in \bar{P}} F_{ij} \\ &= \textstyle\sum_{j \in P} \sum_{i \in \bar{P}} F_{ji} - \sum_{j \in P} \sum_{i \in \bar{P}} F_{ij} \\ &\leq \textstyle\sum_{j \in P} \sum_{i \in \bar{P}} F_{ji} \\ &\leq \textstyle\sum_{j \in P} \sum_{i \in \bar{P}} C_{ji}. \ \blacksquare \end{aligned}$$

Example 3.3.7 *In Figure* 3.7, $4 = F \leq C = 13$.

Definition 3.3.8 *A **minimal cut** is a cut having minimum capacity.*

Theorem 3.3.9 (Max Flow Min Cut) *Let F be a flow in G and let $(P, \bar{P})$ be a cut in G. If equality holds in (3.2), then the flow is maximal and the cut is minimal. Furthermore, equality holds in (3.2) if and only if*
(i) $F_{ij} = C_{ij}$ for $i \in P$, $j \in \bar{P}$ and
(ii) $F_{ij} = 0$ for $i \in \bar{P}$, $j \in P$.

Proof. The first statement is immediate.
From the proof of Theorem 3.3.6, it follows that equality holds when

$$\sum_{j \in P} \sum_{i \in \bar{P}} F_{ij} = 0 \text{ and } \sum_{j \in P} \sum_{i \in \bar{P}} F_{ji} = \sum_{j \in P} \sum_{i \in \bar{P}} C_{ji}. \blacksquare$$

Example 3.3.10 *In Figure 3.8, since the value of flow and the capacity of the cut are equal, namely, both equal 7, the flow is maximal and the cut is minimal.*

Theorem 3.3.9 can be used to show that Algorithm max-flow produces a maximal flow.

Theorem 3.3.11 *At termination, Algorithm max-flow produces a maximal flow. Furthermore, if P is the set of labeled vertices and $\bar{P}$ is the set of unlabeled vertices at the termination of Algorithm max-flow, then the cut $(P, \bar{P})$ is minimal.*

Proof. Consider an edge (i, j), where $i \in P$, $j \in \bar{P}$. Since i is labeled,

$$F_{ij} = C_{ij};$$

else, j would have been labeled at lines 23 and 24. Now consider an edge (j, i), where $j \in \bar{P}$, $i \in P$. Since i is labeled,

$$F_{ji} = 0;$$

else, j would have been labeled at lines 30 and 31. By Theorem 3.3.9, the flow at the termination of Algorithm max-flow is maximal and the cut $(P, \bar{P})$ is minimal. ■

3.4 Maximum Flow in a Network with Fuzzy Arc Capacities

The material in this section is based on [11, 12]. A fuzzy approach to decision tasks has allowed for better mathematical models of those tasks. In particular, fuzzy linear programming allows for tolerances in satisfying constraints of a decision task [22, 31]. In practice, a decision maker usually has several possibilities of violating beforehand fixed constraints of resources if it is required by the goal associated with a given situation. Parameters in problems of decision models often are fixed constraints accompanied by a certain margin. They may be violated only in the range of the margin. Thus a better realization of the goal is possible, but at the same time it worsens conditions for realization of a chosen decision variant. Thus a common problem for decision makers is to find an agreement between goal requirements and the necessity to preserve conditions allowing for an easy decision realization. Fuzzy programming is useful in such a situation.

This section considers the classic problem of maximum flow under the assumption that violation of capacity constraints of network arcs is acceptable in a certain range. Fuzzy linear programming techniques are not efficient for solving these kinds of problems.

In this section, we analyze the problem of maximum flow with fuzzy network arc capacities. We prove results which may be regarded as equivalent to the classic Ford and Fulkerson theorem concerning the minimum cut and maximum flow value. These results enable the formulation of an efficient algorithm for the solution of an integer value network flow.

We formulate, using different notation as in the previous sections, the classic task of the maximum flow in a network $S = \langle N, A, C\rangle$ with a fixed source $s \in N$ and a destination $t \in N$. N denotes a set of nodes, $A \subseteq N \times N$ denotes a set of arcs, and $C : A \to \mathbb{R}^+$ with $C(i,j) = c_{ij}$ denotes the capacity of an arc $(i,j) \in A$. A v-value, x_v, from source s to sink t is a function $x_v : A \to \mathbb{R}^+$, $(i,j) \in A$, $x_v(i,j) = x^v_{ij} \geq 0$ such that:

$$\sum_{i\in\Gamma_j^-} x^v_{ij} - \sum_{k\in\Gamma_j} x^v_{jk} = \begin{cases} -v & \text{for } j = s \\ 0 & \text{for } j \neq s,t \\ v & \text{for } j = t \end{cases} \tag{3.3}$$

$$b_{ij} \leq x^v_{ij} \leq c_{ij}, \tag{3.4}$$

where x^v_{ij} denotes the flow value on an arc $(i,j) \in A, \Gamma_j^-$ and Γ_j are sets of nodes preceding the node $j \in N$ and following the node $j \in N$, respectively, and $b_{ij} \geq 0$ for all i and j.

If no confusion arises, we will denote x^v_{ij} simply by x_{ij}. Also, we denote the flow x_v by $< x^v_{ij} : (i,j) \in A >$ or simply by $< x_{ij} : (i,j) \in A >$

The problem is to find the flow x_v which maximizes the value v:

$$\max v. \tag{3.5}$$

The following are the well-known Ford-Fulkerson theorems which relate the maximal (minimal) flow value with the capacities of cuts in the network [9].

Definition 3.4.1 *Let $(X, \overline{X})$ be a cut in the network S. Let $\overrightarrow{(X, \overline{X})}$ denote the set of those arcs of the cut $(X, \overline{X})$ that are directed from X to $\overline{X}$ and*

$\overleftarrow{(X,\overline{X})}$ *denote the set of those arcs of the cut* $(X,\overline{X})$ *that are directed from* $\overline{X}$ *to* X.

Theorem 3.4.2 *Under one-sided, upper arc capacity constraints (all* $b_{ij} = 0$*), the value* v *of the maximal flow is equal to the capacity of the minimal cut:*

$$v = \wedge_{(X,\overline{X})} \sum_{(i,j)\in\overrightarrow{(X,\overline{X})}} c_{ij}. \tag{3.6}$$

On the arcs of the minimal cut, the maximal flow satisfies the conditions $x_{ij} = c_{ij}$ *for* $(i,j) \in \overrightarrow{(X,\overline{X})}$ *and* $x_{ij} = 0$ *for* $(i,j) \in \overleftarrow{(X,\overline{X})}$. ■

Theorem 3.4.3 *If there are admissible flows (satisfying condition* (3.4)*), then the value of the maximal flow* x_v *satisfies the condition*

$$v = \wedge_{(X,\overline{X})}\Big(\sum_{(i,j)\in\overrightarrow{(X,\overline{X})}} c_{ij} - \sum_{(i,j)\in\overleftarrow{(X,\overline{X})}} b_{ij}\Big). \tag{3.7}$$

On the arcs of the minimal cut, the maximal flow satisfies $x_{ij} = c_{ij}$ *for* $(i,j) \in \overrightarrow{(X,\overline{X})}$ *and* $x_{ij} = b_{ij}$ *for* $(i,j) \in \overleftarrow{(X,\overline{X})}$. ■

Theorem 3.4.4 *If there are admissible flows (satisfying condition* (3.4)*), then the value of the minimal flow* x_v *satisfies the condition*

$$v = \vee_{(X,\overline{X})}\Big(\sum_{(i,j)\in\overrightarrow{(X,\overline{X})}} b_{ij} - \sum_{(i,j)\in\overleftarrow{(X,\overline{X})}} c_{ij}\Big). \tag{3.8}$$

On the arcs of the cut maximizing the value of the expression (3.8)*, the minimal flow satisfies* $x_{ij} = b_{ij}$ *for* $(i,j) \in \overrightarrow{(X,\overline{X})}$ *and* $x_{ij} = c_{ij}$ *for* $(i,j) \in \overleftarrow{(X,\overline{X})}$. ■

Suppose that $b_{ij} = 0$ for all i, j. Then (3.4) becomes

$$0 \le x_{ij}^v \le c_{ij}. \tag{3.9}$$

That is, the capacity constraints are one sided. Next we shall generalize the max flow network problem under one-sided capacity constraints, i.e., (3.9), to the fuzzy case. We will study two-sided capacity constraints separately.

We assume that the decision maker is interested in the flow value at least v_0, but he allows smaller values with a decreasing degree of satisfaction to the limit of $v_1 \le v_0$ beyond which the values are not at all satisfactory to him. Assume that the decision maker allows the capacity to increase on the arc (i,j) to the limit of $\bar{c}_{ij} \ge c_{ij}$. The problem is to find such a flow which has a 'possible large' value and which simultaneously satisfy the capacity constraints in the 'possible best way'. The formulation of the problem is not sharp and can be treated as a fuzzy decision problem.

We may associate criterion (3.5) with a fuzzy set G in the set of alternatives which satisfies (3.3) with the following membership function:

$$\mu_G(x_v) = \begin{cases} 1 & \text{if } v > v_0 \\ L(v_0, v_1; v) & \text{if } v_1 \leq v \leq v_0 \\ 0 & \text{if } v < v_1, \end{cases} \tag{3.10}$$

where $L(v_0, v_1; v)$ is a linear function in the interval $[v_1, v_0]$ such that $L(v_0, v_1; v_1) = 0$ and $L(v_0, v_1; v_0) = 1$.. The value of the function $\mu_G(x_v)$ is the decision maker's degree of satisfaction with respect to the satisfying of goal G by the flow x_v. Since that value depends on the value v only, we use the symbols $\mu_G(x_v)$ and $\mu_G(v)$ interchangeably.

However, each capacity constraint $x_{ij}^v \leq c_{ij}$ can be associated with a fuzzy set C_{ij} of the membership function:

$$\mu_{ij}(x_v) = \begin{cases} 1 & \text{if } x_{ij}^v < c_{ij} \\ \overline{L}(c_{ij}, \overline{c}_{ij}; x_{ij}^v) & \text{if } c_{ij} \leq x_{ij}^v \leq \overline{c}_{ij} \\ 0 & \text{if } x_{ij}^v > \overline{c}_{ij}, \end{cases}$$

where $\overline{L}(c_{ij}, \overline{c}_{ij}; x_{ij}^v)$ is a linear function such that $\overline{L}(c_{ij}, \overline{c}_{ij}; c_{ij}) = 1$ and $\overline{L}(c_{ij}, \overline{c}_{ij}; \overline{c}_{ij}) = 0$. The $\mu_{ij}(x_v)$-feasibility degree of x_v with respect to the capacity constraint on arc (i, j) can also be interpreted as a measure of truth of the fuzzy statement "x_v satisfies $x_{ij}^v \leq c_{ij}$ as well as possible". Since $\mu_{ij}(x_v)$ depends only on the value x_{ij}^v, we often use the symbol $\mu_{ij}(x_{ij}^v)$ or simply $\mu_{ij}(x_{ij})$.

Clearly the functions L and $\overline{L}$ need not be linear in general.

The problem can be reduced to finding a flow which maximizes the value of the membership function $\mu_D(x_v)$ in a fuzzy decision D:

$$D = \cap_{(i,j)\in A} C_{ij} \cap G,$$

$$\mu_D(x_v) = \mu_C(x_v) \wedge \mu_G(x_v) \to \max,$$

where $\mu_C(x_v) = \wedge_{(i,j)\in A} \mu_{ij}(x_{ij})$, [4]. $\mu_C(x_v)$ denotes the degree of simultaneous realization of the capacity constraints on all network arcs.

Definition 3.4.5 *A flow x_v^* is called the **maximum flow** if*

$$\mu_D(x_v^*) = max_{x_v} \mu_D(x_v).$$

Definition 3.4.6 *A flow x_v with the greatest value v among other maximum flows is called an **optimal flow**.*

For a path r from the source s to the sink t in the network $S = \langle N, A, C \rangle$, $\overrightarrow{r}$ denotes the set of directed arcs on the path r and $\overleftarrow{r}$ denotes the set of reverse arcs on the path r. Clearly, $r = \overrightarrow{r} \cup \overleftarrow{r}$.

Lemma 3.4.7 *If x_v is an optimal flow in the network $S = \langle N, A, C\rangle$ and $C(i,j) = c_{ij}$, then for any path r from the source s to the sink t either*
(i) there exists a directed arc $(i,j) \in r$ such that $\mu_{ij}(x_{ij}) = \mu_D(x_v)$, or
(ii) there exists a reverse arc $(j,i) \in r$ for which $x_{ji} = 0$.

Proof. Suppose there exists a path r from the source s to the destination t such that
(i′) for each directed arc $(i,j) \in r, \mu_{ij}(x_{ij}^v) > \mu_D(x_v)$, and
(ii′) for each reverse arc $(j,i) \in r$, $x_{ij} > 0$.

Since each μ_{ij} is continuous and decreasing, there exists a positive real number ε_{ij} such that $\mu_{ij}(x_{ij} + \varepsilon_{ij}) = \mu_D(x_v)$ for $(i,j) \in \overrightarrow{r}$ or $\varepsilon_{ij} = x_{ij}$ for $(i,j) \in \overleftarrow{r}$. Let $\varepsilon = \wedge_{(i,j)\in r}\varepsilon_{ij}$. Then $\varepsilon > 0$. Let x'_w be the flow obtained from the flow x_v by the following transformations:

$$x'_{ij} = \begin{cases} x_{ij} + \varepsilon & \text{for } (i,j) \in \overrightarrow{r} \\ x_{ij} - \varepsilon & \text{for } (i,j) \in \overleftarrow{r} \\ x_{ij} & \text{for other arcs.} \end{cases}$$

Then $\mu_D(x'_w) = \mu_D(x_v)$ and $w = v + \varepsilon$, which contradicts the optimality of the flow x_v. ∎

Theorem 3.4.8 *If x_v is an optimal flow in the network $S = \langle N, A, C\rangle$, there exists a cut $(X, \overline{X})$ in that network satisfying the following conditions:*
(i) $(i,j) \in A,\ i \in X,\ j \in \overline{X} \Rightarrow \mu_{ij}(x_{ij}) = \mu_D(x_v)$,
(ii) $(i,j) \in A,\ i \in \overline{X},\ j \in X \Rightarrow x_{ij} = 0$.

Proof. Let X be a set of nodes such that
(i′) $s \in X$,
(ii′) $i \in X$ and $\mu_{ij}(x_{ij}) > \mu_D(x_v) \Rightarrow j \in X$,
(iii′) $i \in X$ and $x_{ji} > 0 \Rightarrow j \in X$.

Let $\overline{X} = \{v \in N \mid v \notin X\}$. From the construction of the set X it follows that if $t \in \overline{X}$, then $(X, \overline{X})$ is a cut separating s from t and the cut satisfies conditions (i) and (ii) of the theorem. If $t \in X$, then there exists a path r connecting s and t such that for each arc $(i,j) \in \overrightarrow{r}$, $\mu_{ij}(x_{ij}) > \mu_D(x_v)$ and for each arc $(i,j) \in \overleftarrow{r}$, $x_{ji} > 0$. By the way X was constructed, this is a contradiction of Lemma 3.4.7. ∎

The cut $(X, \overline{X})$ appearing in Theorem 3.4.8 will be called a minimal cut because it forms a ‘bottle neck’ for the optimal flow. For optimal flow x_v,

$$v = \sum_{(X,\overline{X})} x_{ij}.$$

Theorem 3.4.8 is equivalent to the theorem of Ford and Fulkerson concerning the maximum flow and the minimal cut in the case of the classic problem [18, 19].

There is an important difference between the idea of the minimal cut in the classic case and in the case discussed in this chapter. In the fuzzy case, the minimal cut in the network cannot be determined without the knowledge of the optimal flow. Both the optimal flow and the respective minimal cut depend on the accepted scope of the criterion G.

The above theorem is not true for integer value flows. That case will be discussed in the next section.

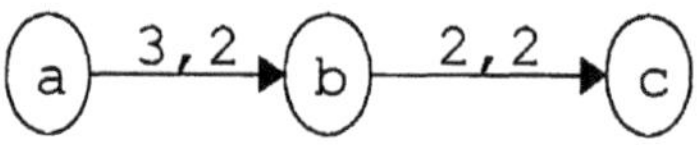

Figure 3.9

Example 3.4.9 *Consider the network in Figure 3.9. Here a is the source and c is the sink. Then $c_{ab} = 3$, $c_{bc} = 2$, and $v = 2$. Let $\bar{c}_{ab} = 4$, $\bar{c}_{bc} = 3$, and $v_1 = 2$, and $v_0 = 4$. Let*

$$L(v_0, v_1; v) = \frac{1}{v_0 - v_1} v - \frac{v_1}{v_0 - v_1}, \qquad v_1 \leq v \leq v_0,$$

and

$$\bar{L}(c_{ij}, \bar{c}_{ij}; x_{ij}) = \frac{1}{c_{ij} - \bar{c}_{ij}} x_{ij} + \frac{-\bar{c}_{ij}}{c_{ij} - \bar{c}_{ij}}, \qquad c_{ij} \leq x_{ij} \leq \bar{c}_{ij}.$$

Then

$$L(4, 2, v) = \frac{1}{2} v - 1, \;\; 2 \leq v \leq 4$$

and

$$\bar{L}(c_{ab}, \bar{c}_{ab}; x_{ab}) = \bar{L}(3, 4; x_{ab}) = -x_{ab} + 4, \qquad 3 \leq x_{ab} \leq 4,$$

$$\bar{L}(c_{bc}, \bar{c}_{bc}; x_{bc}) = \bar{L}(3, 4; x_{bc}) = -x_{bc} + 3, \qquad 2 \leq x_{bc} \leq 3.$$

Since the flow into b must equal the flow out of b, we have $v = x_{ab} = x_{bc}$. Now

$$\mu_{ab}(x_v) = \begin{cases} 1 & x_{ab} \leq 3 \\ -x_{ab} + 4 & 3 \leq x_{ab} \leq 4 \\ 0 & x_{ab} \geq 4 \end{cases}$$

$$\begin{aligned} &v = 2 \Rightarrow x_v = \; < x_{ab}, x_{bc} > \; = \; < 2, 2 >, \\ &\mu_{ab}(\{2, 2\}) = 1, \\ &\mu_{bc}(\{2, 2\}) = 1, \\ &\mu_G(\{2, 2\}) = 0, \end{aligned}$$

$$\mu_{bc}(x_v) = \begin{cases} 1 & x_{bc} \leq 2 \\ -x_{bc} + 3 & 2 \leq x_{bc} \leq 3 \\ 0 & x_{bc} \geq 3 \end{cases}$$

$v = 3 \Rightarrow x_v = \ < x_{ab}, x_{bc} > \ = \ < 3, 3 >,$
$\mu_{ab}(\{3,3\}) = 1,$
$\mu_{bc}(\{3,3\}) = 0,$
$\mu_G(\{3,3\}) = .5,$

$$\mu_C(x_v) = \mu_{ab}(x_v) \wedge \mu_{bc}(x_v)$$

$$\mu_G(x_v) = \begin{cases} 1 & v \geq 4 \\ \frac{1}{2}v - 1 & 2 \leq v \leq 4 \\ 0 & v \leq 2 \end{cases}$$

$v = 2.5 \Rightarrow x_v = \ < x_{ab}, x_{bc} > \ = \ < 2.5, 2.5 >,$
$\mu_{ab}(\{2.5, 2.5\}) = 1,$
$\mu_{bc}(\{2.5, 2.5\}) = .5,$
$\mu_G(\{2.5, 2.5\}) = .25.$

Since $v = x_{ab} = x_{bc}$, *we see that* $\mu_{bc}(x_v) \leq \mu_{ab}(x_v)$ *for all* $v \geq 0$. *Thus* $\mu_C(x_v) = \mu_C(v) = \mu_{bc}(v)$. *Hence* $\mu_D(x_v) = \mu_{bc}(x_v) \wedge \mu_G(x_v)$. *Now* $\max\{\min\{\mu_{bc}(x_v), \mu_G(x_v)\} \mid v \geq 0\} = \max\{\min\{-v+3, \frac{1}{2}v - 1\} \mid 2 \leq v \leq 3\} = \frac{1}{3} = \mu_D(x_v^*)$. *(See Figure 3.10.)* $X = \{a, b\}$, $\bar{X} = \{c\}$ *is the minimal cut.*

$$x_v = \sum_{(X,\bar{X})} x_{ij} = x_{bc} = \frac{8}{3}.$$

See Theorem 3.4.8. *We have* $v = \frac{8}{3} = x_{ab} = x_{bc}$. *Now* $\mu_{ab}(\frac{8}{3}) = 1 > \frac{1}{3} = \mu_D(\frac{8}{3})$ *and* $\mu_{bc}(\frac{8}{3}) = \frac{-8}{3} + 3 = \frac{1}{3} = \mu_D(\frac{8}{3})$. *Thus* $X = \{a, b\}$ *and* $\bar{X}$ *is the cut of Theorem 3.4.8.*

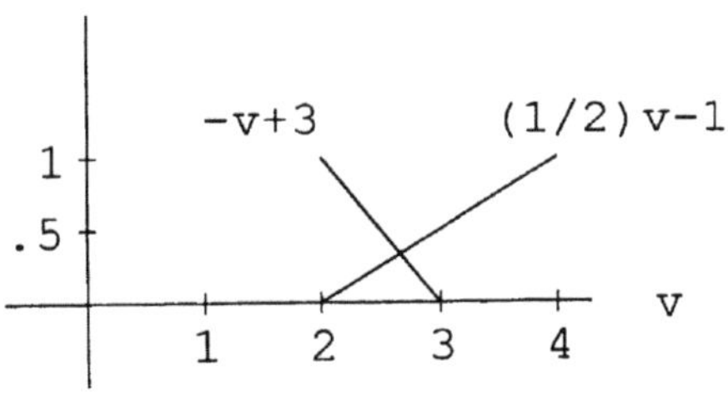

Figure 3.10

3.5 The Maximum Flow with Integer Values

In this section, we discuss network flows with integer values. For these networks, Lemma 3.5.1 and Theorem 3.5.2 proved below correspond to Lemma 3.4.7 and Theorem 3.4.8, respectively.

Lemma 3.5.1 *If x_v is the optimal flow, then for any path from the source s to the sink t either*
(i) there exists an arc $(i,j) \in \overrightarrow{r}$ such that $\mu_{ij}(x_{ij}+1) < \mu_D(x_v)$, or
(ii) there exists an arc $(j,i) \in \overleftarrow{r}$ for which $x_{ji} = 0$.

Proof. Suppose there exists a path r from the source s to the sink t such that
(i′) $(i,j) \in \overrightarrow{r} \Rightarrow \mu_{ij}(x_{ij}+1) \geq \mu_D(x_v)$, and
(ii′) $(j,i) \in \overleftarrow{r} \Rightarrow x_{ij} \geq 1$.
Let x'_w be the flow obtained from the flow x_v by the following transformations:

$$x'_{ij} = \begin{cases} x_{ij}+1 & \text{for } (i,j) \in \overrightarrow{r} \\ x_{ij}-1 & \text{for } (i,j) \in \overleftarrow{r} \\ x_{ij} & \text{otherwise.} \end{cases}$$

Then $\mu_D(x'_w) = \mu_D(x_v)$ and $w = v+1$, which contradicts the optimality of the flow x_v. ■

Theorem 3.5.2 *If x_v is an optimal flow, then there exists in the network $S = \langle N, A, C\rangle$ and a cut $(X, \overline{X})$ such that the following properties hold:*
(i) $(i,j) \in A$, $i \in X$, $j \in \overline{X} \Rightarrow \mu_{ij}(x_{ij}+1) < \mu_D(x_v)$,
(ii) $(i,j) \in A$, $i \in \overline{X}$, $j \in X \Rightarrow x_{ij} = 0$.

Proof. The proof is analogous to the proof of Theorem 3.4.8. Here X is obtained by imposing the following requirements:
(i′) $s \in X$,
(ii′) $i \in X$ and $\mu_{ij}(x_{ij}+1) \geq \mu_D(x_v) \Rightarrow j \in X$,
(iii′) $i \in X$ and $x_{ij} \geq 1 \Rightarrow j \in X$.
The assumption that $t \in X$ leads to a contradiction of Lemma 3.5.1. ■

Lemma 3.5.3 *The following relation holds:*

$$\vee_{x_v} \mu_D(x_v) = \vee_{z \in \mathrm{N}} \{\mu_C(\overline{x}_z) \wedge \mu_G(z)\},$$

where $\mu_C(\overline{x}_z) = \vee_{x_z \in V(z)} \mu_C(x_z)$ and $V(z) = \{x_v \mid v = z\}$.

Proof.

$$\begin{aligned} \vee_{x_v} \mu_D(x_v) &= \vee_{x_v} \{\mu_C(x_v) \wedge \mu_G(x_v)\} \\ &= \vee_{x_v \in \cup_{z \in \mathrm{N}} V(z)} \{\mu_C(x_v) \wedge \mu_G(x_v)\} \\ &= \vee_{z \in \mathrm{N}} \{\vee_{x_z \in V(z)} [\mu_C(x_z) \wedge \mu_G(x_v)]\} \\ &= \vee_{z \in \mathrm{N}} \{[\vee_{x_z \in V(z)} \mu_C(x_z)] \wedge \mu_G(z)\} \\ &= \vee_{z \in \mathrm{N}} \{\mu_C(\overline{x}_z) \wedge \mu_G(z)\}. \ ■ \end{aligned}$$

It follows from Lemma 3.5.3 that in order to find an optimal flow, it suffices to consider for each $V(z)$, $z \in \mathbb{N}$, only that flow $\overline{x}_v \in V(z)$ which satisfies the capacity constraints in the best way. The above property is now used in the construction of an algorithm for the determination of optimal flow.

Algorithm [14]

Step 1. Find a maximum flow x_w in the classic sense in the network S assuming arc capacities equal to c_{ij}, $(i,j) \in A$. Evidently, $\mu_C(x_v) = 1$. If $\mu_G(v) = 1$, then STOP, x_v is optimal. Otherwise go to Step 2.

Step 2. Let x_v be a current flow. Determine a path r leading from s to t and maximizing the value of the expression

$$\mu^r(x_v) = (\wedge_{(i,j)\in \overrightarrow{r}} \mu_{ij}(x_{ij}+1)) \wedge (\wedge_{(i,j)\in \overleftarrow{r}} \mu_{ij}(x_{ij}-1)) \rightarrow \max$$

where $\overrightarrow{r}$ and $\overleftarrow{r}$ are sets of the forward and backward arcs in the path r, respectively. Additionally, we assume that if $x_{ij} = 0$, then $\mu_{ij}(x_{ij}-1) = -1$. If $\mu_D(x_v) > \mu^r(x_v)$, then STOP; x_v is optimal. Otherwise, go to Step 3.

Step 3. Set $x_v := x_{v+1}$, where x_{v+1} is the $v+1$-value flow obtained by increase of the flow x_v with a unit on the path r (adding a unit flow to the forward arcs and subtracting it from the backward arcs.) Go to Step 2.

To determine the path r in Step 2 the shortest path algorithm of Dijkstra (see, e.g., [5]) may be easily adapted.

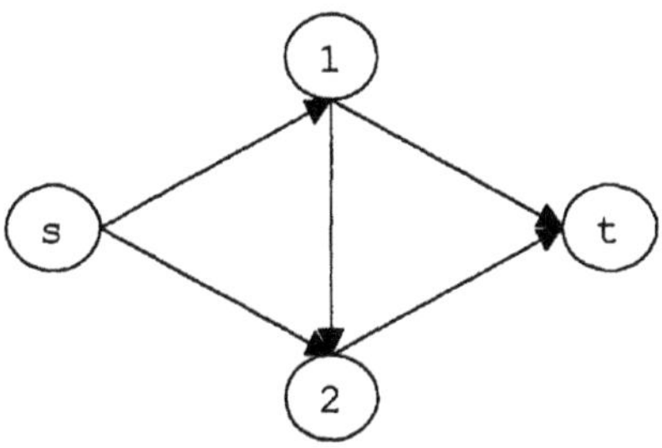

Figure 3.11

Example 3.5.4 *Consider the network of Figure 3.11.*

Suppose that μ_{ij} *and* μ_G *are defined as follows:*

$\mu_{s1}(x_{s1}) = 1$ *for* $x_{s1} = 0, 1, 2,$ *and* $\mu_{s1}(3) = 0.7$, $\mu_{s1}(4) = 0.4$, $\mu_{s1}(5) = 0.2$, $\mu_{s1}(x_{s1}) = 0$ *for* $x_{s1} \geq 6$;

$\mu_{s2}(x_{s2}) = 1$ *for* $x_{s2} = 0, \ldots, 5,$ *and* $\mu_{s2}(6) = 0.8$, $\mu_{s2}(7) = 0.5$, $\mu_{s2}(8) = 0.1$, $\mu_{s2}(x_{s2}) = 0$ *for* $x_{s2} \geq 9$;

$\mu_{1t}(x_{1t}) = 1$ *for* $x_{1t} = 0, 1,$ *and* $\mu_{1t}(2) = 0.9$, $\mu_{1t}(3) = 0.4$, $\mu_{1t}(x_{1t}) = 0$ *for* $x_{1t} \geq 4$;

$\mu_{2t}(x_{2t}) = 1$ *for* $x_{2t} = 0, \ldots, 6,$ *and* $\mu_{2t}(7) = 0.7$, $\mu_{2t}(8) = 0.4, \mu_{2t}(9) = 0.1$, $\mu_{2t}(x_{2t}) = 0$ *for* $x_{2t} \geq 10$;

$\mu_{12}(x_{12}) = 1$ *for* $x_{12} = 0,\ldots,2$, *and* $\mu_{12}(3) = 0.7$, $\mu_{12}(4) = 0.3$, $\mu_{12}(x_{12}) = 0$ *for* $x_{12} \geq 5$;

$\mu_G(v) = 1$ *for* $v \geq 10$, $\mu_G(9) = 0.8$, $\mu_G(8) = 0.6$, $\mu_G(7) = 0.2$, $\mu_G(v) = 0$ *for* $v \leq 6$.

The algorithm runs as follows:

Step 1. $x_v : x_{s1} = 1$, $x_{s2} = 5$, $x_{12} = x_{1t} = 1$, $x_{2t} = 6$. $v = 7$, $\mu_G(7) = 0.2 \neq 1$. *Go to Step 2.*

Step 2. $r = (s,2,1,t)$, $\mu^r(x_v) = 0.8 > \mu_D(x_v) = 0.2$. *Go to Step 3.*

Step 3. $x_v : x_{s1} = x_{1t} = 2$, $x_{s2} = x_{2t} = 6$, $x_{12} = 0$. *Go to step 2.*

Step 2. $r = (s,1,2,t)$, $\mu^r(x_v) = 0.7 > \mu_D(x_v) = 0.6$. *Go to Step 3.*

Step 3. $x_v : x_{s1} = 3$, $x_{s2} = 6$, $x_{12} = 1$, $x_{1t} = 2$, $x_{2t} = 7$. *Go to step 2.*

Step 2. $r = (s,2,1,t)$, $\mu^r(x_v) = 0.4 > \mu_D(x_v) = 0.7$. *STOP.*

The current flow x_v *is optimal.*

3.6 Integer Flows in Network with Two-Sided Fuzzy Capacity Constraints

In this section, we consider integer flows in networks with two sided fuzzy capacity constraints, i.e.,

(iii)′

$$b_{ij} \leq x_{ij} \leq c_{ij},$$

with b_{ij}, x_{ij}, and c_{ij} nonnegative integers for all i, j. Many of the results are from [14].

For each of the constraints in (iii′), we can assign a fuzzy set C_{ij} with the membership function

$$\mu_{ij}(x_v) = \begin{cases} 0 & \text{if } x_{ij} < \underline{b}_{ij} \\ L_l(\underline{b}_{ij}, b_{ij}; x_{ij}) & \text{if } \underline{b}_{ij} \leq x_{ij} \leq b_{ij} \\ 1 & b_{ij} < x_{ij} < c_{ij} \\ L_r(c_{ij}, \overline{c}_{ij}; x_{ij}) & \text{if } c_{ij} \geq x_{ij} \geq \overline{c}_{ij} \\ 0 & \text{if } x_{ij} > \overline{c}_{ij}, \end{cases} \tag{3.11}$$

where $\underline{b}_{ij} < b_{ij}$ and $\overline{c}_{ij} > c_{ij}$ are maximal ranges of admissible tolerance for the lower and upper arc capacities, respectively.

$L_l(\underline{b}_{ij}, b_{ij}; x_{ij})$ is an increasing function, $L_l(\underline{b}_{ij}, b_{ij}; \underline{b}_{ij}) = 0$, and $L_l(\underline{b}_{ij}, b_{ij}; b_{ij}) = 1$. $L_r(c_{ij}, \overline{c}_{ij}; x_{ij})$ is a decreasing function, $L_r(c_{ij}, \overline{c}_{ij}; c_{ij}) = 1$, and $L_r(c_{ij}, \overline{c}_{ij}; \overline{c}_{ij}) = 0$. In particular, L_l and L_r may be linear functions. (In

Figure 3.12, $a = \underline{b}_{ij}$, $b = b_{ij}$, $c = c_{ij}$, $d = \overline{c}_{ij}$, and $U = \mu$.)

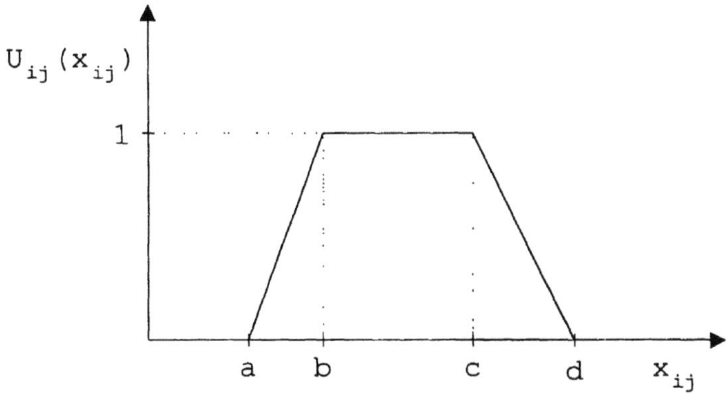

Figure 3.12

Since the functions in (3.11) depend on x_{ij} only, we use either of the notations $\mu_{ij}(x_{ij})$ or $\mu_{ij}(x_v)$.

For easy reference, we call integer flows in networks with one sided constraints (that we discussed in the previous section), Problem I and integer flows in networks with two sided fuzzy constraints, Problem II.

As before, Problem II can be reduced to finding a flow which maximizes the value of the membership function $\mu_D(x_v)$ in a fuzzy decision D

$$D = \bigcap_{(i,j)\in A} C_{ij} \cap G,$$

$$\mu_D(x_v) = \mu_C(x_v) \wedge \mu_G(x_v) \to \max, \tag{3.12}$$

where

$$\mu_C(x_v) = \bigwedge_{(i,j)\in A} \mu_{ij}(x_{ij})$$

and $\mu_{ij}(x_{ij})$ is of the form (3.11).

Before we analyze the formal model formulated above, we discuss the possible interpretation of the membership functions μ_{ij} and μ_G. The interpretation is closely related to the problem of selection of the proper operation in defining the fuzzy decision D.

The first possible interpretation is closely connected with the notion of tolerance. The arc capacities and the flow value are considered to be standards established beforehand which may be violated in given ranges of tolerance. However, a decision maker wishes to minimize these deviations. The values $1 - \mu_{ij}(x_{ij})$ and $1 - \mu_G(v)$ may be regarded as degrees of deviations from the respective standards and problem (3.12) is reduced to the determination of a flow for which the largest of the deviations is minimal. With such an interpretation, it seems reasonable to assume the linear form of the membership functions μ_{ij} and μ_G. The min operations used in (3.12) may be replaced by other operators, e.g., t-norms.

Regarding the membership functions μ_{ij} and μ_G as possibility distributions of fuzzy variables in the sense of [43] yields an approach which is closer to fuzzy set theory. The $\mu_{ij}(x_{ij})$ would mean the possibility that the (i,j) arc capacity is at least equal to or greater than x_{ij}. Similarly, $\mu_G(v)$ would mean the possibility that a demand for product in the sink is not greater than v.

There are two completely different interpretations of the possibility notion. The first is a physical interpretation. Here possibility is interpreted as a degree of easiness of a system performance when a concrete value of a parameter, described by a fuzzy variable, is assumed. The second interpretation is an epistemic one. Here, possibility is a measure of uncertainty connected with an event. Usually the possibility distribution is treated as a subjective characterization resulting from the meaning of fuzzy propositions, e.g., expert opinions expressed in a natural language, [43].

The functions μ_{ij} and μ_G may be defined subjectively in the case of the physical interpretation. However, in some situations they may be determined by a strict dependence on the measurable physical parameters. The min operation used in (3.12) is acceptable in this case and can be interpreted.

In the case of the epistemic interpretation of the possibility distribution, the problem is more complicated and the form of the membership functions μ_{ij} and μ_G as well as the operation in (3.12) should be selected separately according to the nature and structure of information used to define the membership functions.

In [9], a special case of problem (3.12) with the product operation, for real flows with fixed value v, has been considered.

Properties of the Optimal Flows

For real flows (i.e., b_{ij}, x_{ij}, and c_{ij} are real numbers), the following theorems, which correspond to the Ford and Fulkerson max-flow min-cut theorem, are valid. In Theorem 3.6.1, we assume that the functions L, L_l, and L_r in (3.11) to be continuous.

Theorem 3.6.1 *If x_v is an optimal flow for Problem II (i.e., networks with two-sided fuzzy capacity constraints), then there exists a cut $(X, \overline{X})$ in the network such that the following implication holds:*

$$(i,j) \in (X, \overline{X}) \Rightarrow \mu_{ij}(x_{ij}) = \mu_D(x_v),$$

where for an arbitrary $\varepsilon > 0$ the following conditions hold:

$$(i,j) \in \overrightarrow{(X, \overline{X})} \Rightarrow \mu_{ij}(x_{ij} + \varepsilon) < \mu_D(x_v),$$

$$(i,j) \in \overleftarrow{(X,\overline{X})} \Rightarrow \mu_{ij}(x_{ij} - \varepsilon) < \mu_D(x_v).$$

Proof. We assume that the problem is admissible, i.e., the support of the fuzzy decision is not empty. By the assumption on the membership functions μ_{ij}, it follows that r-cuts of fuzzy constraint C_{ij} are closed intervals, i.e., $C_{ij}^r = [b_{ij}^r, c_{ij}^r]$ $\forall r \in (0,1]$ and $(i,j) \in A$. Now

$$\begin{array}{l} \mu_{ij}(b_{ij}^r) = \mu_{ij}(c_{ij}^r) = r \\ \underline{b}_{ij} < b_{ij}^r \leq b_{ij},\ c_{ij} \leq c_{ij}^r < \overline{c}_{ij}. \end{array} \tag{3.13}$$

Let $\mu_D(x_v) = \lambda$. Since the problem is admissibe, it follows that $\lambda > 0$. Thus we can determine a network $\underline{S} = \langle N, A, B, C \rangle$ with two-sided capacity constraints on arcs, $B(i,j) = b_{ij}^\lambda$ and $C(i,j) = c_{ij}^\lambda$, where x_v is feasible. Then x_v is also a maximum flow in $\underline{S}$ since the existence of $\underline{S}$ of a flow x_w with a greater value would contradict the optimality of the flow x_v in the previous network S since the flow x_w satisfies the condition $w > v$ and $\mu_D(x_w) \geq \mu_D(x_v)$. By the Ford and Fulkerson max-flow min-cut theorem, there exists a cut $(X, \overline{X})$ in $\underline{S}$ satisfying the following conditions:

(1) $(i,j) \in \overrightarrow{(X,\overline{X})} \Rightarrow x_{ij} = c_{ij}^\lambda$,

(2) $(i,j) \in \overleftarrow{(X,\overline{X})} \Rightarrow x_{ij} = b_{ij}^\lambda$.

This cut satisfies the conditions formulated in the theorem in view of relations (3.13) and the properties of the μ_{ij}. ■

Lemma 3.6.2 *Let x_v be an integer optimal flow for Problem II. Then for any path r from the source s to the sink t either*

(i) $\exists$ an arc $(i,j) \in \overrightarrow{r}$ such that $\mu_{ij}(x_{ij}+1) < \mu_D(x_v)$ or

(ii) $\exists$ an arc $(i,j) \in \overleftarrow{r}$ such that $\mu_{ij}(x_{ij}-1) < \mu_D(x_v)$,

where $\overrightarrow{r}$ and $\overleftarrow{r}$ are the sets of forward and backward arcs in the path r, respectively.

Proof. Suppose that there is a path r from s to t such that

(i) for every arc $(i,j) \in \overrightarrow{r}$, $\mu_{ij}(x_{ij}+1) \geq \mu_D(x_v)$ and

(ii) for every arc $(i,j) \in \overleftarrow{r}$, $\mu_{ij}(x_{ij}-1) \geq \mu_D(x_v)$.

Let $\widehat{x}_w$ be the flow obtained from the flow x_v by the following transformation:

$$\widehat{x}_{ij} = \begin{cases} x_{ij}+1 & \text{for } (i,j) \in \overrightarrow{r} \\ x_{ij}-1 & \text{for } (i,j) \in \overleftarrow{r} \\ x_{ij} & \text{for other arcs.} \end{cases}$$

Hence $\mu_D(\widehat{x}_w) = \mu_D(x_v)$ and $w = v+1$, which contradicts the optimality of the flow x_v. ■

Theorem 3.6.3 *If x_v is an integer optimal flow for Problem II, then there exists a cut $(X, \overline{X})$ for which the following implications are true:*

$$(i,j) \in \overrightarrow{(X,\overline{X})} \Rightarrow \mu_{ij}(x_{ij}+1) < \mu_D(x_v),$$
$$(i,j) \in \overleftarrow{(X,\overline{X})} \Rightarrow \mu_{ij}(x_{ij}-1) < \mu_D(x_v).$$

Proof. Let X be a set of vertices of the network such that
(i) $s \in X$,
(ii) $i \in X$ and $\mu_{ij}(x_{ij}+1) \geq \mu_D(x_v) \Rightarrow j \in X$,
(iii) $i \in X$ and $\mu_{ij}(x_{ij}-1) \geq \mu_D(x_v) \Rightarrow j \in X$.
Let $\overline{X}$ denote the set of vertices not in X. Then $(X, \overline{X})$ is a cut satisfying the conditions formulated in the theorem since $t \in \overline{X}$. If $t \in X$, then there exists a path r from vertex s to vertex t such that
(i′) for every $(i,j) \in \overrightarrow{r}$, $\mu_{ij}(x_{ij}+1) \geq \mu_D(x_v)$ and
(ii′) for every $(i,j) \in \overleftarrow{r}$, $\mu_{ij}(x_{ij}-1) \geq \mu_D(x_v)$.
However, this is a contradiction of Lemma 3.6.2. ■

The cut $(X, \overline{X})$ appearing in Theorems 3.4.8, 3.6.1, 3.5.2, and 3.6.3 could be called the minimal cut as it determines the bottleneck of the network for the optimal flow.

There is an important difference between the idea of the minimal cut in the classic case and in the case discussed above. In the fuzzy case, the minimal cut in the network cannot be determined without knowing the optimal flow. Both the optimal flow and the respective minimal cut depend on the accepted form of the criterion G. In the next section, we will give a direct generalization of the max-flow min-cut theorem for real-valued flows.

Algorithm for Determination of Integer Flows (for Problem II)

In the algorithm presented below, subsequently constructed flows $\overline{x}_r$, $\overline{x}_{r+1}$, $\overline{x}_{r+2}, \ldots$, are compared and each satisfies in maximum degree the capacity constraints in its class of flows with the same value, i.e.,

$$\mu_C(\overline{x}_k) = \vee_{x_k} \mu_C(x_k) \text{ for } k = v,\ v+1,\ v+2, \ldots .$$

An inspection of a finite number of flows is executed. The procedure allows also for the identification of flows close to the optimal flow.

Algorithm 3.6.4 (for Problem II) [14]

Step 1. Determine the maximum flow x_v^1 in the network assuming arc flows $x_{ij} \in [\underline{b}_{ij}, \overline{c}_{ij}]$. Substitute $\overline{x}_v^1 := x_v^1$ and go to step 2.

Step 2. Determine the path r from the source s to the destination t of the greatest ability with respect to $\overline{x}_v^1$. Generate the flow x_{v+1}^2 by increasing the value of the flow $\overline{x}_v^1$ by a unit on the path r. If $\mu_C(\overline{x}_v^1) \geq \mu_C^r(\overline{x}_v^1)$, then substitute $\overline{x}_{v+1}^2 := x_{v+1}^2$ and go to step 5. Otherwise go to step 3.

Step 3. Determine in the network an arc (k,l) for which $\mu_{kl}(x_{kl}^2) = \mu_C(x_{v+1}^2) = \wedge_{(i,j)} \mu_{ij}(x_{ij}^2)$.

If $x_{kl}^2 < b_{kl}$, then determine a path r from l to k of the greatest ability with respect to x_{v+1}^2 and go to step 4.

If $b_{kl} \leq x_{kl}^2 \leq c_{kl}$, then substitute $\overline{x}_{v+1}^2 := x_{v+1}^2$ and go to step 5.

If $x_{kl}^2 > c_{kl}$, then determine a path r from vertex k to vertex l of greatest ability with respect to x_{v+1}^2 and go to step 4.

Step 4. If $\mu_{kl}(x_{kl}^2) \geq \mu_C^r(x_{v+1}^2)$, *then substitute* $\overline{x}_{v+1}^2 := x_{v+1}^2$ *and go to step 5. Otherwise, transform the flow* x_{v+1}^2 *on the cycle formed by the arc* (k,l) *and the path as follows:*

$$x_{ij}^2 := \begin{cases} x_{ij}^2 + 1 & \text{if } (i,j) \in \overrightarrow{r} \\ x_{ij}^2 - 1 & \text{if } (i,j) \in \overleftarrow{r}, \end{cases} \tag{3.14}$$

$$x_{kl}^2 := \begin{cases} x_{kl}^2 + 1 & \text{if } x_{kl}^2 < b_{kl} \\ x_{kl}^2 - 1 & \text{if } x_{kl}^2 > c_{kl} \end{cases}$$

and go to step 3.

Step 5. Check whether the condition $\mu_D(\overline{x}_v^1) \geq \mu_D(x_{v+1}^2)$ *is valid. If so, then stop and the flow* $\overline{x}_v^1$ *remains optimal. Otherwise substitute* $\overline{x}_v^1 := \overline{x}_{v+1}^2$ *and go to step 2.*

The reader should be aware of the segment in the algorithm which is the cause of a greater complexity of Algorithm 3.6.4 compared with the algorithm given previously for Problem I. It occurs in the case when $\mu_C(\overline{x}_{v+1}^2) \geq \mu_C(\overline{x}_v^1)$. An increase of the flow $\overline{x}_v^1$ by a unit on the path of the greatest ability from s to t may not result in the flow $\overline{x}_{v+1}^2$. Thus in that case one has to "improve" the flow x_{v+1}^2 which is done in step 4 of the algorithm. We note that this case does not occur in Problem I.

The following theorem is a basis for the procedure of improving the admissibility of a flow in the class of flows with a fixed value. The theorem justifies both algorithms. It is proved for Problem II, but it is also valid for Problem I.

Before stating and proving the theorem, we explain the notion of "**improving cycles**". A cycle $(k,l) \cup r$, where r is a path from l to k if $x_{kl} < b_{kl}$ either from k or l if $x_{kl} > c_{kl}$, is called an improving cycle with respect to arc (k,l) if and only if $\mu_{kl}(x_v) < \mu^r(x_v)$ (see [20]). The transformation of the flow x_v in the improving cycle as executed in Step 4 of Algorithm 3.6.4 results in a flow x_v^1 such that $\mu_{kl}(x_{kl}^1) > \mu_{kl}(x_{kl})$ and $\mu_C(x_v^1) \geq \mu_C(x_v)$.

Theorem 3.6.5 *Let* x_v *and* $\overline{x}_v$ *be flows for Problem II (*$\overline{x}_v$ *is a flow admissible to the highest degree among the flows with a fixed value* v*, i.e.,* $\mu_C(\overline{x}_v) = \vee_{x_v}\mu_C(x_v)$*). If there exists no improving cycle for* x_v *with respect to arc* (k,l)*, where* (k,l) *is any arc such that* $\mu_{kl}(x_v) = \mu_C(x_v)$*, then* $\mu_C(x_v) = \mu_C(\overline{x}_v)$.

Proof. Suppose that $\mu_C(x_v) < \mu_C(\overline{x}_v)$. We show the existence of an improving cycle with respect to the arc (k,l) for flow x_v. The construction of such a cycle follows.

We say that an arc $(i,j) \in A$ has a type if and only if $\overline{x}_{ij} \neq x_{ij}$. If $\overline{x}_{ij} < x_{ij}$, then (i,j) is called an arc of type p. If $\overline{x}_{ij} > x_{ij}$, then (i,j) is called an arc of type q.

Since $\mu_C(\overline{x}_v) > \mu_C(x_v)$, it follows that for all $(i,j) \in A$

$$\mu_{ij}(\overline{x}_v) > \mu_{kl}(x_v). \tag{3.15}$$

Thus, $\mu_{kl}(\overline{x}_v) > \mu_{kl}(x_v)$. Hence, $\overline{x}_{kl} \neq x_{kl}$ and the type of arc (k,l) is determined. One can assume that (k,l) is an arc of type q. If (k,l) is an arc of type p, then the argument is analogous.

It is easy to show that $\forall i \in N$,

$$\sum_{j \in \Gamma_i^-} (\overline{x}_{ji} - x_{ji}) = \sum_{k \in \Gamma_i} (\overline{x}_{ik} - x_{ik}). \tag{3.16}$$

Let $\overline{S} = \langle N, \overline{A}, d \rangle$ be the network obtained from S by changing the direction of all arcs of type p. Define the function $d : \overline{A} \to \mathbb{Z}$ as follows:

$$d_{ij} = \begin{cases} -(\overline{x}_{ji} - x_{ji}) & \text{if } (j,i) \in A \text{ and } (j,i) \text{ is of type } p, \\ \overline{x}_{ij} - x_{ij} & \text{otherwise.} \end{cases}$$

Then the change of direction of an arc (i,j) of type p results in the change of sign of the respective value $\overline{x}_{ij} - x_{ij}$. The defined transformation can be performed by taking the term $\overline{x}_{ij} - x_{ij}$ on the other side of equality (3.16). It follows that the values of the function d satisfy the following conditions:

$$\sum_{j \in \Gamma_i^-} d_{ji} = \sum_{k \in \Gamma_i} d_{ik} \ \forall i \in N.$$

Thus we have

(i) $\{d_{ij}\}$ is a non-zero flow round in the network $\overline{S}$ (flow without source and destination),

(ii) the arc $(k,l) \in \overline{A}$.

Hence there exists a network $\overline{S}$ and a contour K (a cycle with all forward arcs) such that $(k,l) \in K$ and $d_{ij} > 0$ for every $(i,j) \in K$. That contour determines a path $\underline{K}$ in the original network S linking every vertex l with vertex k. Each forward arc of the path $\underline{K}$ is of type q and each backward arc is of type p. By condition (3.15), we have

$$\mu_{ij}(\overline{x}_{ij}) > \mu_{kl}(x_{kl}),\ \overline{x}_{ij} > x_{ij}\ \forall (i,j) \in \overrightarrow{\underline{K}}$$

and

$$\mu_{ij}(\overline{x}_{ij}) > \mu_{kl}(x_{kl}),\ \overline{x}_{ij} < x_{ij}\ \forall (i,j) \in \overleftarrow{\underline{K}}.$$

From the properties stated in the definition of μ_{ij}, we have that

$$\mu_{kl}(x_{kl}) < (\wedge_{(i,j) \in \overrightarrow{\underline{K}}} \mu_{ij}(x_{ij}+1)) \wedge (\wedge_{(i,j) \in \overleftarrow{\underline{K}}} \mu_{ij}(x_{ij}-1))$$

holds. Thus it follows that the cycle formed by the arc (k,l) and the path $\underline{K}$ is the required improving cycle with respect the arc (k,l) for the flow x_v. Finally note that if all arcs in the considered network have no type, then the statement is obvious. ∎

A numerical example can be found in [14, p.28].

3.7 Real-Valued Flows in a Network with Fuzzy Arc Capacities

In the previous sections, we presented a generalization of classical problems of the maximal flows in a network with one-sided and two-sided arc capacity constraints. In particular, we studied integer-valued flow in a network. In the present section, we study real-valued flows in networks with fuzzy capacities constraints, i.e., for all i and j, b_{ij}, x_{ij}, and c_{ij} are real numbers. The results of this section are from [13].

In order to study networks with real-valued flows, we need to introduce the notion of fuzzy numbers since the arc capacities are fuzzy numbers. The reader can find the relevant material in [10,16,17,32].

For ease of notation, we sometimes speak of a fuzzy set A on a set X and use the notation μ_A for its membership function.

Definition 3.7.1 *Let A be a fuzzy set on a set X. A is called* ***normalized*** *if $\mu_A(x) = 1$ for some $x \in X$.*

Definition 3.7.2 *Let M be a fuzzy set on $\mathbb{R}$. M is called a* ***fuzzy number*** *if μ_M is continuous and*
(i) $\mu_M(x) = 0\ \forall x \in (-\infty, c] \cup [d, \infty)$,
(ii) μ_M is strictly increasing on $[c, a]$ and strictly decreasing on $[b, d]$,
(iii) $\mu_M(x) = 1\ \forall x \in [a, b]$.

In the above definition, $c = -\infty$ or $a = b$ or $c = a$ or $b = d$ or $d = \infty$ is allowable.

A fuzzy number M can also be represented as

$$\mu_M(x) = \begin{cases} L(\frac{m_1 - x}{\alpha}) & \text{for } x \leq m_1,\ \alpha > 0 \\ 1 & \text{for } x \in [m_1, m_2] \\ R(\frac{x - m_2}{\beta}) & \text{for } x \geq m_2,\ \beta > 0 \end{cases} \tag{3.17}$$

where L and R are functions such that L (as well as R) satisfy $L(x) = L(-x)$, $L(0) = 1$, L is non-increasing on $[0, \infty)$, but strictly decreasing on that part of the $[0, \infty)$ interval for which $L(x) > 0$. This representation is called the $L - R$ **representation** of M (see [17]). The numbers α and β are called the **left** and **right spread** of M, respectively. For instance, one can assume $L(x) = \max\{0, 1 - x^p\}$, $p \geq 0$ or $L(x) = \exp(-|x|^p)$.

Let M be a fuzzy number with an $L - R$ representation. Then we write $M = (m_1, m_2, \alpha, \beta)_{LR}$.

Definition 3.7.3 *Let M and N be fuzzy numbers. Define the fuzzy subset $M * N$ by*

$$\mu_{M*N}(x) = \vee_{x_1, x_2}(\mu_M(x_1) \wedge \mu_N(x_2)) \tag{3.18}$$

for all $x \in \mathbb{R}$, where $x = x_1 * x_2$ and $*$ is a binary operation on the real numbers.

Let M and N be $L-R$ fuzzy numbers such that $M = (m_1, m_2, \alpha, \beta)_{LR}$, $N = (n_1, n_2, \gamma, \sigma)_{LR}$. Then

$$M + N = (m_1 + n_1, m_2 + n_2, \alpha + \gamma, \beta + \sigma)_{LR} \tag{3.19}$$

If M is an $L-R$ fuzzy number and N is an $R-L$ fuzzy number, then

$$M - N = (m_1 - n_2, m_2 - n_1, \alpha + \sigma, \beta + \gamma)_{LR} \tag{3.20}$$

For details see [17].

For subtraction, the minued must have the inverse representation with respect to the type of subtrahend. However, for a set of fuzzy numbers whose left and right sides are represented by a function of the same type, this inconvenience no longer exists. Such a set is closed with respect to subtraction.

It is the case that the r-cuts of a fuzzy number N are closed intervals, i.e., $N^r = [n_1^r, n_2^r]$.

Fuzzy numbers M and N are equal to each other, written $M = N$, if and only if $\mu_M(x) = \mu_N(x)\ \forall x \in \mathbb{R}$.

Theorem 3.7.4 $N = M \Leftrightarrow N^r = M^r\ \forall r \in (0,1]$. ■

Theorem 3.7.5 *Let N and M be fuzzy numbers. Then*
(i) $[N + M]^r = N^r + M^r = [n_1^r + m_1^r, n_2^r + m_2^r]$,
(ii) $[N - M]^r = N^r - M^r = [n_1^r - m_2^r, n_2^r - m_1^r]$, *and*
(iii) $[N \cap M]^r = N^r \cap M^r = [n_1^r \vee m_1^r, n_2^r \wedge m_2^r]$. ■

We recall the formulation of the classical problem of the maximal (minimal) flow in a network before we generalize it to the fuzzy case.

Let $S = \langle N, A \rangle$ be a network. The classical problem of the maximal (minimal) flow in the network consists in the determination of a flow which satisfies the arc capacity constraints,

$$b_{ij} \le x_{ij} \le c_{ij},\ (i,j) \in A, \tag{3.21}$$

and which is characterized by the maximal (minimal) value

$$v \longrightarrow \max\ (\min). \tag{3.22}$$

The minimization problem is not trivial only if not all $b_{ij} = 0$. Otherwise the minimal flow is the zero one.

We now generalize the flow problem in the network by allowing fuzziness in constraints (3.4) and goal (3.22). Consequently, a range of admissible flows in the arc $(i,j) \in A$ is not given now in the form of the precisely determined interval $[b_{ij}, c_{ij}]$, but in the form of a flat fuzzy number which we denote by C_{ij}. The fuzzy number may be considered as a fuzzy interval. Let μ_{ij} denote the membership function of the fuzzy interval C_{ij}. The value

$\mu_{ij}(x_{ij})$ for a given x_{ij} may be considered as the degree of satisfaction of the fuzzy capacity constraint by the arc flow x_{ij} as the possibility ('easeness') of the x_{ij} flow realization in the arc (i, j). C_{ij} may be also considered as a verbally defined fuzzy variables, for example $C_{ij} \equiv$ 'approximately between b_{ij} and c_{ij}', and μ_{ij} is then a possibility distribution relevant to this variable [43], $\mu_{ij}(x_{ij}) = \text{Poss}(x_{ij} \text{ is } C_{ij})$.

We formulate the goal (3.22) in another way. A desirable value v of the flow is given by a decision-maker in the form of a fuzzy number, which we denote by G. The membership value $\mu_G(v)$ determines the degree of the goal realization at the flow value v, the degree of the decision-maker's satisfaction with the obtained flow quantity.

We introduce the following terminology: Let $\mu_{ij}(x_v) \triangleq \mu_{ij}(x_{ij})$ denote the degree of satisfying the (i, j) arc capacity constraint by the flow x_v, and let $\mu_G(x_v) \triangleq \mu_G(v)$ denote the degree of realization of the fuzzy goal by x_v, where $x_v = \{x_{ij} \mid x_{ij} \geq 0,\ (i, j) \in A\}$ satisfies the condition (3.3). The fuzzy constraints and goal as fuzzy sets on the set of all flows generated by means of the above definitions will be denoted by the same symbols: C_{ij} and G.

From the concept of decision making in fuzzy conditions given in [4], the problem reduces now to the fuzzy decision D:

$$\begin{aligned} \mu_D(x_v) &= \mu_G(x_v) \wedge \mu_C(x_v) \quad \rightarrow \quad \max \\ D &= \cap_{(i,j)\in A} C_{ij} \cap G, \\ \mu_C(x_v) &= \wedge_{(i,j)\in A} \mu_{ij}(x_v). \end{aligned} \tag{3.23}$$

$\mu_C(x_v)$ denotes, according to the definition, the degree of simultaneous satisfaction of all arc capacity constraints in the network by the flow x_v.

Properties of the Optimal Flows

We assume that problem (3.23) is admissible, i.e., $S(D) \neq \emptyset$.

The theorems mentioned below are a consequence of the use of the Ford-Fulkerson theorems on the crisp r-cuts of the fuzzy capacity constraints.

Theorem 3.7.6 *Let $x_v^* = \{x_{ij} \mid (i, j) \in A\}$ be the optimal flow in the network S (with two-sided capacity constraints). Then there exists a cut $(X, \bar{X})$ in the network such that*

$$(i, j) \in (X, \bar{X}) \Rightarrow \mu_{ij}(x_{ij}^*) = \mu_D(x_v^*).$$

Also for $\varepsilon > 0$, the following conditions hold.

(i) For the maximization problem

$$\begin{aligned} (i, j) \in \overrightarrow{(X, \bar{X})} &\Rightarrow \mu_{ij}(x_{ij}^* + \varepsilon) < \mu_D(x_v^*), \\ (i, j) \in \overleftarrow{(X, \bar{X})} &\Rightarrow \mu_{ij}(x_{ij}^* - \varepsilon) < \mu_D(x_v^*). \end{aligned}$$

(ii) For the minimization problem

$$(i,j) \in \overrightarrow{(X,\bar{X})} \Rightarrow \mu_{ij}(x^*_{ij} - \varepsilon) < \mu_D(x^*_v),$$
$$(i,j) \in \overleftarrow{(X,\bar{X})} \Rightarrow \mu_{ij}(x^*_{ij} + \varepsilon) < \mu_D(x^*_v).$$

Proof. The proof for the maximization problem is given in Theorem 3.6.1. Next, we prove the part concerning the minimization problem.

From the definition of a fuzzy number, it follows that the r-cuts of C_{ij} are closed intervals, $C^r_{ij} = [b^r_{ij}, c^r_{ij}]$ for all $r \in (0,1]$ and $(i,j) \in A$. By the admissibility of problem (3.23), it follows that $\mu_D(x^*_v) = s > 0$. We set the lower and upper arc capacities equal to b^s_{ij} and c^s_{ij}, respectively, the ends of interval $C^s_{ij} = [b^s_{ij}, c^s_{ij}]$. The flow x^*_v is admissible under arc capacities determined this way, i.e., conditions $b^s_{ij} \leq x^*_{ij} \leq c^s_{ij}$ are satisfied for all $(i,j) \in A$. The flow x^*_v is the flow with the minimal value, else there would exist a flow $x_w = \{x_{ij} \mid (i,j) \in A\}$ such that $b^s_{ij} \leq x_{ij} \leq c^s_{ij}$ for all $(i,j) \in A$, $w < v$, $\mu_G(w) \geq \mu_G(v)$. However, we would have that $\mu_D(x_w) \geq \mu_D(x^*_v)$. Since x^*_v is the minimal flow, we must have equality. Thus, the existence of x_w would contradict the optimality of the flow x^*_v. Hence, this proves the minimality of x^*_v in the network S with arc capacities b^s_{ij} and c^s_{ij}.

By the Ford-Fulkerson theorem (Theorem 3.4.4), we have that in the network S there exists a cut, $(X, \bar{X})$, such that

$$(i,j) \in \overrightarrow{(X,\bar{X})} \Rightarrow x^*_{ij} = b^s_{ij}$$
$$(i,j) \in \overleftarrow{(X,\bar{X})} \Rightarrow x^*_{ij} = c^s_{ij}.$$

Thus, $\mu^*_{ij}(x_{ij}) = s = \mu_D(x_v)$ for all $(i,j) \in (X,\bar{X})$. Hence for the remaining arcs, we have $\mu^*_{ij}(x_{ij}) \geq s$. From the above facts and the form of the membership functions μ_{ij}, the main statement as well as conditions (ii) follow. Note that

$$\mu_C(x^*_v) = \wedge_{(i,j)\in A}\mu_{ij}(x^*_{ij}) = s = \mu_D(x^*_v) \ \blacksquare.$$

Theorem 3.7.7 *Let x^*_v be the optimal flow in the network S with one-sided capacity constraints. Then there exists in the network a cut $(X, \bar{X})$ such that*

$$(i,j) \in \overrightarrow{(X,\bar{X})} \Rightarrow \mu_{ij}(x^*_{ij}) = \mu_D(x^*_v),$$
$$(i,j) \in \overleftarrow{(X,\bar{X})} \Rightarrow x^*_{ij} = 0.$$

Proof. The proof is similar to that of Theorem 3.7.6. It suffices to note that in this case $b^s_{ij} = 0$ for all $s \in (0,1]$ and $(i,j) \in A$ and to apply Theorem 3.4.2. ■

The above theorems can be regarded as parallels to the Ford-Fulkerson theorems. They indicate the existence of a 'bottleneck' in the network, that is, the existence of a cut limiting directly the optimal flow value.

We next formulate and prove the theorem which is a more natural and immediate generalization of the Ford-Fulkerson theorems.

Determination of Optimal Real Flows

Definition 3.7.8 *A fuzzy set V on $\mathbb{R}$ with the membership function $\mu_V(v) = \vee_{x_v} \mu_C(x_v)$ is called a **fuzzy capacity** of the network S.*

The value $\mu_V(v)$ can be interpreted as the possibility of reaching the flow value v in the network S, taking into account the existence of fuzzy arc capacity constraints.

Definition 3.7.9 *The fuzzy capacity of the cut $(X, \overline{X})$ is the fuzzy number $C(X, \overline{X})$ defined as follows:*

(i) for a network with two-sided capacity constraints

$$C(X,\overline{X}) = \sum_{(i,j)\in\overrightarrow{(X,\overline{X})}} C_{ij} - \sum_{(i,j)\in\overleftarrow{(X,\overline{X})}} C_{ij}; \tag{3.24}$$

(ii) for a network with one-sided capacity constraints

$$C(X,\overline{X}) = \sum_{(i,j)\in\overrightarrow{(X,\overline{X})}} C_{ij}. \tag{3.25}$$

The summation and subtraction signs appearing in (3.24) *and* (3.25) *denote respective operations on fuzzy numbers (see* (3.18)).

Theorem 3.7.10 *Let W denote the set of all cuts in the network S and I be the fuzzy set such that $\mu_I(x) = \vee_v \mu_V(v)$ $\forall x \in \mathbb{R}$. Then the following relation holds:*

$$V = I \cap \cap_{(X,\overline{X})\in W} C(X,\overline{X}). \tag{3.26}$$

Proof. We prove that the corresponding r-cuts of the right and left-side of (3.26) are equal for every $r \in (0,1]$. For $r \in (\vee_v \mu_V(v), 1]$, we have $V^r = \emptyset$ and $I^r = \emptyset$. Thus the condition is trivially satisfied. Hence we only need to consider $r \in (0, \vee_v \mu_V(v)]$. We note that $V^r = [v^r_{\min}, v^r_{\max}]$, where $v^r_{\min}$ and $v^r_{\max}$ are values of the minimal and maximal flows, respectively, in the network S with arc capacities b^r_{ij} and c^r_{ij}. In fact,

$$\begin{aligned} V^r &= \{v \mid \mu_V(v) \geq r\} \\ &= \{v \mid x_v \text{ exists such that } \mu_C(x_v) \geq r\} \\ &= \{v \mid x_v \text{ exists such that } \mu_{ij}(x_{ij}) \geq r \ \forall (i,j) \in A\} \\ &= \{v \mid \text{a flow } x_v \text{ of value } v \text{ exists in the network } S \text{ with} \\ &\qquad \text{arc capacities } b^r_{ij} \leq x_{ij} \leq c^r_{ij}\} \\ &= [v^r_{\min}, v^r_{\max}]. \end{aligned}$$

From Theorems 3.4.3 and 3.4.4, it follows that

$$v^r_{\min} = \vee_{(X,\overline{X})\in W}\Big(\sum_{(i,j)\in\overrightarrow{(X,\overline{X})}} b^r_{ij} - \sum_{(i,j)\in\overleftarrow{(X,\overline{X})}} c^r_{ij}\Big)$$

and

$$v^r_{\max} = \wedge_{(X,\overline{X})\in W}\Big(\sum_{(i,j)\in\overrightarrow{(X,\overline{X})}} c^r_{ij} - \sum_{(i,j)\in\overleftarrow{(X,\overline{X})}} b^r_{ij}\Big).$$

We now evaluate the r-cut of the right side of (3.26) by using Theorem 3.7.5:

$$\begin{aligned}[C(X,\overline{X})]^r &= \textstyle\sum_{(i,j)\in\overrightarrow{(X,\overline{X})}} C^r_{ij} - \sum_{(i,j)\in\overleftarrow{(X,\overline{X})}} C^r_{ij} \\ &= [\textstyle\sum_{(i,j)\in\overrightarrow{(X,\overline{X})}} b^r_{ij}, \sum_{(i,j)\in\overrightarrow{(X,\overline{X})}} c^r_{ij}] \\ &\quad -[\textstyle\sum_{(i,j)\in\overleftarrow{(X,\overline{X})}} b^r_{ij}, \sum_{(i,j)\in\overleftarrow{(X,\overline{X})}} c^r_{ij}] \\ &= [\textstyle\sum_{(i,j)\in\overrightarrow{(X,\overline{X})}} b^r_{ij} - \sum_{(i,j)\in\overleftarrow{(X,\overline{X})}} c^r_{ij}, \\ &\quad \textstyle\sum_{(i,j)\in\overrightarrow{(X,\overline{X})}} c^r_{ij} - \sum_{(i,j)\in\overleftarrow{(X,\overline{X})}} b^r_{ij}] \\ &\triangleq [b^r, c^r],\end{aligned}$$

$$\begin{aligned}[I \cap \cap_{(X,\overline{X})\in W} C(X,\overline{X})]^r &= I^r \cap [\vee_{(X,\overline{X})\in W} b^r, \wedge_{(X,\overline{X})\in W} c^r] \\ &= \mathbb{R} \cap [v^r_{\min}, v^r_{\max}] \\ &= [v^r_{\min}, v^r_{\max}] \\ &= V^r.\end{aligned}$$

Since the r-cuts of the left and right side of (3.26) are equal, equation (3.26) holds as well as the theorem. Equality (3.26) can be proved similarly for a network with one-sided capacity constraints. ■

Theorem 3.7.10 is an analogue to the Ford-Fulkerson theorems relating the possible flow value in the network (that is the network capacity) with the cut capacities. The analog is clear in the case of maximizing the flow under one-sided capacity constraints because the fuzzy set I, appearing in equality (3.26) can be left out since $\vee_v \mu_V(v) = 1$. Under the assumptions on the form of the numbers C_{ij}, the intersection operation in (3.26) is the same in this case as the fuzzy minimum operation and can be replaced by

$$V = \widetilde{\wedge_{(X,\overline{X})\in W}} C(X,\overline{X}). \tag{3.27}$$

For two-sided capacity constraints, both the minimization and maximization problems are treated jointly in the theorem. The left, increasing part of μ_V provides information on the possibility of achieving 'small' flow values in the network (determining the 'minimal' fuzzy network capacity) whereas the right, decreasing part of μ_V, characterizes the 'maximal' network capacity.

In the following algorithm it is assumed that the C_{ij} are numbers of the same type, that is $C_{ij} = (0, c_{ij}, 0, \beta_{ij})_{LR}$, when determining the optimal flow in the network with one-sided arc capacities conatraints. Only in that case, is it possible to calculate efficiently and exactly the fuzzy arc capacities according to (3.25). The algorithm seeks the intersection point of functions μ_V and μ_G, which determines the value and estimate of the optimal flow.

Algorithm 3.7.11 (12) *Step 1. Determine the maximal real flow x_w and the respective minimal cut $(X, \overline{X})$ in the network with arc capacities c_{ij}. If $\mu_G(w) = 1$, then x_w is optimal. Otherwise go to Step 2.*

Step 2. Determine the fuzzy number $C(X, \overline{X})$ and coordinates (v, r) of the intersection point of the functions $\mu_{C(X,\overline{X})}$ and μ_G. If $r = 0$, then STOP (the problem is inadmissible). Otherwise, go to Step 3

Step 3. Determine the maximum flow x_w and the respective minimal cut $(X, \overline{X})$ in S with arc capacities c_{ij}^r. If $\mu_D(x_w) = r$ and $w = v$, then the flow x_w is optimal. Otherwise go to Step 2.

The algorithm terminates in a finite number of steps from the fact that there are finite number of cuts in the network and from the fact that in Step 3 of the algorithm, different cuts are generated in succession (except the last two).

It follows from equality (3.26) and properties of the fuzzy summations, that the membership function μ_V is constructed by 'segments' of the membership functions of the fuzzy capacities of a group of cuts in the network. These cuts are called **essential**. The minimal cut characterized by Theorem 3.7.7 is one of the essential cuts.

Every cut generated in Step 3 is essential, otherwise it could not be a minimal cut in the network S with arc capacities $0 \leq x_{ij} \leq c_{ij}^r$.

In the case of linear fuzzy constraints and goal, that is, when C_{ij} and G are numbers of the form

$$\begin{aligned} C_{ij} &= (0, c_{ij}, 0, d_{ij} - c_{ij})_{LR} \\ G &= (a_G, \infty, a_G - c_G, 0)_{LR}, \end{aligned} \tag{3.28}$$

where $L(x) = R(x) = (1 - |x|) \vee 0$, the right parts of $\mu_{C(X,\overline{X})}$, $(X, \overline{X}) \in W$ are linear functions and the right side of μ_V is a piece-wise linear function. Thus in seeking the intersection point of $\mu_{C(X,\overline{X})}$ and μ_G, a system of two linear equations with two unknowns is solved in Step 2 of the algorithm. Hence under the assumption that c_{ij}, d_{ij}, a_G and c_G are rational, the coordinates of (v, r) generated in Step 2 are also rational. Consequently, the flow generated in Step 3 is rational. This seems essential for the numerical realization of the algorithm when we must restrict our consideration to the case of rational numbers.

Example 3.7.12 (13) *Consider the network as given in Figure* 3.11. *Assume that $C_{ij} = (c_{ij}, \overline{c}_{ij} - c_{ij})_R$, where $R(x) = 0 \vee (1 - |x|)$, i.e., $\mu_{ij}(x_{ij})$*

is linear on $[c_{ij}, \bar{c}_{ij}]$, $(i,j) \in A$. *Let it be :* $C_{S1} = (4,16)_R$, $C_{S2} = (5,40)_R$, $C_{12} = (4,4.5)_R$, $C_{1t} = (7.5,1.5)_R$, $C_{2t} = (8.5,4)_R$. *Let* μ_G *be linear on* $[v_1, v_0]$ *and* $v_1 = 18$, $v_0 = 20$.
The algorithm runs as follows:

Step 1. $x_w = x_9$ *and* $x_{s1} = 4$, $x_{s2} = 5$, $x_{12} = 0$, $x_{1t} = 4$, $x_{2t} = 5$, $(X, \overline{X}) = (\{s\}, \{1,2,t\})$.

$\mu_G(9) = 0 \neq 1$. *Go to Step 2.*

Step 2. $C(\{s\},\{1,2,t\}) = (9,56)_R$, $(v,r) = (19.62,\ 0.81)$. $r = 0.81 \neq 0$. *Go to Step 3.*

Step 3. c^r_{ij} *s :* $c^r_{s1} = 7.04$, $c^r_{s2} = 12.6$, $c^r_{12} = 4.86$, $c^r_{1t} = 7.78$, $c^r_{2t} = 9.26$. $w = 16.3$ *and* x_w : $x_{s1} = 7.04$, $x_{s2} = 9.26$, $x_{12} = 0$, $x_{1t} = 7.04$, $x_{2t} = 9.26$. $(X, \overline{X}) = (\{s,2\},\{1,t\})$.

$\mu_D(x_w) = 0 \neq r = 0.81$ *and* $w = 16.3 \neq v = 19.62$. *Go to Step 2*

Step 2. $C(\{s,2\},\{1,t\}) = (12.5,20)_R$, $(v,r) = (19.32,\ 0.66)$. $r = 0.66 \neq 0$. *Go to Step 3.*

Step 3. c^r_{ij} *s :* $c^r_{s1} = 9.44$, $c^r_{s2} = 13.6$, $c^r_{12} = 5.53$, $c^r_{1t} = 8.01$, $c^r_{2t} = 9.86$. $w = 17.87$ *and* x_w : $x_{s1} = 8.01$, $x_{s2} = 9.86$, $x_{12} = 0$, $x_{1t} = 8.01$, $x_{2t} = 9.86$. $(X, \overline{X}) = (\{s,1,2\},\{t\})$.

$\mu_D(x_w) = 0 \neq r = 0.66$ *and* $w = 17.87 \neq v = 19.32$. *Go to Step 2.*

Step 2. $C(\{s,1,2\},\{t\}) = (16,5.5)_R$, $(v,r) = (18.93,\ 0.467)$. $r = 0.467 \neq 0$. *Go to Step 3.*

Step 3. c^r_{ij} *s :* $c^r_{s1} = 12.54$, $c^r_{s2} = 26.32$, $c^r_{12} = 6.4$, $c^r_{1t} = 8.3$, $c^r_{2t} = 10.63$. $w = 18.93$ *and* x_w : $x_{s1} = 8.3$, $x_{s2} = 10.63$, $x_{12} = 0$, $x_{1t} = 8.3$, $x_{2t} = 10.83$. $(X, \overline{X}) = (\{s,1,2\},\{t\})$.

$\mu_D(x_w) = 0.467 = r$ *and* $w = 18.93 = v$.

STOP. The current x_w *is optimal.*

Algorithm 3.7.11 can be adopted to the 'maximization' of the flow in a network with two-sided capacity constraints. In this case, the maximal flow x_w in the network with two-sided constraints b^r_{ij} and c^r_{ij} should be tetermined in Step 3. It would be more convenient to use numbers C_{ij} of $L-R$ type, where $L = R$. This allows for the easy execution of subtraction in (3.24) when evaluating fuzzy cut capacities. Still some problems can arise in Step 1 of the algorithm since the flow problem in the network with capacities b_{ij} and c_{ij} may be inadmissible. In that case, V is not a normal fuzzy number and the problem of determining the modal value of μ_V arises, that is v^* such that $\mu_V(v^*) = r^* = \vee_v \mu_V(v)$. The point (v^*, r^*) would be a starting point in the algorithm. It is worth noticing that if $\mu_V(v^*) \geq r^*$, then r^* is just the estimate of the optimal flow.

Another interesting problem arises in Algorithm 3.7.11 since not all essential cuts are identified. Therefore, the complete form of V is not identified. This knowledge could be of some importance for the decision-maker. The problem consists in the construction of an efficient algorithm determining all essential cuts. An analysis of the dual maximal flow problem may allow for a positive solution to the problem.

3.8 Petri Nets

In this section, we briefly introduce the notion of a (crisp) Petri Net. The instructions in certain computer programs would be processed sequentially. However, often there is no logical reason that prevents some instructions being processed in any order or concurrently. Due to the decline of the cost of computer hardware, processors in particular, there is increasing interest in concurrent processing to achieve greater speed and efficiency. The use of **Petri nets** is one method of modeling and studying concurrent processing.

Definition 3.8.1 *A* ***Petri net*** *is directed graph* $G = (V, E)$, *where* $V = P \cup T$ *and* $P \cap T = \emptyset$. *Any edge e in E is incident on one member of P and one member of T. The set P is called the set of* ***places*** *and the set T is called the set of* ***transitions.***

A Petri net is a directed, bipartite graph, where the two classes of vertices are called places and transitions. In general, parallel edges are allowed in Petri nets. However, for simplicity, we do not permit parallel edges. Thus $E \subseteq (P \times T) \cup (T \times P)$ in Definition 3.8.1.

Example 3.8.2 *The directed graph of Figure 3.13 is an example of a Petri net. Circles represents places and bars (or rectangles) represents transitions.*

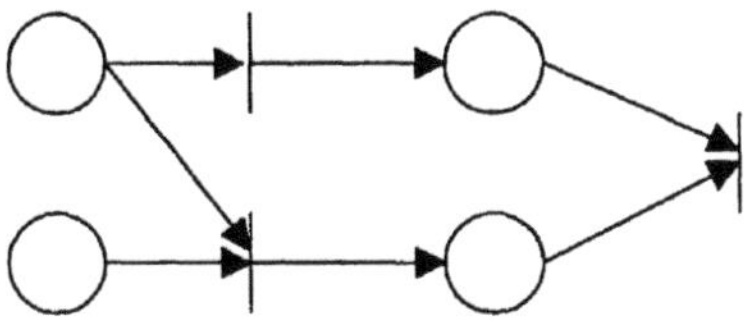

Figure 3.13

Definition 3.8.3 *A* ***marking*** *of a Petri net is a function from the set of places to the nonnegative integers. It assigns each place a nonnegative integer. A Petri net with a marking is called a* ***marked Petri net*** *(or sometimes just a Petri net).*

If a marking assigns the nonnegative integer n to place p, we say that there are n **tokens** on p. The tokens are represented as black dots.

Example 3.8.4 *An example of a marked Petri net is given in Figure 3.14.*

Figure 3.16. *Transition* t_3 *is now enabled. If we then fire transition* t_3, *we obtain the net shown. At this point no transition is enabled and thus none may be fired.*

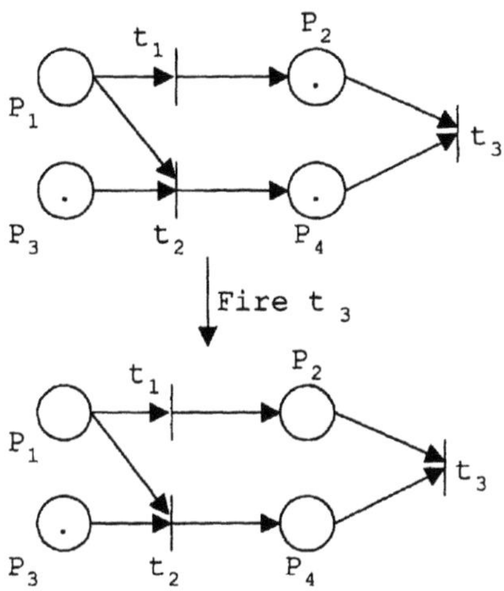

Figure 3.16

Definition 3.8.8 *If a sequence of firings transforms a marking* M *to a marking* M', *we say that* M' *is* **reachable** *from* M.

In Figure 3.17, M'' is reachable from M by first firing transition t_1 and then firing t_2.

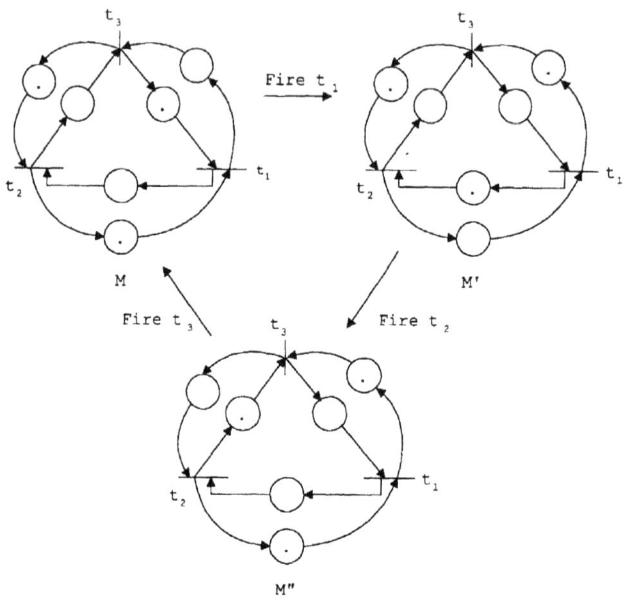

Figure 3.17

The notion of **liveness** and **safeness** are among the most important properties studied in Petri net theory. Liveness is related to the absence of deadlocks and safeness is related to bounded memory capacity.

Definition 3.8.9 *A marking M for a Petri net is* ***live*** *if, beginning from M, no matter what sequence of firings has occurred, it is possible to fire any given transition by proceeding through some additional firing sequence.*

If a marking M is live for a Petri net G, then no matter what sequence of transition is fired, G will never deadlock. Indeed, any transition can be fired by proceeding through some additional firing sequence.

If a place is regarded as having limited capacity, **boundedness** assures us that no place will overflow.

Definition 3.8.10 *A marking M for a Petri net is* ***bounded*** *if there is some positive integer n such that in any firing sequence, no place ever receives more that n tokens. If a marking M is bounded and in any firing sequence no place ever receives more that one token, we call M a* ***safe*** *marking.*

For example, if each place represents a register capable of holding one computer word and if an initial marking is safe, we are guaranteed that the memory capacity of the register will not be exceeded.

3.9 Fuzzy Petri Nets for Rule-Based Decisionmaking

The material of this section is from [28].

In this section, we extend to Petri nets the technique of fuzzy reasoning by way of transformations of fuzzy truth state vectors by fuzzy rule matrices. A new type of neural network, where the transition bars serve as the neurons and the nodes are conditions is thus presented. Conditions may be conjuncted and disjuncted in order to allow the firing of the neurons. Fuzzy values are allowed for rules and truths of conditions that appear in rules. Then, multiple copies, rather than the original, of the fuzzy truth tokens are passed along all arrows that leave a node or transition bar, where the truth token is present. We present an algorithm for reasoning by way of these networks. Abduction may be done analogously by reversing all arrows and propagating truth tokens backwards.

In [30], a compact and efficient method for reasoning by means of propositional logic was developed. In [29], it was extended to fuzzy logic. These methods have possible applications to control theory and pattern recognition. These algorithms execute quickly without human interaction. They also are appropriate for many real-time control situations. However, their structure is not general enough for predicate logic to work in them. We modify Petri nets to model fuzzy reasoning with propositional logic.

A compact technique for reasoning forward and backward using transformations of truth state vectors by a rule matrix was devised in [30]. A truth state vector u contains a Boolean-valued component for k conditions, $C_1, \ldots, C_k$. The truth of C_i is denoted by $u_i = 1$, and its falsity is denoted by $a_i = 0$. Each condition C_i is a string that forms a set of English words. However, conditions in processing are not used because it is more efficient to process their truth states instead. The conditions are used only for communicating with users, e.g., for explaining or editing.

A rule matrix R associated with a set of k conditions is a $k \times k$ matrix, where the entry $R_{ij} = 1$ if Condition $(j) \Rightarrow$ Condition (i). The entry $R_{ij} = 0$ when there is no known implication. Hence all elements on the diagonal equal 1. When a rule matrix R transforms a truth state vector u, the row-by-column dot product is performed by replacing multiplication and addition with the logical operations AND and OR, respectively.

Example 3.9.1 *Suppose that we have four conditions, C_1, C_2, C_3, and C_4. C_1 and C_2 might represent symptoms of a patient, while C_3 and C_4 might represent diagnoses. Suppose also that our experience is represented by the following rules:*

$$R_{31} = 1 : [C_1 \Rightarrow C_3] \tag{3.29}$$

$$R_{42} = 1 : [C_2 \Rightarrow C_4]. \tag{3.30}$$

We also have that the diagonal elements are all 1's and the remaining entries are all 0's. Suppose that the given known truth of C_1, is $u_1 = 1$. Then the matrix transformation of all the appropriate truth states is

$$Ru = \begin{bmatrix} 1 & 0 & 0 & 0 \\ 0 & 1 & 0 & 0 \\ 1 & 0 & 1 & 0 \\ 0 & 1 & 0 & 1 \end{bmatrix} \begin{bmatrix} 1 \\ 0 \\ 0 \\ 0 \end{bmatrix} = \begin{bmatrix} 1 \\ 0 \\ 1 \\ 0 \end{bmatrix} = v. \tag{3.31}$$

The resultant or deduced truth state v shows the truth of C_3 and is given by the rule $R_{31} = 1$.

In [39] it has been shown for binary matrices that the transpose of the rule matrix takes a truth state of given symptoms backwards into an abduced state of logical causes [30]. For example, if the truth of C_4 is given as true by the symptomatic truth state component $v_4 = 1$, then the transpose of the rule matrix transforms v as follows:

$$R^T v = \begin{bmatrix} 1 & 0 & 1 & 0 \\ 0 & 1 & 0 & 1 \\ 0 & 0 & 1 & 0 \\ 0 & 0 & 0 & 1 \end{bmatrix} \begin{bmatrix} 0 \\ 0 \\ 0 \\ 1 \end{bmatrix} = \begin{bmatrix} 0 \\ 1 \\ 0 \\ 1 \end{bmatrix} = u. \tag{3.32}$$

This yields the logically causal truth state u with $u_2 = 1$. Thus C_2 is a candidate for the logical cause of the truth of C_4 by the rule $R_{42} = 1$.

We extend the above method from reasoning with Boolean values to reasoning with fuzzy values [29], i.e., values from $[0, 1]$. These values designate the degree of belief or subjective certainty in the truths. A fuzzy truth state component value u_i denotes the degree of belief in the truth of the corresponding condition C_i. The degree could be given by humans or by the output of a pattern recognition system to which sensor signals are input. A fuzzy value in a rule matrix entry R_{ij} gives the degree of belief that Condition (j) implies Condition (i). Such beliefs are based on human experience or relative success counts, both of which are based on feedback.

The fuzzy row-by-column dot product is analogous to the Boolean case, except that $\wedge$ and $\vee$ replace and generalize AND and OR, respectively.

Example 3.9.2 *Let the rules in Example 3.9.1 be given the following fuzzy values:*

$$R_{31} = 0.7 \;\; [C_1 \Rightarrow (0.7)C_3] \tag{3.33}$$

$$R_{42} = 0.6 \;\; [C_2 \Rightarrow (0.6)C_4]. \tag{3.34}$$

Let the degree of belief or certainty in the truth of Condition (1) *be* 0.4. *Then we have*

$$\begin{bmatrix} 1 & 0 & 0 & 0 \\ 0 & 1 & 0 & 0 \\ 0.7 & 0 & 1 & 0 \\ 0 & 0.6 & 0 & 1 \end{bmatrix} \begin{bmatrix} 0.4 \\ 0 \\ 0 \\ 0 \end{bmatrix} = \begin{bmatrix} 0.4 \\ 0 \\ 0.4 \\ 0 \end{bmatrix}. \tag{3.35}$$

Next, letting Condition (4) *be given as true with degree of belief* 0.8, *the transpose of* R *reasons backwards as follows:*

$$\begin{bmatrix} 1 & 0 & 0.7 & 0 \\ 0 & 1 & 0 & 0.6 \\ 0 & 0 & 1 & 0 \\ 0 & 0 & 0 & 1 \end{bmatrix} \begin{bmatrix} 0 \\ 0 \\ 0 \\ 0.8 \end{bmatrix} = \begin{bmatrix} 0 \\ 0.6 \\ 0 \\ 0.8 \end{bmatrix}. \tag{3.36}$$

Rules whose conditions are the conjunctions of other conditions are managed by the use of extra conditions $C_{k+1}, \dots, C_n$ and their associated truth state components $u_{k+1}, \dots, u_n$. Each of these components is associated with a conjunctive string that gives the number of conjunctive conditions and their component indices from among the previous K conditions. Suppose the truths of $C_1, \dots, C_k$ are given. Then the truths of $C_{k+1}, \dots, C_n$ are computed by mining the truths of the appropriate conjuncted conditions. Thus an extended truth state u (or v) is completed. It is then transformed into the extended deduced (or abduced) truth state v (or u).

Petri nets are natural for logical arrays of MIN and MAX (AND and OR) gates, as long as certain modifications are made.

We denote the nodes by $C_1, \dots, C_k$ of the Petri net in order to represent a set of k conditions in our rule-based system. The truths of the conditions (the tokens of Petri nets) will be designated by $T_1, \dots, T_k$. The T_i make up a truth state T for the conditions, $i = 1, 2, \dots, k$. Each component of the truth state T will be either 0 (false or disabling), or 1 (true or enabling). The truths of the conditions activate the transition bars, called **neurons.** The truth values for the neurons make up a neuron state vector denoted by $N = (N_1, \dots, N_l)$, for appropriate l.

Figure 3.18 depictes the properties of concurrency and conflict.

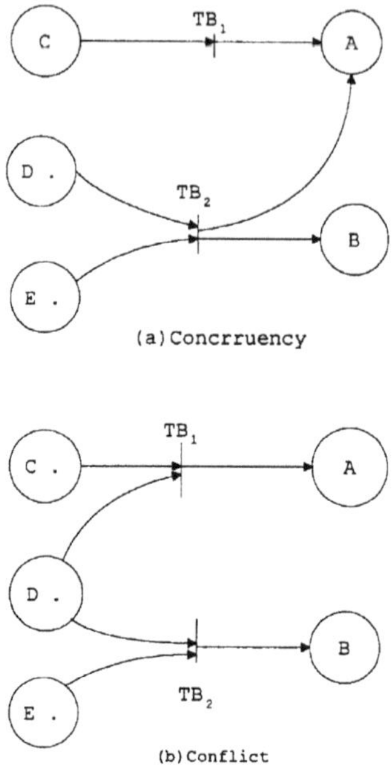

Figure 3.18

In Figure 3.18(a), the order of activation of node A from transition bar TB_1 or TB_2 is important in logic, where token passing represents truth implication. A token follows a single arrow in the usual Petri nets. A token in the Petri net shown in Figure 3.18(b) could go from node D to either TB_1 or TB_2. However, in logical implication, the original token would remain at node D and copies would be sent to both TB_1 and TB_2. This is necessary in logic since the truth of a condition may imply the truth of multiple conditions simultaneously.

Condition 1 In a (Petri) logic network, an activated node passes copies of its truth token along all arrows that leave that node. An activated neuron (one that fires) passes copies of its token along all arrows that depart from it.

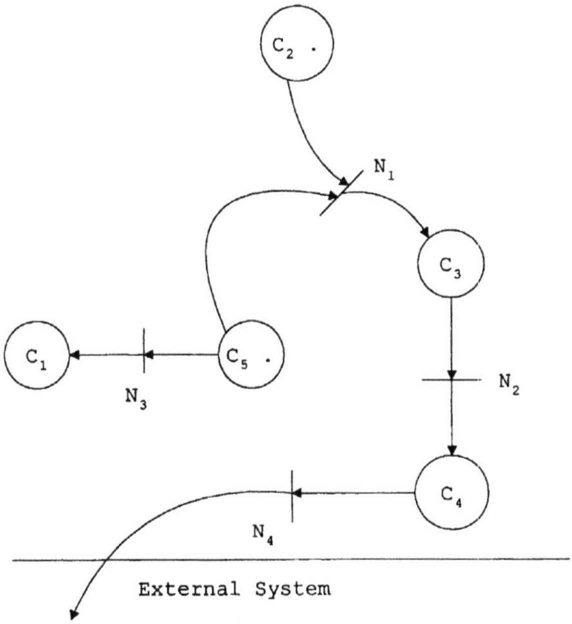

Figure 3.19

Figure 3.19 depicts a simple rule-based system which has the following rules:

$R_1 : (C_2 \text{ AND } C_5) \Longrightarrow C_3$,
$R_2 : C_5 \Longrightarrow C_1$,
$R_3 : C_3 \Longrightarrow C_4$,
$R_4 : C_4 \Longrightarrow C'$ (external node).

Suppose now that the truths of C_2 and C_5 are given, i.e., that node C_2 and node C_5 are given enabling tokens as shown in Figure 3.19. Then neurons N_1 and N_3 are fired. Hence C_1 and C_3 become true. Next, the truth of C_3 causes neuron N_2 to fire so that C_4 becomes true. C_4 sends a truth token copy to C' in an external system.

We now allow the truth tokens $T_1, \ldots, T_k$ associated with the conditions $C_1, \ldots, C_k$ to take on fuzzy values. This corresponds to the state (marking) for a Petri net, which uses only 1's and 0's. We associate with each arrow leaving a node a fuzzy value for multivalued logical implication rules (fuzzy rules). If we do not want to allow a neuron to fire at low levels, we include a threshold (decision) D_i for each neuron state component N_i. We call such models, with Condition 1 holding and with no cycles, a fuzzy Petri net or fuzzy logic network (FLN).

Consider Figure 3.19. Let C_5 and C_2 be given the fuzzy truth token values $T_5 = 0.6$ and $T_2 = 0.7$, respectively. Then neuron N_1 would have the fuzzy value $0.6 \wedge 0.7 = 0.6$. If we use a decision value $D_1 = 0.5$, then

$N_1 > D_1$ and the neuron would fire and transfer its token value of 0.6 across the neuron to node C_3, i.e., $T_3 = 0.6$.

A fuzzy logic network executes when one or more neurons fire [33, p. 23]. Given an FLN $\tilde{N}$ and a fuzzy truth token state T, the reachability set of $(\tilde{N}, T)$ is the set of all fuzzy truth token states that can be reached by a sequence of executions of $\tilde{N}$. Standard problems in Petri nets are reachability, deadlock, and optimization. Deadlock occurs when every tranisition bar is waiting to be activated and there are no firings. By optimization, we mean the minimal sequence of executions to reach a given final state from a given initial state. Since states must be reached to control a process, we equate reachability in a FLN.

Actually, the problems of reachability, deadlock, and optimization are not important in FLNs as shown by the following the algorithm. It is well-known and used in [30] that a digraph with no cycles has one or more maximal paths. In an FLN, a maximal path may not be an executable path since tokens from other nodes may be needed to fire a neuron. However, there still is a longest string of nodes. We call the length of this the **maximal span** of $\tilde{N}$.

Proposition 3.9.3 *Let M denote the maximal span of the FLN $\tilde{N}$. Then the number of executions needed to reach any final state T' from an initial state T does not exceed $[M/2]$, where $[M/2]$ is the greatest integer less than or equal to $M/2$.*

Proof. There is an input node and an output node that attach to every neuron that executes. When a neuron fires, tokens pass through the output conditional node on to the next neuron. This latter neuron must be fired for further traversing of the implication chain. Hence every firing traverses two arrows. Recall that an FLN contains no cycles. There are $n-1$ neurons between n conditional nodes. Consequently, the total number of conditions and transition bars in a maximal chain is $M = 2n - 1$. Thus $n - 1 = [M/2]$. If a neuron had to wait for a token from a different partial chain that fed into it at some point and that partial chain together with the remaining part of $\tilde{N}$ (from that point forward) were longer than M, then the maximality of $\tilde{N}$ would be contradicted. ■

We now present an algorithm for implementing the Petri model.

There are two fuzzy states, namely, T and N, to be updated in sequence in any algorithm that implements the fuzzy logic network described above. Multiple subsystems may be updating asynchronously in a distributed system, with each one executing its own algorithm (which could all be the same). Consider a single subsystem. First, the conditions, or nodes, $C_1, \ldots, C_k$, represent the conditions. Associated with these is the fuzzy truth token state $T = (T_1, \ldots, T_k)$ that at any instant gives the fuzzy truth tokens for these conditions. There is also the neuron truth state defined by $N = (N_1, \dots, N_l)$, where the N_i are the fuzzy truths at the neurons (transition bars). There is also the decision vector $D = (D_1, \dots, D_l)$ of thresholds.

Next, a fuzzy partial rule matrix gives the certainty or degree-of-belief values of the fuzzy rules $R = [R_{ij}]$. The partial rule matrix for fuzzy logic networks is different from the rule matrix first described in this section. The columns represent condition truths and the rows represent neuron values. For example, $R_{ij} = 0.6$ means that the truth token value of Condition (j) implies (transmits) a truth value of $[0.6 \wedge T_j]$ to the neuron N_i. If other arrows enter N_i from other nodes, then all implied fuzzy values arriving there must be MINed to obtain the new fuzzy value for N_i. Hence the weakest evidence of conjuncted conditions prevails. This is in agreement with the fact that if the weakest one were 0, then the conjoined condition would not be true. On the other hand, truth values coming into a node from more than one neuron are maxed. This is in accordance with the rules of fuzzy logic.

The higher level algorithm for deduction starts with fetched truth token values (T) and updates the neuron state component values (N). It then updates the truth token values (T) and so on until all implication paths have been traversed.

Algorithm [28]

1. Initialize (reset) all fuzzy truth tokens T_i and neuron state component fuzzy values N_i to 0. Then read in new fuzzy truth tokens from the sensor/recognition input system for the initialized fuzzy truth state T.

2. Set all neuron state components, N_i that are 0 to 1 (for MINing). Propagate all fuzzy truth tokens for the conditions, as given in the fuzzy truth state T, by the fuzzy rules in the fuzzy partial rule matrix R to obtain a resultant fuzzy neuron state vector N of fuzzy implication values by MINing with N_i.

3. Compare each neuron state component, N_i with its decision (threshold) value D_i: if $N_i < D_i$, then set the neuron value to $N_i = 0$; else leave it at its current value.

4. Update the truth token state T of the conditions. Transfer all fuzzy neuron values N_i along their departing arrows to the appropriate conditional nodes by MAXing each incoming N_i in turn with the fuzzy truth value T_j for the corresponding receiving node C_j.

5. If the maximal implication chains have been traversed (in which case no further changes will occur), then transfer the fuzzy truth state T to another subsystem (for selecting control commands, responses, etc.); else go to Step 2.

6. Get stop condition from external subsystem, and if set, then stop; else go to Step 1.

Step 1 starts with all fuzzy truth components equal to 0 in preparation for later MAXing. Then the fuzzy truth values are fetched from a sensor/pattern-recognition system. (This system may use a fuzzy logic network to make decisions on data from sensor signals). The neuron state is then set to all 1's (maximum value) in preparation for the MINing to be done late. Next, the fuzzy truth state T is propagated by the fuzzy rules to obtain a set of implication values for each neuron state component N_i.

Step 2 starts with either the current value for each N_i if it is nonzero from a previous MINing, or if N_i is 0, it is set to 1 for MINing. Each N_i is MINed with the values coming into it, which are themselves results of MINing truth tokens and rule matrix values.

In Step 3, each neuron value N_i above that neuron's threshold value D_i is kept for firing; otherwise that neuron is disabled by setting N_i to 0. Step 4 fires the neuron by transferring each nonzero fuzzy value N_i present at the ith neuron along all arrows that depart that neuron to the appropriate fuzzy truth state components T_j. The rule values for these implication arrows are always 1's. Since a fuzzy truth (token) value T_j for node C_j may be affected by more than a single neuron that feeds into it, T_j is determined by MAXing the incoming values with the current value of T_j. In Step 5, if the fuzzy truth state T has reached a final value, it is output to another subsystem for some particular use. If not the inner iteration loop is repeated). In Step 6, the process is either stopped or iterated again from Step 1.

The preceding algorithm is for deductive reasoning. All arrows can be reversed since there are no constraints on the number of arrows into or out of either a node or a neuron. Hence abductive reasoning can be done analogously by reversing all arrows.

Consider again the situation given by Figure 3.19 and apply the foregoing algorithm. Let all threshold values be 0, i.e., disable the thresholding. First, all fuzzy truth state components (tokens) T_i are set to 0, $1 \leq i \leq k$, and all neuron state components N_j are reset, $1 \leq j \leq l$. Then the nonzero fuzzy truth values are fetched for the T_i. For this example, let be $T_2 = 0.8$ and $T_5 = 0.5$. Let the rules denoted by the arrows in the figure each have fuzzy value 1 for simplicity. Next update the truth tokens and neuron state components (see Figure 3.19):

$T_1 = 0$ [Step 1: get initial truths]

$T_2 = 0.8$

$T_3 = 0$

$T_4 = 0$

$T_5 = 0.5$

(reset all N_i that are 0 to 1 for MINing)

$N_1 = ((N_1 \wedge T_2) \wedge T_5) = ((1 \wedge 0.8) \wedge 0.5) = 0.5$ [Step 2: update neurons]

$N_2 = (N_2 \wedge T_3) = (1 \wedge 0) = 0$

$N_3 = (N_3 \wedge T_5) = (1 \wedge 0.5) = 0.5$

$N_4 = (N_4 \wedge T_4) = (1 \wedge 0) = 0$

— [Step 3: threshold-not used]

$T_1 = (T_1 \vee N_3) = (0 \vee 0.5) = 0.5$ [Step 4: update truth state tokens]

$T_2 = 0.8$

$T_3 = (T_3 \vee N_1) = (0 \vee 0.5) = 0.5$

$T_4 = (T_4 \vee N_2) = (0 \vee 0) = 0$

$T_5 = 0.5$

(reset all N_i that are 0 to 1 for MINing)

$N_1 = ((N_1 \wedge T_2) \wedge T_5) = ((0.5 \wedge 0.8) \wedge 0.5) = 0.5$ [Step 2: update neurons]

$N_2 = (N_2 \wedge T_3) = (1 \wedge 0.5) = 0.5$

$N_3 = (N_3 \wedge T_5) = (0.5 \wedge 0.5) = 0.5$

$N_4 = (N_4 \wedge T_4) = (1 \wedge 0) = 0$

— [Step 3: thresholding-not used]

$T_1 = (T_1 \vee N_3) = (0.5 \vee 0.5) = 0.5$ [Step 4: update truth state]

$T_2 = 0.8$

$T_3 = (T_3 \vee N_1) = (0.5 \vee 0.5) = 0.5$

$T_4 = (T_4 \vee N_2) = (0.5 \vee 0.5) = 0.5$

$T_5 = 0.5$

(reset all N_i that are 0 to 1 for MINing)

On the next iteration, N_4 will become 0.5. Hence a fuzzy truth token of 0.5 will be transmitted to the external system which could then be interpreted as a command. However, no values would change on the succeeding iterations. Thus the process would be reset for another set of iterations or else stopped entirely.

The reader may consult [28] for a further discussion of the advantages of fuzzy logic networks.

3.10 References

1. Agerwala, T., Putting Petri to work, *IEEE Computer,* (1979) 85-94.

2. Alsina, C., Trillas, E. and Valverde, L, On non-distributed logical connectivity for fuzzy set theory, *Busefal,* 2 (1980) 18-29.

3. Atabakhche, H., Barhalho, D., Valette, R., and Courvoisier, M., From Petri net based PLC's to knowledge based control, in Proc. IECON'86, 817-822.

4. Bellman, R.E. and Zadeh, L.A., Decision making in a fuzzy environment, *Managemant Sci.,* 17 (1970) 141-164.

5. Boffey, T.B., *Graph Theory in Operations Research,* Macmillan, London, 1982.

6. Burr, D. J., A neural netswork digital recognizer, in *Proc. 1986 Conf. Syst. Man Cybern.,* Atlanta, GA, 2 (1986) 1621-1625.

7. Cardoso, J. and Camargo, H., Editors, *Fuzziness in Petri Nets*, Studies in Fuzziness and Soft Computing 22, Physicia-Verlag, 1999.

8. Carpenter, G. and Grossberg, S., A massively parallel architecture for a self-organizing neural pattern recognition machine, *Comput. Vision, Graphics, Image Processing,* 37 (1987) 54-115.

9. Chanas, S., Reliable flows in network, Institute of Management, Technical University of Wrocław, Report 36, 1984.

10. Chanas, S. and Kamburowski, J., The use of fuzzy variables in PERT, *Fuzzy Sets and Systems,* 1 (1981) 11-19.

11. Chanas, S. and Kołodziejczynk, W., Maximum flow in a network with fuzzy arc capacities, *Fuzzy Sets and Systems,* 8 (1982) 165-173.

12. Chanas, S. and Kołodziejczynk, W., Integer flows in networks with fuzzy capacity constraints, Inst. of Management, Tech. Univ. of Wrocław, Report 368 (1982).

13. Chanas, S. and Kołodziejczynk, W., Real-valued flows in a network with fuzzy arc capacities, *Fuzzy Sets and Systems,* 13 (1984) 139-151.

14. Chanas, S. and Kołodziejczynk, W., Integer flows in network with fuzzy capacity constraints, *Networks,* 16 (1986) 17-31.

15. Dijkstra, E.W., A note on two problems in connection with graphs, *Numer. Math.,* 1 (1959) 269-271.

16. Dubois, D. and Prade, H., Operations on fuzzy numbers, *Internat. J. Systems Sci.,* 9 (1978) 613-626.

17. Dubois, D. and Prade, H., *Fuzzy Sets and System,* Academic Press, New York 1980.

18. Ford, L.R. Jr., and Fulkerson, D.R., Maximal flow through a network, *Canad. J. Math.,* 8 (1956) 339-404.

19. Ford, L.R. Jr., and Fulkerson, D.R, A simple algorithm for finding maximal network flows and an application to Hitchcock problem, *Canad. J. Math.*, 9 (1957) 210-218.

20. Ford, L.R. Jr., and Fulkerson, D.R, Flows in Networks, A RAND Corporation Research Study (Princeton Univ. Press 1962).

21. Gaines, B. R. and Shaw, M. L., From fuzzy logic to expert systems, *Inform. Sci.,* 36 (1985) 5-15.

22. Hamacher, H., Leberling H., and Zimmermann, H.-J., Sensitive analysis in fuzzy linear programming, *Fuzzy Sets and Systems,* 1 (1978) 269-281.

23. Holt, A. W. and Commoner, F., *Events and Condition.* New York: Applied Data Research, 1970; also in *Rec. Project MAC Conf. Concurrent Syst. and Parallel Computation.* New York: ACM 1970, pp. 1-52.

24. Hopfield, J., Neural networks and physical systems with emergent collective computational abilities, *Proc. Nat. Acad. Sci. USA.* 17 (1987).

25. Johnsonbaugh, R., *Discrete Mathematics*, 4th edition, Prentice Hall, Inc, 1997.

26. Klement, E.P., Fuzzy sigma algebra and fuzzy measures with respect to t-norms, Round table on Fuzzy Sets, Lyon 1980.

27. Kołodziejczynk, W., Flows in transportion networks with non-sharply defined parematers analysis, Ph.D. Thesis, Inst. of Management, Techn. Univ. of Wrocław, Report 347 (1981) (in Polish).

28. Looney, C.G., Fuzzy petri nets for rule-based decision making, *IEEE Trans. Sys. Man and Cyb.,* 18 (1988) 178-183.

29. Looney, C.G., Expert control design with fuzzy rule matrices, *Int. J. Expert Syst.: Res. Appl.,* in press.

30. Looney, C.G. and Alfize, A. R., Logical controls via Boolean rule matrix transformations, *IEEE Trans. Syst. Man Cybern.,* 17 (1987).

31. Negoita, C.V. and Sularia, M., Fuzzy linear programming and tolerance in planning, *Econ. Comput. Econ. Cyberte. Stud. Res.,* 1 (1975) 3-15.

32. Negoita, C.V., Flonder, P. and Sularia, M., On fuzzy environment in optimization problems. Proc. Third International Congress of Cybernetics and Systems, Bucharest August 1975, (Abacus Press, London 1976).

33. Peterson, J. L., *Petri nets. Theory and the Modelling of Systems.* Englewood Cliffs, NJ: Prentice-Hall, 1981.

34. Petri, C. A., Kommunikation mit Automaten, Schriften des Rheinish-Westfalischen Institutes fur Instrumentelle Mathematik an der Universtät Bonn, Heft 2, Bonn, W. Germany, 1962.

35. Sugeno, M., An introductory survey of fuzzy control, *Inform. Sci.,* vol. 36, pp. 59-83, 1985.

36. Tabak, D., Petri net representation of decision models, *IEEE Trans. Syst. Man Cybern.,* 15 (1985) 812-818.

37. Thole, U., Zimmermann, H.-J., and Zysno, P., On the suitability of minimum and product operators for the intersection of fuzzy sets, *Fuzzy Sets and Systems,* 2 (1979) 167-180.

38. Thomason, M. G., Convergence of powers of a fuzzy matrix, *J. Math. Anal. Appl.,* 57 (1977) 476-480.

39. Warefield, J., Binary matrices in system modelling, IEEE Trans. Syst. Man Cybern., 3 (1973) 441-449.

40. Zimmermann, H.-J., Fuzzy programming and linear programming with several objective functions, *Fuzzy Sets and Systems,* 1 (1978) 45-55.

41. Zimmermann, H.-J. and Zysno, P., Latent connectives and human decision making, *Fuzzy Sets and Systems,* 4 (1980) 37-51.

42. Zadeh, L.A., Fuzzy sets, *Information and Control,* 8(1965) 338-353.

43. Zadeh, L.A., Fuzzy sets as a basis for theory of possibility, *Fuzzy Sets and Systems,* 1 (1978) 3-28.

44. Zisman, M. D., Use of production systems for modelling asynchronous, concurrent processes, in *Pattern Directed Inference Systems,* D. A. Waterman and F. Hayes-Roth, Eds. New York: Academic Press. 1978, pp. 53-68.

4

Fuzzy Graphs and Shortest Paths

4.1 Fuzzy Shortest Paths

The material in this section is taken from [36].

Posing problems on networks serves as a means for visualizing a problem and for developing a better understanding of the problem. It also has certain computational advantages. In fact, it is easier for a decision maker to draw a picture of what he wants than it is to write down constaints.

There are a wide variety of network type problems. These include such models as location, transportation, flow, reliability, and shortest path to name a few. The majority of the time, these models deal with deterministic data and a single objective. For fuzzy counterparts, see [1, 9-13, 19, 28, 29, 44, 48].

In this chapter, we concentrate on the most basic network problem, the shortest path problem. The fuzzy shortest path problem was first analyzed in [19]. However, a fuzzy shortest path length can be found, but it may not correspond to an actual path in the network. This problem is circumvented with some more recent models based on fuzzy shortest paths and multiple objective functions presented in this and the next section. Algorithms for these models based on classical dynamic programming procedures and multi-criteria optimization are also presented. Submodular functions are used to analyze these models.

A network is generally depicted by a graph and the terms will be used interchangeably. Recall that a graph G is a pair (V, E), where V is a set of points and E is a set of edges. In this chapter, an edge is denoted by a pair of vertices $\{i, j\}$. If E is changed to a set of ordered pairs of distinct elements of V, (V, E) is a directed graph and $E \subseteq V \times V$. The ordered pairs $(i, j) \in E$ are referred to as arcs or edges and an arc goes from vertex i to vertex j. An arc (i, i) is referred to as a **loop**.

A path from s to t is called an **(s, t)-path**. An (s, t)-path is called **open** if $s \neq t$ and is called **closed** if $s = t$. A **cycle** is a closed (s, s)-path in which no vertices are repeated except s and there exists at least one arc. A graph that contains no cycles is called **acyclic**.

In an acyclic directed graph with n vertices, the source can be labeled 1 and the sink n and all other vertices can be labeled such that for any arc (i, j), $i < j$.

Definition 4.1.1 *Let (V, E) be an acyclic directed graph. Then (V, E) is called **layered** if $\exists$ a partition $\{V_1, \dots, V_m\}$ of V such that if $|V_k| > 1$, the vertices in V_k are sequentially numbered and there does not exist an arc (i, j) for $i, j \in V_k$ for $k = 1, 2, \dots, n$. For a layered graph, generally V_1 contains only the source vertex and V_m contains only the sink vertex.*

Suppose each arc (i, j) has an associated weight or length c_{ij}. Then an (s, t)-path has an associated weight or length equal to the sum of the weights of the arcs in the path. This gives rise to the shortest path problem, i.e., is to find the path with minimal weight between two vertices s and t.

There are many ways to find the shortest part for a network [42]. Some of the more general methods such as the labeling algorithm follow from dynamic programming. We assume that the graphs for the models presented in this chapter are directed acyclic graphs. Any graph that has no cycles of negative weight can easily be converted to a directed acyclic graph [42]. Consequently, this assumption is not a major restriction.

Let (V, E) be an acyclic directed graph with n vertices numbered from 1 to n such that 1 is the source and n is the sink. A **dynamic programming (DP)** shortest path problem can be formulated as follows:

$$\begin{aligned} f(n) &= 0 \\ f(i) &= \wedge_{i<j}\{c_{ij} + f(j) \mid (i, j) \in E\}. \end{aligned} \tag{4.1}$$

Here c_{ij} is the length of an arc (i, j) and $f(i)$ is the length of the shortest path or paths from vertex i to vertex n. The algorithm labels vertex i with $f(i)$, and the labels allow the determination of the path.

We note that the sum in (4.1) can be replaced by almost any binary operator and the recursion will hold.

For a variety of fuzzy network problems, see [1, 9-13, 19, 28, 29, 44, 48]. Generally fuzziness is introduced into the network through arc capacities, arc lengths, or vertex restrictions.

Fuzziness can be introduced in a network in a variety of ways. We shall limit ourselves to two specific categories. The first is the classical fuzzy shortest path problem in which the length of each arc is a fuzzy number and the second is the length of the path is a fuzzy number and each arc in the graph has a membership value.

The classical fuzzy shortest path problem seems to have been first introduced by Dubois and Prade [19]. In [19], the solution of this problem is discussed through the use of the extended sum and also the extended min and max, $\widetilde{\min}$ and $\widetilde{\max}$, respectively. To solve the problem, Floyd's Algorithm and Ford's Algorithm were used, [30, 42]. However, although a fuzzy shortest path length can be found, a shortest path with that length may not exist since the extended min of several fuzzy numbers may not be one of those numbers.

This fuzzy shortest path problem can also be viewed in terms of the formulation given in equation (4.1). This point of view is very close to Ford's Algorithm and the formulation is easily extended to fuzzy numbers. The fuzzy DP recursion is given by

$$\begin{aligned} \widetilde{f}(n) &= \{0/1\}, \\ \widetilde{f}(i) &= \widetilde{\bigwedge}_{i<j}\{\widetilde{c_{ij}} + \widetilde{f}(j) \mid (i,j) \in E\}. \end{aligned} \tag{4.2}$$

where $\widetilde{c_{ij}}$ is the fuzzy arc length, x/y represents the element x and its membership grade y, and $\widetilde{f}(i)$ is the fuzzy shortest path distance from i to n. This can essentially be viewed as a labelling routine, where each vertex i is given the fuzzy label $\widetilde{f}(i)$. In a directed acyclic network, $\widetilde{f}(1)$ yields the fuzzy shortest distance from the source, node 1, to the sink, node n. As Dubios and Prade noted there may not be a path corresponding to the fuzzy number $\widetilde{f}(1)$. We illustrate this with the following example.

Example 4.1.2 *Consider a fuzzy network on four vertices* $1, 2, 3,$ *and* 4 *having directed edges* $(1,2)$, $(1,3)$, $(2,4)$, *and* $(3,4)$. *Thus* 1 *is the source node and* 4 *is the sink node. Let*

$$\begin{aligned} \widetilde{c_{12}} &= \{2/0.8,\ 3/0.3\}, \\ \widetilde{c_{13}} &= \{1/0.5,\ 2/0.6,\ 3/0.2\}, \\ \widetilde{c_{24}} &= \{1/0.2,\ 2/0.3\},\ \textit{and} \\ \widetilde{c_{34}} &= \{1/0.2,\ 2/0.8\}. \end{aligned}$$

The DP algorithm begins by setting

$$\begin{aligned} \widetilde{f}(4) &= \{0/1\}, \\ \widetilde{f}(3) &= \{\widetilde{c_{34}} + \widetilde{f}(4)\} \\ &= \{1/0.2,\ 2/0.8\} + \{0/1\} \\ &= \{1/0.2,\ 2/0.8\}. \end{aligned}$$

Similarly,

$$\begin{aligned}
\widetilde{f}(2) &= \{1/0.2,\ 2/0.3\} \\
\widetilde{f}(1) &= \widetilde{min}\{\widetilde{c_{12}} + \widetilde{f}(2),\ \widetilde{c_{13}} + \widetilde{f}(3)\} \\
&= (\{2/0.8, 3/0.3\} + \{1/0.2, 2/0.3\})\widetilde{\wedge} \\
&\quad (\{1/0.5, 2/0.6, 3/0.2\} + \{1/0.2, 2/0.8\}) \\
&= \{5/0.2, 6/0.3, 7/0.3\}\widetilde{\wedge}\{4/0.2, 5/0.5, 6/0.6, 7/0.2\} \\
&= \{4/0.2, 5/0.3,\ 6/0.3,\ 7/0.2\}.
\end{aligned}$$

Note that $4/0.2$ *is determined as follows:* $(0.2 \wedge 0.2)$, $(0.3 \wedge 0.2)$, *and* $(0.3 \wedge 0.2)$ *correspond to* $5 \wedge 4$, $6 \wedge 4$, *and* $7 \wedge 4$, *respectively, and* $0.2 = \vee\{0.2 \wedge 0.2, 0.3 \wedge 0.2, 0.3 \wedge 0.2\}$. *The path* $1-2-4$ *is of length* $\{5/0.2, 6/0.3, 7/0.3\}$ *and the path* $1-3-4$ *is of length* $\{4/0.2, 5/0.5, 6/0.6, 7/0.2\}$. *Hence there is no path of length* $\{4/0.2, 5/0.3,\ 6/0.3,\ 7/0.2\}$ *from source to sink. Note also that no path has membership value* 0.3 *for* 5.

We now present a very simple example.

Example 4.1.3 *Consider a fuzzy network on four vertices* 1, 2, 3, *and* 4 *having directed edges* $(1,2)$, $(1,3)$, $(2,4)$, *and* $(3,4)$. *Hence* 1 *is the source and* 4 *is the sink. Let* $\widetilde{c_{12}} = \{1/1\}$, $\widetilde{c_{24}} = \{1/1\}$, $\widetilde{c_{13}} = \{1/1\}$, *and* $\widetilde{c_{34}} = \{2/0.5\}$. *Then* $\widetilde{f}(4) = \{0/1\}$, $\widetilde{f}(3) = \{2/0.5\}$, $\widetilde{f}(2) = \{1/1\}$, *and* $\widetilde{f}(1) = \widetilde{\min}(\{1/1\} + \{1/1\}, \{2/0.5\} + \{1/1\}) = \{2/1\}\widetilde{\wedge}\{3/0.5\} = \{2/\vee\{1 \wedge 0, 1 \wedge 0.5, 0 \wedge 0\}, 3/0 \wedge 0.5\} = \{2/0.5\}$. *The path* $1-2-4$ *is of length* $\{2/1]$ *and the path* $1-3-4$ *is of length* $\{3/0.5\}$. *Thus there is no path of length* $\{2/0.5\}$.

To circumvent the problem of the fuzzy shortest path distance not corresponding to an actual path, a different category of fuzzy path problems is considered. Rather than viewing each arc as a fuzzy number, let each possible arc length and path length be a fuzzy set. Then each arc has a membership grade in each fuzzy set corresponding to a length. Then through a DP recursion, it is possible to find the shortest fuzzy path length. This length will be a fuzzy number that may not correspond to an actual path. However, each value in the fuzzy set with a positive membership grade will correspond to a path in the network.

We assume that the network we are dealing with is layered with m layers. Any network without negative cycles can be converted to a layered network by the Bellman-Ford method [42]. We also assume that each arc can take an integer value for length between 1 and R. Let the integer values 1 through R be fuzzy sets and let the grade of membership of each arc (i,j) in each fuzzy set I be denoted by $\mu_I(i,j)$, $I = 1, 2, \ldots, R$. Hence each arc has associated with it an r-tuple of membership values. In order to solve this problem, we use a hybrid DP approach.

This approach is based on the concept of multi-criteria programming. It is incoporated into the dynamic programming recursion. The membership grades are associated to the arcs and 'nondominated' paths are found.

In order to use the nondominated concept, we show that the principle of optimality holds for the concept of domination. This principle has been established for general multi-criteria in [7]. It is stated here in terms of paths for clarity.

A nondominated policy has the property that it contains nondominated subpolicies: A policy is a path that possesses such attributes as length and membership. The path consists of subpolicies that are subpaths, i.e., if a path is from 1 to n, a subpath would be the path from 1 to k. The ordering procedure is based on a domination operator since the ordering determines the inclusion or the exclusion of an edge to be added to the path. That is, if an edge is added to the path and with that addition another possible path is dominated, it is excluded from the candidate list. This can be stated as follows. Let π be an optimal policy. π has subpolicies and can be written $\pi = \pi_1\pi_2$. Let $h(s, \pi)$ be the consequence of applying π to state s. We say the policy π' is said to be dominated by π if $h(s, \pi') < h(s, \pi)$ for a given binary function $<$. If we let Ω^* be the set of nondominated policies, then the claim is equivalent to: if $\pi = \pi_1\pi_2 \in \Omega^*$ and $\pi_1 \in \Omega(s)$, then $\pi_1 \in \Omega^*(s)$, where $\Omega(s)$ is the set of feasible decision leading from state 1 to state s and $\Omega^*(s)$ is the set of nondominated decisions leading from state 1 to state s. Suppose that $\pi_1 \notin \Omega^*(s)$. Then there exists a $\pi^* \in \Omega(s)$ such that $h(1, \pi_1) < h(1, \pi^*)$. Since $\pi_1, \pi^* \in \Omega(s)$ and π_2 leads from s to n, it follows that $h(1, \pi_1\pi_2) < h(1, \pi^*\pi_2)$. Hence $\pi = \pi_1\pi_2 \notin \Omega^*$, a contradiction.

We also assume the following conditions in order to use the hybrid dynamic programming method for the shortest route problem and other network problems:

(1) the network is directed and acyclic;

(2) the network is layered;

(3) $d(e) \geq 0 \; \forall e$, where $d(e)$ is an attribute of the edge e other than membership.

These assumptions are not restrictive for the following reasons. If a graph is not directed, it can be converted to a directed graph by inclusion of an additional arc for each arc in the network. Then if a directed graph has no negative cycles (i.e., a path from i to itself with negative length), the Bellman-Ford method can be used to reconstruct it into an acyclic network. This new network is acyclic and layered. The original graph can be transformed into a layered directed acyclic graph by use of a heuristic method.

Many applications take on a layered network form. Activity networks such as PERT are always directed and acyclic. It is generally true that edges have positive lengths. This is especially true in terms of transportation related problems. Therefore assumption (3) above may not be truly binding. Also, depending on the type of network model, some or all of these assumptions can be relaxed. However, they are used here for the shortest path problem.

We use the following hybrid multi-criteria DP recursion:

$$\begin{aligned} f(n) &= (1, 1, \ldots, 1) \\ f(i) &= \text{dom}_{(i,j)\in E}(e_{ij} \tilde{+} f(j)), \end{aligned} \tag{4.3}$$

where e_{ij} is an r-tuple associated with each arc or edge (i,j) or path from i to j. This r-tuple consists of the membership grades of arc (i,j) or the membership grades of the paths from i to j in the respective fuzzy sets associated with the possible lengths, 1 through R. Thus

$$e_{ij} = (\mu_1(i,j), \mu_2(i,j), \ldots, \mu_R(i,j)).$$

The operator $\tilde{+}$ represents the combinatorial sum and dom is the domination operator based on bigger is better.

The combinatorial sum for fuzzy shortest paths is defined as follows. Recall that we assume there are m layers and the possible lengths of an edge are 1 through R. Hence the shortest a path could be in length is $m-1$ and the longest a path could be in length is $(m-1)\cdot r$. To find the paths of possible lengths, combinations of the possible lengths must be considered. To find the possible length I of a path, the lengths that can be used in combination are $1, 2, \ldots, i-m+1$. The combinatorial sum of two tuples is then defined as follows. Let

$$z = \min\{\mu_x(j,k), \mu_y(k,q)\}.$$

Then $e_{j,q} = e_{j,k} \tilde{+} e_{k,q}$ where the i-th element of the r-tuple $e_{j,q}$ is given by

$$(e_{j,q})^i = \max_{x+y=i}(z) = \mu_I(j,q).$$

The recursion in equation (4.3) yields the set of nondominated paths from source 1 to n. We define the fuzzy shortest path length as

$$\widetilde{P}_{1,n} = \{1/\max_i\{\mu_1^i(1,n)\}, \ldots, k/\max_i\{\mu_K^i(1,n)\}, \ldots, r/\max_i\{\mu_R^i(1,n)\}\},$$

where $\mu_K^i(s,t)$ represents the membership grade in the fuzzy set K of the path from vertex s to vertex t given by the ith nondominated r-tuple.

The fuzzy number found for the fuzzy shortest path length is different than that for the classical method. However, there is a path associated with each fuzzy set I and its membership function. Consider the example in Figure 4.1.

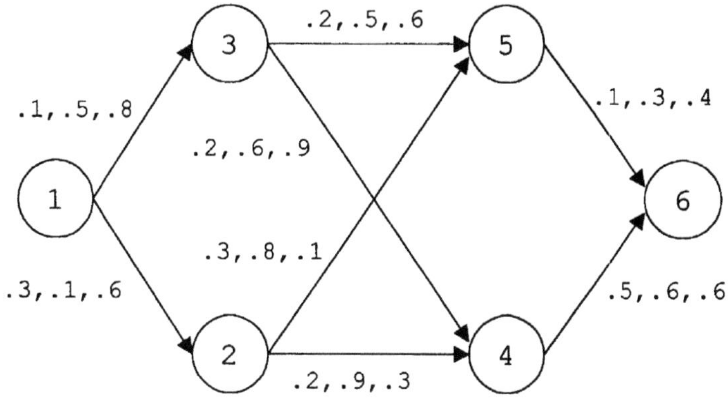

Figure 4.1

Assume that each edge can take a length of 1, 2, or 3. The membership value of the edge in each of the fuzzy sets 1, 2, and 3 is given by the 3-tuples associated with each edge. There are 4 layers in the graph. Hence $m-1=3$ and the shortest possible path is 3 units and the most a path could be is 9 units since $r = 3$. Applying the recursion given by equation (4.3), we obtain

$$\begin{aligned} f(6) &= (1,1,1,\ldots,1) \\ f(5) &= (0.1,0.3,0.4,0,0,0,0,0,0) \\ f(4) &= (0.5,0.6,0.6,0,0,0,0,0,0) \end{aligned}$$

$$\begin{aligned} f(3) &= \text{dom}\{e_{35}\tilde{+}f(5),\ e_{34}\tilde{+}f(4)\} \\ &= \text{dom}\{(0,0.1,0.2,0.3,0.4,0.4,0,0,0), \\ &\qquad (0,0.2,0.5,0.6,0.6,0.6,0,0,0)\} \\ &= (0,0.2,0.5,0.6,0.6,0.6,0,0,0) \end{aligned}$$

$$\begin{aligned} f(2) &= \text{dom}\{e_{24}\tilde{+}f(4),\ e_{25}\tilde{+}f(5)\} \\ &= \text{dom}\{(0,0.2,0.5,0.6,0.6,0.3,0,0,0), \\ &\qquad (0,0.1,0.3,0.3,0.4,0.1,0,0,0)\} \\ &= (0,0.2,0.5,0.6,0.6,0.3,0,0,0) \end{aligned}$$

$$\begin{aligned} f(1) &= \text{dom}\{e_{13}\tilde{+}f(3),\ e_{12}\tilde{+}f(2)\} \\ &= \text{dom}\{(0,0,0.1,0.2,0.5,0.5,0.6,0.6,0.6), \\ &\qquad (0,0,0.2,0.3,0.3,0.5,0.6,0.6,0.3)\} \\ &= (0,0,0.2,0.3,0.5,0.5,0.6,0.6.0.6). \end{aligned}$$

The fuzzy shortest path length is then given by

$$\widetilde{P}_{1,6} = (1/0, 2/0, 3/0.2, 4/0.3, 5/0.5, 6/0.5, 7/0.6, 8/0.6, 9/0.6).$$

Thus the shortest possible path has length 3 with a grade of membership of 0.2. The length of the path 1–2–4–6 is $(0, 0, 0.2, 0.3, 0.3, 0.5, 0.6, 0.6, 0.3)$ and the path length of the path $1-3-4-6$ is $(0, 0, 0.1, 0.2, 0.5, 0.5, 0.5, 0.6, 0.6)$. Thus the path with the shortest fuzzy path length corresponds to the path $1-2-4-6$ in the original network and is formed by the general backtracking techniques of DP. If the decision maker has a threshold of membership of 0.5, then the path of length 5 which corresponds to the path $1-3-4-6$ could be choosen.

We note that in Example 4.1.3,

$$\text{dom}\{(2/1, 3/0), (2/0, 3/0.5)\} = \{2/1, 3/0.5\}.$$

The model yields a fuzzy number representing the fuzzy shortest path length, but it also associates an actual path or paths to that number. This procedure is a naive DP approach. It can be easily made more efficient.

We now consider other fuzzy path models.

This model presented above can be extended to incorporate other network concepts. In particular, we can consider each arc itself as a fuzzy proposition. That is to say, an arc with known nonfuzzy parameters may be present with a fuzzy membership grade.

There are a variety of applications for this particular type of model. Having the arcs as members of fuzzy sets allows flexibility and interactive input from a decision maker. Finding the most 'reliable' path in a communications network is one motivation for this model. In order to find the most reliable path, a probability of being operative is generally associated with each arc and the probability a path is operative in the product of the arc probabilities in the path. This notion of an arc having an associated property other than length or capacity leads to the models presented in this section.

The models presented below extend this idea. They maintain the length of an arc, but also associate other properties to the arc by the inclusion of the arc in a fuzzy set or sets. This allows the modeling of a variety of situations, e.g., wastesite management (see [36, p. 104]).

Suppose that each edge in (V, E) has associated attributes, $d(e)$, $\mu_Z(e)$, where $d(e)$ is the length (distance) of edge $e \in E$ and $\mu_Z(e)$ is the membership grade of edge e in some fuzzy set Z. Let s be the source vertex and t be the sink vertex. Also, let P be the set of all paths from s to t.

Let $p_i \in P$. Then p_i is a set of edges that connect s to t. The length of p_i is defined as in classical (nonfuzzy) network theory:

$$d(p_i) = \sum_{e \in p_i} d(e).$$

Let μ_Z be the fuzzy subset of P which gives the membership grade of the paths associated with the fuzzy set Z. Much flexibility is afforded by the relationship of the membership of the path to the memberships of the individual arcs. There are a large number of possible relationships. However, only a few of these relationships are considered here. See also [33]. The relations to be considered are the following:

$$\mu_Z(p) = \Pi_{e \in p} \mu_s(e), \tag{4.4}$$

$$\eta_Z(p) = \sum_{e \in p} \mu_s(e), \tag{4.5}$$

where $p \in P$ and $\eta_Z(p)$ represents the nonnormalized membership grade of p. That is, the membership grade denoted by $\eta_Z(p)$ is not restricted to the interval $[0, 1]$. A normalized membership function can be easily obtained by equation (4.6).

$$\mu_Z(p) = \sum_{e \in p} \mu_s(e) / |p| . \tag{4.6}$$

Other possible path membership functions are

$$\mu_Z(p) = \sum_{e \in p} \mu_s(e) d(e), \tag{4.7}$$

$$\mu_Z(p) = \Pi_{e \in p} \mu_s(e) d(e), \tag{4.8}$$

$$\mu_Z(p) = \vee_{e \in p} (\wedge) \{\mu_s(e)\}. \tag{4.9}$$

A variety of objectives for the model can be considered with the defined relationship between paths and their respective edges. For example, the path that has maximum or minimum membership in Z or the shortest path from s to t such that $\mu_Z(p) \leq \alpha$ or $\mu_Z(p) \geq \alpha$ might be desired, where $\alpha \in [0, 1]$.

These models can be solved by use of hybrid multi-criteria DP. Suppose that each edge or arc has a nonfuzzy attribute, say, length. Then the appropriate DP strategy is developed as follows.

Let $d_{ij} \in \mathbb{R}^+$ be the length or distance from vertex i to vertex j if edge $(i, j) \in E$. If $(i, j) \notin E$, then let $d_{ij} = -\infty$. For each edge $(i, j) \in E$, define an n-tuple, e_{ij} by

$$e_{ij} = (d_{ij}, f_{S_2}(i, j), f_{S_3}(i, j), \dots, f_{S_n}(i, j)),$$

where $S_2, S_3, \dots, S_n$ are $n - 1$ fuzzy sets associated with the edge (i, j). For the n-tuple e_{ij}, define the operator $*$ as follows:

$$e_{im} = e_{ij} * e_{km} = (d_{ij} + d_{km}, f_{S_2}(P), f_{S_3}(P), \dots, f_{S_n}(P)).$$

where P is the path formed by edges (i,j) and (k,m) and $f_{S_i}(P)$, $i = 2,3,\dots,n$, is the path membership function given by one of the equations (4.4) through (4.9). If the edges do not have a common vertex, then they do not form a path. In this case, they are given an associated value of $-\infty$. An edge is called **nondominated** if there does not exist an n-tuple e' such that each element of e' dominates the corresponding element in e. We define the domination operator, dom, as

$$\text{dom}(e_{ij}, e_{km}, \dots, e_{xy}) = \{e \mid e \text{ is a nondominated } n\text{-tuple}\},$$

The concept of domination is dependent on the decision maker and can be mixed within the n-tuple. That is, for one element of e_{ij}, more is better, but for another element, less is better.

Let $f(i)$ be an n-tuple which represents the shortest nondominated path distance and path membership from vertex i to a sink vertex t. Then $f(i)$ can be found by DP with the following functional equation:

$$f(t) = (0, 0, \dots, 0),$$

$$f(i) = \text{dom}_{(i,j)\in E}(e_{ij} * f(j)).$$

Specific examples and applications of these models can be found in [38, 40, 37].

4.2 Analysis of the Fuzzy Path Models

Network or operations research models with fuzzy components are analyzed in terms of fuzzy concepts. However, when using fuzzy numbers, set functions are used. In this section, the fuzzy models discussed in Section 4.1 are analyzed, in terms of the set-function concept of submodularity.

Submodularity

Submodularity gives a unifying structure for set functions similar to the way convexity does for continuous functions. The minimization of a submodular function can be done in polynomial time. Likewise, submodularity is the structure that allows the greedy algorithm to work for matroids.

Let E be a finite set and recall that $\mathcal{P}(E)$ denotes the power set of E. Let $f : \mathcal{P}(E) \to \mathbb{R}$ be a set function that associates a real number with each subset of E.

Definition 4.2.1 *The function $f : \mathcal{P}(E) \to \mathbb{R}$ is called **submodular** if $\forall A, B \subseteq E$,*

$$f(A \cup B) + f(A \cap B) \le f(A) + f(B). \tag{4.10}$$

*Likewise **supermodular** and **modular** can be defined by replacing $\le$ with $\ge$ and $=$, respectively.*

Note that if f is submodular, $-f$ is supermodular.

Information additional to what is given below can be found in [43], [39], or [21].

Let f and g be submodular functions on the set E and $X \subseteq E$. Then
(1) $f + c$ is submodular, where $(f + c)(X) = f(X) + c$, $c \in \mathbb{R}$,
(2) $f + g$ is submodular, and
(3) $\overline{f}$ is submodular, where $\overline{f}(X) = f(E \backslash X)$.

Definition 4.2.2 *A set function f is said to be* ***monotone nonincreasing*** *if $\forall X, Y \subseteq E$ such that $X \subseteq Y$, $f(X) \geq f(Y)$.*

It is often useful to know the marginal value of adding a value to a set. This **marginal value** is given by

$$\rho_j(X) = f(X \cup \{j\}) - f(X).$$

The concept of marginal value yields an alternative characterization for submodularity.

Definition 4.2.3 *Let f be a set function defined on a set E. Then f is* ***submodular*** *if*

$$\rho_j(X) \geq \rho_j(Y)$$

$\forall X, Y \subseteq E$ with $X \subseteq Y$ and $\forall \{j\} \in (E \backslash Y)$. An analogous characterization holds for ***supermodular*** *and* ***modular*** *set functions by replacing $\leq$ with $\geq$ and $=$, respectively.*

Discrete problems that have the additional structure of submodularity appear to be inherently easier to solve. In particular, the ellipsoid algorithm has been used to solve the minimization of an unconstrained submodular function in polynomial time, [43]. The technique used in [43] is derived by the use of extensions of submodular functions to convex functions. This extension is carried out as follows. Let f be any set function defined on the subsets of S and let χ be the characteristic vector of a set $T \subseteq S$. Set $f(\chi) = f(T)$. Hence f is defined for $0-1$ vectors. Let b be any nonnegative vector. The vector b can be written uniquely as

$$b = \sum_{i=1}^{k} a_i \chi_i$$

where $a_i > 0$ and $\chi_1 \geq \chi_2 \geq \ldots \geq \chi_k$ are distinct $0-1$ vectors. The Lovász-extension is defined as

$$\widehat{f}(b) = \sum_{i=1}^{k} a_i (f \chi_i).$$

The Lovász-extension of a submodular function f yields the following results from [43].

Proposition 4.2.4 *Let f be a set function defined on the subsets of S. Let $\widehat{f}$ be its Lovász-extension to nonnegative vectors. Then $\widehat{f}$ is convex (concave) if and only if f is submodular (supermodular).* ■

Lemma 4.2.5 *Let f be a set function defined on the subsets of S such that $f(\emptyset) = 0$. Then*

$$\wedge\{f(X) \mid X \subseteq S\} = \wedge\{\widehat{f}(x) \mid x \in [0,1]^s\}. \blacksquare$$

From Proposition 4.2.4 and Lemma 4.2.5, it is possible to minimize a submodular function in polynomial time with the use of the ellipsoid method. This polynomial solvability will be utilized in analyzing the fuzzy path problems.

We analyze now the fuzzy path problems discussed previously in terms of their complexity. We use submodularity and view the membership functions as classic set-functions.

As noted previously the classical fuzzy shortest path problem is basėd on the DP recursion given in equation (4.2) and need not yield an associated path. We next analyze the reformulated shortest path problem whose solution is given by equation (4.3).

Equation (4.3) finds the membership value of a path P for each of the possible lengths 1 through $(m-1)\cdot r$. These values are found simultaneously through a multi-criteria approach. It is however, possible to find the values one at a time. In order to find the membership value for a path of length l, $1 \leq l \leq r$, the following problem formulation can be used.

$$\text{(SP)} \qquad \vee_{P \subseteq E} \mu_l(P)$$

subject to

$$|P| = m - 1, \tag{4.11}$$

$$v_i(P) = 2, \ i = 1, 2, \ldots, n-1, \tag{4.12}$$

$$h(P) = 1, \tag{4.13}$$

$$g(P) = 1, \tag{4.14}$$

where E is the set of all edges in the network, m is the number of layers, n is the number of vertices, and

$$v_i(P) = \sum_{(j,k) \in P} |\{j,k\} \cap \{i\}|,$$

$$h(P) = \sum_{(j,k) \in P} |\{j,k\} \cap \{1\}|,$$

$$g(P) = \sum_{(j,k)\in P} |\{j,k\} \cap \{n\}|,$$

$$\mu_l(P) = \wedge_{(j,k)\in P} \mu_T(i,j).$$

Since the network is assumed to be directed, acyclic and layered, the constraints (4.11)-(4.14) guarantee that the set $P \subseteq E$ is a path from vertex 1 to vertex n.

Lemma 4.2.6 *$\mu_l(P)$ is supermodular.*

Proof. Let $A, B \subseteq E$. Then

$$\mu_l(A \cap B) = \mu_l(A) \wedge \mu_l(B)$$

and

$$\mu_l(A \cup B) \geq \mu_l(A) \vee \mu_l(B).$$

Hence $\mu_l(A \cup B) + \mu_l(A \cap B) \geq \mu_l(A) + \mu_l(B)$. ■

Lemma 4.2.7 *$v_i(P)$ is modular.*

Proof. Let $f_a(X) = (v_i(X \cup \{a\}) - v_i(X))$, where $\{a\} \subseteq (E \backslash X)$. Then

$$f_a(X) = \begin{cases} 1 & \text{if } \{a\} \cap \{i\} \neq \phi \\ 0 & \text{if } \{a\} \cap \{i\} = \phi, \end{cases}$$

for any $X \subseteq E$. Let $X \subseteq Y$. Then for $f_a(Y)$ and $f_a(X)$, $\{a\} \subseteq E \backslash Y$. Hence $f_a(X)$ and $f_a(Y)$ are equal and by Definition 4.2.3, v_i is modular. ■

Lemma 4.2.8 *$h(P)$ and $g(P)$ are modular.*

Proof. The proof follows as the proof of Lemma 4.2.7. ■

Since the constraints are equalities and modular, we can form the following Lagrangian problem:

$$\text{(LSP)} \qquad \vee_{P \subseteq E} L(a, P) = \mu_l(P) - a(|P| - m + 1) - \textstyle\sum_{i=2}^{n-1} (a_i v_i(P) - 2) - a_n(h(P) - 1) - a_{n+1}(g(P) - 1).$$

The function $L(a, P)$ is supermodular. Consequently, this problem can be solved in polynomial time since maximizing a supermodular function is equivalent to minimizing a submodular function (see [43] and [21]). Therefore, the fuzzy shortest path problem given can be solved in polynomial time for a value I. Hence the new model of the fuzzy shortest path length with corresponding paths can be solved in polynomial time also. Further discussion on problem formulation can be found in [36].

4.3 On Valuation and Optimization Problems

The material in the remainder of this chapter is from [17].

Imprecise problems that can be represented by graphs have been approached in the following two major ways depending upon how the imprecision appears:

(1) Suppose that the problem structure is well defined and can be modeled by means of a classical graph. However, suppose some values are expressed by arcs or nodes that stand for costs, capacities, etc. that are imprecise in nature. Then optimization problems can be defined and solved using a classical graph that includes fuzzy quantities. The papers [8, 16, 18, 19, 35, 44] concern such problems and also contain pertinent bibliographies.

(2) Suppose that the problems involve finite sets of items whose association structure is not clear. Then they can be modeled by means of a fuzzy set of nodes and arcs, that is by fuzzy graphs. Most of the work with fuzzy graphs has had one of the following two objectives in mind:

(a) To add definitions and properties to the classical concepts concerning the fuzzy graph framework, such as paths and fuzzy trees [48, 14].

(b) To consider a fuzzy graph as a representation of imprecise relations between items and to use connectedness properties to do a cluster analysis of them, [31, 32], [41], [45], [46], [49], and [52].

These papers use only values from $[0, 1]$ for the nodes and arcs that express an imprecise structure. However, these values could be replaced by fuzzy numbers, [45]. The development of a search procedure for implicit fuzzy graphs where the imprecision of the nodes and arcs can be expressed by $[0, 1]$ real values or by fuzzy numbers can be found in [51].

Not many papers use a more general approach to treat such problems, i.e., to consider additional calculations (usually fuzzy quantities) that can be defined on fuzzy graphs. Dubois and Prade [18] examine the problem of finding the minimal path on a graph where the arcs are fuzzy and fuzzy numbers represent their costs. In [15], the minimal spanning tree problem on a fuzzy graph with additional costs associated with its edges is studied.

In this section, we consider similar problems, i.e., those with fuzzy graphs and additional valuations for cost capacities, which may be real numbers, fuzzy quantities or linguistic values.

We consider the following two problems:

(1) finding the minimal path between two nodes in a fuzzy graph where the costs are linguistic values, and

(2) finding the minimal path where times are given by unnormalized fuzzy numbers.

The general algorithm is particularized in both cases. Several examples show that good results are obtained.

We first recall the definition of a fuzzy graph.

Definition 4.3.1 (48) *Let X be a finite set which is assumed equal to $\{1, 2, \ldots, n\}$. The triple (X, σ, μ) is called a* ***fuzzy graph*** *on X, where*

$$\sigma : X \to [0, 1]$$

and

$$\mu : X \times X \to [0, 1].$$

are such that

$$\mu(i, j) \leq \sigma(i) \wedge \sigma(j), \qquad \forall i, j \in X.$$

A fuzzy graph is useful in representing structures of relationships between items where the existence of a concrete item (node) and relationship between two items (arc) are matters of degree.

Example 4.3.2 *Consider the items as the propositions in an expert system with the logical implication: If A, then B. A degree of truth is generally associated with both propositions and implications. This degree may be formulated by means of probabilities, possibilities, or belief measures. They are $[0, 1]$-valued. Hence the implication network of the system may be viewed as a fuzzy graph.*

Example 4.3.3 *Consider the classical model of activity scheduling, where arcs stand for activities and the nodes stand for the beginning and end of these activities. If we assume that no activities are equally significant in accomplishing the whole project, we can associate a degree of importance or necessity with each arc. The activity network then becomes a fuzzy graph.*

Example 4.3.4 *Consider the items as some kind of service points, such as towns or hospitals, which are connected by means of communication routes, such as roads or streets. This problem is often modeled as an undirected graph. By associating an importance level with both points and routes, a fuzzy graph is obtained.*

Much of classical graph theory deals with valued graphs. A real function is defined on the graph arc set that stands for costs, times, capacities, etc. The problem is to find a "special" subgraph such as a path or a tree, that makes up the sum (product) of the arcs that form its optimal (maximum or minimum) value.

Hence many classical problems may be framed, e.g.:

(1) Problems of minimal (maximal) paths.

(2) The problem of minimal spanning trees.

(3) The traveling salesman problem.

In the first approach, a fuzzy graph may be considered as a valued graph. Such graphs were initially studied by solving such problems as finding maximal intensity paths, [34]. However, this approach turned out to be insufficient.

The real values that initially appear in a fuzzy graph are real in a different sense than those considered in classical theory, since they originated either in the uncertainty associated with nodes or arcs (Example 4.3.2) or in additional classification of graph elements (Examples 4.3.3 and 4.3.4). In any case, the model's structure is determined by these values. Therefore, this valuation is not put "on" an underlying structure, but rather it is "a component" of it.

Under these conditions, a valued fuzzy graph should have two kinds of valuation, that associated with the structure ([0, 1]-valued), and some other functions defined on arc or node sets that allows the equation of these components with different approaches.

In Example 4.3.2, we assume that our expert system has a risk factor associated with the inference process. That is, it is possible to find an implication that has a low degree of certainty and a high degree of risk if it is not taken into account because it leads to very important propositions (from a semantic point of view).

For these models, the essential problem may not be to obtain conclusions of maximal certainty, but rather to reach a conclusion of fixed certainty and minimal risk or, alternatively, of fixed risk and maximal certainty.

In Example 4.3.3, we assume that every network activity has a time that can be formulated as a real number, or a real interval, or fuzzy number, or a linguistic value.

In Example 4.3.4, we assume that with every communication route there is associated a length or time to cross it, formulated in the same manner as above.

Let (X, σ, μ) be a fuzzy graph. Then μ is a fuzzy relation on X. Hence we can define μ^n as the composition of μ, n times with itself by means of the classical max-min operation and we define

$$\mu^\infty(i,j) = \vee_{n=1}^{\infty} \mu^n(i,j)$$

$\forall i, j \in X$. This later is defined as the strength of the strongest path between i and j ([34]).

Definition 4.3.5 *Let $G = (X, \sigma, \mu)$ be a fuzzy graph. We call $F = (X, \tau, \nu)$ a **fuzzy subgraph** of G if F is a fuzzy graph and*

$$\tau(i) \leq \sigma(i) \ \forall i \in X$$

$$\nu(i,j) \leq \mu(i,j) \ \forall i,j \in X.$$

Definition 4.3.6 *Let $G = (X, \sigma, \mu)$ be a fuzzy graph. For every $a \in [0,1]$ define the a-**cut** of G to be the graph $G^a = (X^a, U^a)$, where*

$$\sigma_a = X^a = \{i \in X \mid \sigma(i) \geq a\},$$

$$\mu_a = U^a = \{(i,j) \in X \times X \mid \mu(i,j) \geq a\}.$$

Since both the set of nodes and the set of arcs of a fuzzy graph are finite sets, a fuzzy graph will have a finite number of distinct a-cuts. In fact, there exists a finite a-value sequence $\mathcal{A} = (a_1, \ldots, a_k)$ with $a_1 < \ldots < a_k$ such that

$$G^a = G^{a_i}$$

for all $a \in (a_{i-1}, a_i]$, $i = 1, 2, \ldots, k$, where $a_0 = 0$.

We denote this interval sequence $(0, a_1]$, $(a_1, a_2]$, ... , $(a_{k-1}, a_k]$ by $\mathcal{I}$.

Definition 4.3.7 *Let (X, σ, μ) be a fuzzy graph and let*

$$\Gamma = \{(i,j) \in X \times X \mid \mu(i,j) > 0\}.$$

For every $(i,j) \in \Gamma$, define the valuation associated with (i,j) to be a function,

$$c_{ij} : [0, \mu(i,j)] \to \mathcal{H},$$

where $\mathcal{H}$ is a totally ordered set with an interval operation $*$ such that $*$ is associative and compatible with that order relation, i.e., $\forall h_1, h_2, h_3, h_4 \in \mathcal{H}$

$$h_1 \leq h_2, \quad h_3 \leq h_4 \Rightarrow h_1 * h_3 \leq h_2 * h_4.$$

Depending upon the problem under consideration, we assume such functions c_{ij} to be increasing or decreasing on their definition intervals.

Definition 4.3.7 is very general in nature. The particular characteristics of $\mathcal{H}$ are determined from the application being considered.

We list some examples.

(A) $\mathcal{H}$ may be a finite set of linguistic labels such as "very low, low, ... , high, very high" with an internal operation given by a table.

(B) $\mathcal{H}$ may be the set of unnormalized fuzzy numbers with their usual operations and some compatible ranking procedure, [20], [16].

(C) Suppose that no dependence exists between the arcs' membership degree and valuations. Then we can assume constant functions. However, even in this situation, these functions will exist only for the interval defined, since it would be meaningless to consider valuations associated with an arc for membership levels where such an arc does not exist.

General Problem of Optimization on a Fuzzy Graph

A general way of solving optimization problems in a fuzzy environment is by using the a-cut representation. A set of crisp problems dependent on membership degree is thus obtained. These may be solved by classical methods to obtain a set of crisp solutions that are also dependent upon membership degree. Such solutions are called **fuzzy solutions of the problem.**

The solution procedure for such problems as finding the minimal path is summarized in the *Fundamental Algorithm.*

FUNDAMENTAL ALGORITHM
For $l = 1$ to k do:
Calculate minimal cost path in G^{a_l} using
$$\forall (i,j) \in U^{a_l};\ c_{ij}(b);\ \forall b \in I_l$$
as cost function.

The problem is reduced to solving k parametric subproblems on crisp graphs. The set $\mathcal{H}$, its order relation, and the operation $*$ determine if there are classical algorithms to solve the subproblems.

Since the feasible solution set of every subproblem is finite, we have that $\forall l \in \{1, 2, \ldots, k\}$ there is a $S_{l1} \ldots S_{lh}(1)$ solution sequence, in which each solution is optimal on a subinterval of I_1. Each solution S is an arc subset of U^1 satisfying a specific property (path, spanning tree, etc.) and making the composition of valuations associated with their arcs optimal.

Therefore the whole solution for our optimization problem on a fuzzy graph is given as follows:

There exists a sequence $\mathcal{B} = \{b_1, \ldots, b_p\}$ of memberships levels and a sequence of solutions $\mathcal{S} = \{S_1, \ldots, S_p\}$ such that

S_1 is included in G^{b_1} and remains optimal in $J_1 = (0, b_1]$
S_2 is included in G^{b_2} and remains optimal in $J_2 = (b_1, b_2]$
$\vdots$
S_p is included in G^{b_p} and remains optimal in $J_p = (b_{p-1}, b_p]$.

It should be noted that J intervals are usually not equal to the I intervals defined by a-cuts. We can insure this only when the valuations are constant functions. However, clearly $b_p = a_k$.

We see that the optimal objective of the problem does not consist of a single $\mathcal{H}$ value. Rather it consists of a function $d : (0, a_k] \to \mathcal{H}$ defined by $d(a) = d_l(a)$ if $a \in J_l$, where $d_l(a) = *_{(i,j)\in S}(c_{ij}(a))$.

The function d_l satisfies the following properties:

Property 1 d_l is increasing or decreasing on $J_1 \cup J_2 \cup \ldots \cup J_l$ depending on the defined valuations.

Property 2 Let $l, h \in \{1, 2, \ldots, p\}$ with $l \leq h$:

If the problem is to minimize, then $d_l(a) \leq d_h(a)\ \forall a \in J_l$.

If the problem is to maximize, then $d_l(a) \geq d_h(a)\ \forall a \in J_l$.

By a-cut properties, $G^h \subset G^{h-1} \subset \ldots \subset G^1$. Hence if we take into account that d_h is actually defined on $J_1 \cup J_2 \cup \ldots \cup J_h$ being optimal only on I_h, we can easily verify Property 2. However d_l is optimal within I_l.

Property 3 If the problem is to minimize and the valuations are increasing, then d is increasing.

If the problem is to maximize and the valuations are decreasing, then d is decreasing.

Proof. Let $a, a' \in (0,1]$ be such that $a \leq a'$. Assume the associated valuations are increasing and the problem is to minimize. For a and a', two cases occur:

(1) There exists J such that $a, a' \in J$. In this case, the property has been proven by Property 1.

(2) $a \in J_l$ and $a' \in J_h$ with $l \leq h$. By using Properties 1 and 2 and the definition of d, we have

$$d(a) = d_l(a) \leq d_h(a) \leq d_h(a') = d(a'). \blacksquare$$

We next attempt to find a fuzzy solution to our problem. The solution procedure by a-cuts lead to a solution sequence $\mathcal{L}$ associated with the sequence of membership levels $\{b_l\}$. There are two different alternatives for obtaining a fuzzy solution from these sequences. The choice depends on the fuzzy solution concept used.

First consider a fuzzy solution as a fuzzy set of crisp solutions. The solution $\mathcal{C}$ is defined as follows:

$$\mathcal{C} = \{S_l/b_l \mid l = 1, 2, \dots, p\}.$$

With this definition every a-cut of $\mathcal{C}$ will be a set of crisp solutions that is optimal for membership levels greater than or equal to a.

Next consider a fuzzy solution as a fuzzy subgraph. A fuzzy structure (path, tree, ...), a fuzzy subgraph of the initial graph, that can be considered an optimal solution in the following way:

For every $l = 1, 2, \dots, p$, define the fuzzy set of arcs

$$S_l = \{(i,j)_l/\beta_l \mid (i,j) \in S_l\}.$$

From these sets, the fuzzy subgraph $F = (X, \tau, \nu)$ given by

$$\nu(i,j) = \mu_S(i,j), \qquad \text{where } S = \cup_l S_l,$$

$$\tau(i) = \vee_{j \in X}\{\nu(i,j) \vee \nu(j,i)\}$$

is obtained. It can be considered the fuzzy optimal solution.

We now demonstrate how our general procedure is applied by considering two optimization problems. The problems chosen are well known in classical theory. Both have been treated from a fuzzy point of view [18] in a more specialized approach.

Consider a road network with two large towns designated 1 and n. Assume that we want to find a minimal time path from 1 to n. We also assume that our knowledge about roads and towns is imprecise and that we have placed $[0,1]$-valued functions on the towns and roads representing our knowledge level about them. Under these assumptions the problem structure may be modeled as a fuzzy graph $G = (X, \sigma, \mu)$, where $X = \{1, 2, \ldots, n\}$ represents the town set, $\sigma : X \to [0,1]$ gives our knowledge level about these towns, and $\mu : X \times X \to [0,1]$, $\mu(i,j) = \alpha > 0$ indicates that a road runs between i and j, which is known at level α.

Clearly, if precise information about the roads is not known, precise evaluation of the time necessary to travel them is not known. We consider the following two situations.

In the first one, the time is estimated roughly be linguistic labels, "low, high," etc. In the second, unnormalized fuzzy numbers are used to represent the time. More precisely, we assume that for every road (i,j) with knowledge level u_{ij}, there exists a linear fuzzy number with height u_{ij} which may be represented by

$$t_{ij} = (u_{ij}, a_{ij}, b_{ij}, d_{ij}, e_{ij}),$$

where $d_{ij} < a_{ij} \le b_{ij} < e_{ij}$ and defined as follows:

$$t_{ij} = \begin{cases} u_{ij} & \text{if } a_{ij} \le x \le b_{ij} \\ \frac{u_{ij}}{a_{ij}-d_{ij}} x - \frac{u_{ij} d_{ij}}{a_{ij}-d_{ij}} & \text{if } d_{ij} \le x \le a_{ij} \\ \frac{u_{ij}}{b_{ij}-e_{ij}} x - \frac{u_{ij} e_{ij}}{b_{ij}-e_{ij}} & \text{if } b_{ij} \le x \le e_{ij} \\ 0 & \text{if } x \le d_{ij} \\ 0 & \text{if } x \le e_{ij}. \end{cases}$$

Both cases may be solved by using the general approach described above. The first one is a problem of finding the minimal path when the time is expressed linguistically. The second is a similar problem with the time expressed with fuzzy numbers.

The Problem of Finding the Minimal Path when the Time Is Expressed Linguistically

Consider a fuzzy graph $G = (X, \sigma, \mu)$ with a valuation function associated with its arcs given by the definition of fuzzy graph valuations, where $\mathcal{H}$ is an ordered set of linguistic labels with an operation $*$ compatible with the order. Suppose also that $*$ is associative.

Consider two nodes x_1 and x_n for which we want to find a fuzzy solution for a minimal path between x_1 and x_n.

The general procedure given previously is used. Its direct application leads to the solution of k crisp subproblems of minimal path with parametric linguistic times. Unfortunately, there are not appropriate algorithms for these problems.

However, each parametric subproblem may be treated as a finite set of subproblems with constant times since $\mathcal{H}$ is a finite set. There exist classical algorithms suitable for these kinds of problems.

Since $\mathcal{H}$ is finite, the valuations c_{ij} satisfy $\forall i, j$, $\exists \mathcal{D} = \{\delta_{ij1}, \dots, \delta_{ijt(ij)}\}$ such that $\forall l$; $\delta_{ijl} \in (0,1]$, $\delta_{ij1} \leq \delta_{ij2} \leq \dots \leq \delta_{ijt(ij)}\}$ and $\delta_{ijt(ij)} = u_{ij}$. This defines c_{ij} as follows:

$$\begin{aligned} c_{ij}(b) &= h_{ij1} && \text{if } b \in (0, \delta_{ij1}] \\ c_{ij}(b) &= h_{ij2} && \text{if } b \in (\delta_{ij1}, \delta_{ij2}] \\ &\vdots \\ c_{ij}(b) &= h_{ijt(ij)} && \text{if } b \in (\delta_{ijt(ij)-1}, \delta_{ijt(ij)}]. \end{aligned}$$

We now consider the entire sequence $\mathcal{R} = \{r_1, \dots, r_t\}$ such that $r_1 \leq \dots \leq r_t$ is defined by $\forall b \in (0,1]$,

$$b \in \mathcal{R} \Leftrightarrow b \in \mathcal{A} \text{ or } \exists i, j \text{ such that } b \in \mathcal{D}_{ij}$$

and its corresponding interval sequence

$$K_1 = (0, r_1], \dots, K_t = (0, r_t]$$

according to the properties of G and c_{ij} within each interval K_l. The former may be viewed as a graph (G^{r_l}) and the latter are constant functions. Hence the initial problem becomes t problems of minimal path on classical graphs with constant linguistic times.

In order to solve the latter problems, a classical minimal path algorithm may be used with very simple modifications. The Bellman-Kalaba's algorithm is used in the following example.

Example 4.3.8 *Let* $X = \{1, 2, 3, 4, 5, 6, 7, 8, 9\}$. *Let* σ *be defined as follows:*

$$\sigma(x) = \begin{cases} 1 & \textit{if } x = 1, 2, 7, 7, 8, \textit{ or } 9, \\ .8 & \textit{if } x = 4, \\ .7 & \textit{if } x = 6, \\ .6 & \textit{if } x = 3. \end{cases}$$

Let μ *be given by Table* 4.1.

μ	1	2	3	4	5	6	7	8	9
1		1.0	0.6						
2			0.3	1.0	0.6		1.0		
3						0.3		0.7	
4					0.5		0.3		
5						0.6			
6								1.0	
7								1.0	
8									1.0
9									

Table 4.1

Suppose that $\mathcal{H}$ = {null, low, moderately low, moderate, moderately high, high, highest, complete, infinite}, with the obvious order relation. Let "infinite" be the time associated with non-existent arcs of an α-cut. Operation $*$ is given by Table 4.2, where linguistic values have been abbreviated.

In the examples to follow, nu stands for null, mlo stands for moderately low, mod stands for moderate, mdh stands for moderately high, hgh stands for high, hst stands for highest, com stands for complete, and inf stands for infinite.

In this example, the associated times are assumed to be constant functions given by Table 4.3.

Figure 4.2 gives a graphic representation of a fuzzy graph with additional time evaluations.

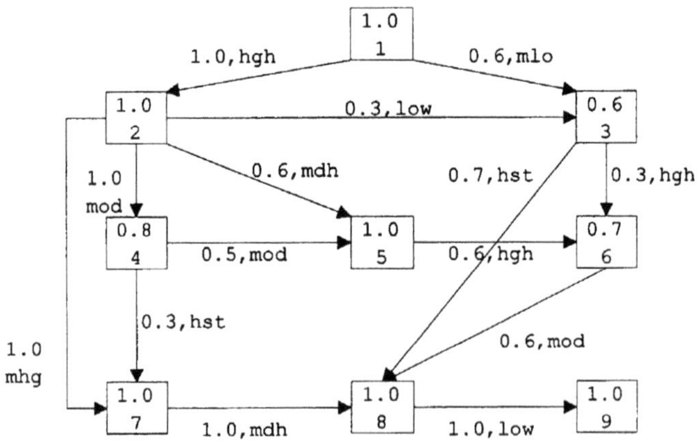

Figure 4.2: Fuzzy graph with additional time evaluations.

We now obtain a minimal path between 1 and 9. By applying a general procedure and by using, in this case, the sequence $\mathcal{A}$, we obtain the solutions:

Membership Degree 0.30
Path Solution
$1 \to 3 \to 6 \to 8 \to 9$
Minimal Time: moderately low

Membership Degree 0.50
Path Solution
$1 \to 3 \to 8 \to 9$
Minimal Time: moderate

Membership Degree 0.60
Path Solution
$1 \to 3 \to 8 \to 9$
Minimal Time: moderate

Membership Degree ≥ 0.70
Path Solution
$1 \to 2 \to 7 \to 8 \to 9$
Minimal Time: high

A fuzzy graph solution is given in Figure 4.3.

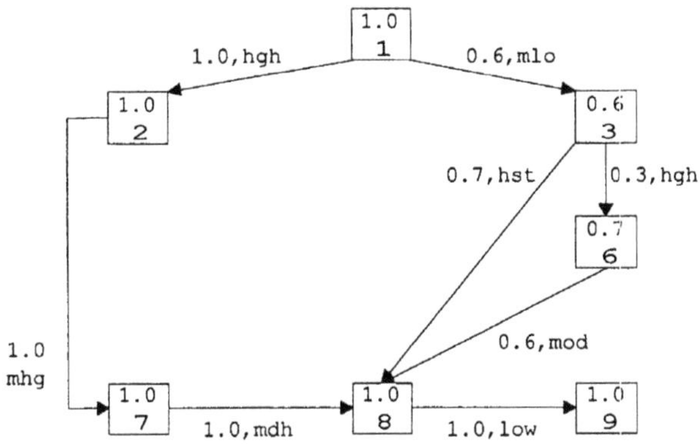

Figure 4.3: Fuzzy graph solution

We can interpret results as follows: If we use roads with a knowledge level $\alpha \geq 0.30$, we obtain a minimal path $1 \to 3 \to 6 \to 8 \to 9$ that takes a moderately low amount of time. If $\alpha \geq 0.50$, the minimal path $1 \to 3 \to 8 \to 9$, which takes a moderate amount of time to compute. The value $\alpha \geq 0.70$ has a minimal path $1 \to 2 \to 7 \to 8 \to 9$ which takes a high amount of time.

	nu	low	mlo	mod	mdh	hgh	hst	com	inf
nu	nu	nu	nu	nu	low	mlo	mod	mdh	inf
low	nu	nu	nu	low	mlo	mod	mdh	hgh	inf
mlo	nu	nu	low	mlo	mod	mdh	hgh	hst	inf
mod	nu	low	mlo	mod	mdh	hgh	hst	com	inf
mhg	low	mlo	mod	mdh	hgh	hst	com	com	inf
hgh	mlo	mod	mdh	hgh	hst	hst	com	com	inf
hst	mod	mdh	hgh	hst	com	com	com	com	inf
com	mdh	hgh	hst	com	com	com	com	com	inf
inf	inf	inf	inf	inf	inf	inf	inf	inf	inf

Table 4.2

	1	2	3	4	5	6	7	8	9
1		hgh	mlo						
2			low	mod	mdh		mdh		
3						hgh	hst		
4					mod		hst		
5						hgh			
6								mod	
7								mdh	
8								low	
9									

Table 4.3

A Problem Where Times Are Unnormalized Linear Fuzzy Numbers

Consider a fuzzy graph $G = (X, \sigma, \mu)$, where the times associated with its arcs are given by the fuzzy numbers t_{ij} as defined above. The functions c_{ij} are obtained from t_{ij} in the following way:

Let $\mathcal{N}$ be the set of unnormalized linear fuzzy numbers.

Let $\mathcal{I} = \{I_l\} = \{(\alpha_{l-1}, \alpha_l)\}$ be the sequence of intervals defined by G.

Define $\forall t_{ij} = (u_{ij}, a_{ij}, b_{ij}, d_{ij}, e_{ij})$ the function $c_{ij}(\alpha)$, given by $\forall \alpha \in (\alpha_{l-1}, \alpha_l]$ with $\alpha \leq u_{ij}$

$$c_{ij}(\alpha) = (u_{ijl}, a_{ijl}, b_{ijl}, d_{ijl}, e_{ijl}),$$

where

$$d_{ijl} = \max\{0, d_{ij} - \delta\};\ e_{ijl} = \max\{0, e_{ij} - \delta\}$$

with

$$\delta = (u_{ij} - \alpha_l)/(u_{ij}\alpha_l).$$

Clearly, $c_{ij} : (0, u_{ij}] \rightarrow \mathcal{N}$ and for each α all existing valuations have equal height.

The reasons for this definition are contained in this last point. Operations with fuzzy numbers and comparisons of fuzzy numbers work better when the numbers have equal heights ([18], [6]). Actually, a simpler flattening effect on t_{ij} could be considered to get this result. But this transformation plus the fuzzy numbers causes too much imprecision in the result. Information provided by $c_{ij}(\alpha)$ keeps constant throughout $(0, u_{ij}]$, [23].

To ensure that the functions c_{ij} are defined properly, we need to prove that they are increasing or decreasing. This depends on the procedure chosen to rank the fuzzy number. There are many procedures ([6]). The following method seems to be appropriate for problems involving linear fuzzy numbers.

Let $c, c' \in \mathcal{N}$ be such that $c = (\alpha, a, b, d, e)$, $c' = (\alpha', a', b', d', e')$ and $t \in (0, 1]$. Suppose

$$\alpha = \alpha \wedge \alpha', \qquad \delta = (\alpha' - \alpha)/(\alpha\alpha')$$

and

$$d'' = 0 \vee (d' - \delta), \qquad e'' = 0 \vee (e' - \delta).$$

Define $>$ on $\mathcal{N}$ as follows:

$$c > c' \Leftrightarrow \begin{cases} a + tb > a' + tb' \\ \text{or} \\ a + tb = a' + tb' \\ \text{and} \\ (a - d) + t(b + e) > (a' - d'') + t(b' + e'') \end{cases}$$

$$c = c' \Leftrightarrow \begin{cases} a + tb = a' + tb' \\ \text{and} \\ (a - d) + t(b + e) = (a' - d'') + t(b' + e'') \end{cases}$$

$$c < c' \text{ otherwise.}$$

The above definitions give a total order relation on $\mathcal{N}$. They also keep c_{ij} constant. This procedure is a particular case of the discrete methods proposed in [24].

If the times associated with the fuzzy graph are represented by the functions c_{ij} given above, we have a crisp graph whose times are fuzzy numbers of equal height within each interval. The fuzzy number sum and the order relation are suitable in this case. Hence classical minimal path algorithms can be used to solve the subproblems that appear in each interval. The Belman-Klalba's algorithm was used to solve the following example.

Example 4.3.9 *Consider the fuzzy graph appearing in Example* 4.3.8, *where the fuzzy numbers* t_{ij} *are given by the matrices shown in Tables* 4.4, 4.5, 4.6, *and* 4.7.

	1	2	3	4	5	6	7	8	9
$a_{ij} = 1$		18	13						
2			18	14	15		13		
3						16		19	
4					14		17		
5						17			
6								11	
7								12	
8									17
9									

Table 4.4

	1	2	3	4	5	6	7	8	9
$b_{ij}=1$		20	15						
2			18	16	17		14		
3					12			16	
4					15		19		
5						18			
6								13	
7								12	
8									17
9									

Table 4.5

	1	2	3	4	5	6	7	8	9
$d_{ij}=1$		3	2						
2			4	4	2		2		
3					3			1	
4					2		3		
5						3			
6								2	
7								2	
8									3
9									

Table 4.6

	1	2	3	4	5	6	7	8	9
$e_{ij} = 1$		4	3						
2			2	5	2		4		
3					2			1	
4					4		4		
5						2			
6								2	
7								2	
8									4
9									

Table 4.7

A minimal time path is obtained between 1 and 9. The application of a general procedure, yields the following solution:

Membership Degree 0.30
Path Solution
$1 \to 3 \to 8 \to 9$
Minimal Time: 49.00, 48.00, 0.10, 2.10

Membership Degree 0.50
Path Solution
$1 \to 3 \to 8 \to 9$
Minimal Time: 49.00, 48.00, 4.09, 6.09

Membership Degree 0.60
Path Solution
$1 \to 3 \to 8 \to 9$
Minimal Time: 49.00, 48.00, 5.09, 7.09

Membership Degree 0.70
Path Solution
$1 \to 2 \to 7 \to 8 \to 9$
Minimal Time: 60.00, 63.00, 8.29, 10.29

Membership Degree 0.80
Path Solution
$1 \to 2 \to 7 \to 8 \to 9$
Minimal Time: 60.00, 63.00, 9.00, 13.00

Membership Degree 0.90
Path Solution
$1 \to 2 \to 7 \to 8 \to 9$
Minimal Time: 60.00, 63.00, 9.55, 13.55

Membership Degree 1.00
Path Solution
$1 \to 2 \to 7 \to 8 \to 9$
Minimal Time: 60.00, 63.00, 10.00, 14.00

The fuzzy graph solution is given in Figure 4.4.

There are three possible minimal paths. If a knowledge level $\alpha \geq 0.30$ is used, the minimal path is $1 \longrightarrow 3 \longrightarrow 8 \longrightarrow 9$ with the time given by the fuzzy number $(49.00, 48.00, 0.10, 2.10)$ and if $\alpha \geq 0.60$, then the minimal path is $1 \to 3 \to 8 \to 9$ with time $(49.00, 48.00, 5.09, 7.09)$. If a knowledge level equal to 1 is used, the minimal path is $1 \to 2 \to 7 \to 8 \to 9$ with time $(60.00, 63.00, 10.00, 14.00)$.

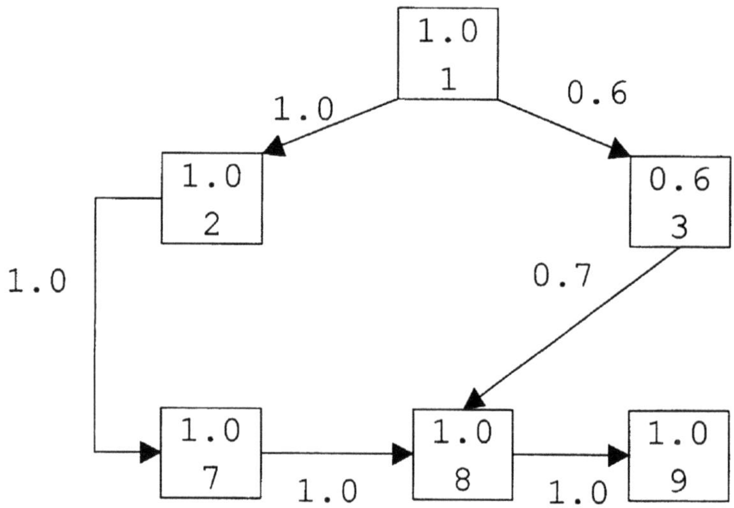

Figure 4.4: Fuzzy graph solution

4.4 References

1. Adamo, J.H., Fuzzy decision trees, *Fuzzy Sets and Systems, 4 (1980) 207-219.*

2. Behzad, M., Chartrand, G., and Lesniak-Foster, L., *Graphs and Digraphs*, Wadsworth International, Belmont, CA, 1979.

3. Bellman, R. and Klaba R., On kth best policies, *J. of Siam.* 8 (1960) 582-585.

4. Bollabas, B., *Random Graphs*, Academic Press, London, 1985.

5. Bonissone, P.P., Summarizing and propogating uncertain information with triangular norms, *Internat. J. Approximate Reasoning*, 1 (1987) 71-101.

6. Bortolan, G. and Degani, R., A review of some methods for ranking fuzz subsets, *Fuzzy Sets and Systems,* 16 (1985), 1-19.

7. Carrayway, R. and Morin, T., *Monotonicity, preference structures and principles of optimality, Computational Combinatorics*, URI Technical Report CC-86-3, Purdue University, West Lafayette, In, 1986.

8. Chanas, S., Fuzzy optimization in networks, In: Operation Models Using Fuzzy Sets and Possibility Theory, Eds.: J. Kacprzyk and S.A. Orlovski, D. Reidel Publishing Co., 1987, 303-327.

9. Chanas, S. and Kamburowski, The use of variables in PERT, *Fuzzy Sets and Systems,* 5 (1981) 11-19.

10. Chanas, S. and Kołodziejczynk, W., Maximum flow in a network with fuzzy arc capacities, *Fuzzy Sets and Systems,* 8 (1982) 165-173.

11. Chanas, S. and Kołodziejczynk, W., Integer flows in networks with fuzzy capacity constraints, Inst. of Management, Tech. Univ. of Wroclaw, Report 368 (1982).

12. Chanas, S. and Kołodziejczynk, W., Real-valued flows in a network with fuzzy arc capacities, *Fuzzy Sets and Systems,* 13 (1984) 139-151.

13. Chanas, S. and Kołodziejczynk, W., Integer flows in network with fuzzy capacity constraints, *Networks,* 16 (1986) 17-32.

14. Delgado, M., Verdegay, J. L., and Vila, M. A., On fuzzy tree definition, *European J. Oper. Res.*, 22 (1985) 243-249.

15. Delgado, M., Verdegay, J.L,. and Vila, M.A., El problema del arbol generador minimal para grafos disusos, *Trabajos de Investigacion Operativa,* 2 (1987) 3-20.

16. Delgado, M., Verdegay, J.L., and Vila, M.A., Fuzzy transportation problems: A general analysis, in J. Kacprzyk and S. Orlovski (eds.), Optimization Models Using Fuzzy Sets and Possibility Theory, P. Reidel, Dordrecht, The Netherlands, 1987, 342-358.

17. Delgado, M., Verdegay, J. L. and Vila, M. A., On valuation and optimization problems in fuzzy graphs: a general approach and some particular cases, *ORSA J. Comput.,* 2 (1990) 74-83.

18. Dubois, D. and Prade, H., Algorithmes de plus court chemins pour trater des donnees flous, *R.A.I.R.O. Recherche Operationelle/Operational Research,* 12 (1978) 214-227.

19. Dubois, D. and Prade, H., *Fuzzy Sets and System,* Academic Press, New York 1980.

20. Dubois, D., Farreny, H. and Prade, H., Combinatorial search with fuzzy estimates, in J. Kacprzyk and S. Orlovski (eds.), Optimization Models Using Fuzzy Sets and Possibility Theory, P. Reidel, Dordrecht, The Netherlands, 1987, 171-185.

21. Fujishige, S., Submodular systems and related topics, *Math. Programming Study*, 22 (1984) 113-131.

22. Garey, M.R. and Johnson, D.S., *Computers and Interactability: A Guide to the Theory of NP-Completeness*, Freeman, San Francisco, CA, 1979.

23. Gonzalez, A., Metodso Subjetivos para la Comparacion de Numeros Difusos, Ph. D. Thesis, University of Granada, 1987.

24. Gonzalez, A. and Vila, M.A., A discrete method to study indifference and order relations between fuzzy numbers, *Inform. Sci.*, 56 (1991) 245-258.

25. Gonzalez, A. and Vila, M.A., Dominance relation on fuzzy numbers, submitted to *Inform. Sci.,* 64 (1992) 1-16.

26. Händler, W., Multiprocessors: Efficiency and fault tolerance, in: Architektur and Betrieb von Rechensystemen (NTG-Fachberichte 92, VDE-Verlag, Berlin-Offenbach, 1986), 7-29 (in German).

27. Hearn, D.W., Lawphongpanich, S., and Ventura J.A., Restricted simplicial decomposition: Computation and extensions, *Math. Programming Study*, 31 (1987) 99-118.

28. Ignizio, J. and Daniels, S., Fuzzy multicriteria integer programming via fuzzy generalized networks, *Fuzzy Sets and Systems*, 10 (1983) 261-270.

29. Itakura, H. and Nishikawa, Y., Fuzzy network technique for technological forecasting, *Fuzzy Sets and Systems*, 14 (1984) 99-113.

30. Jensen, P. and Barnes, J., *Network Flow Programming*, John Wiley and Sons, New York, 1980.

31. Katsumata Y., Tsuda E. and Yamashita H., Fuzzy clustering and ordering. Instructional Analysis Applying Fuzzy Graphs, preprints of International Workshop on Fuzzy Systems Applications, Fukuoka, Japan, 1988.

32. Katsumata, Y., Tsuda, E. and Yamashita, H., The difference of two fuzzy graphs, preprints of International Workshop on Fuzzy Systems Applications, Fukuoka, Japan, 1988.

33. Kaufman, A., *Introduction to the Theroy of Fuzzy Sets*, Vol. 1, Academic Press, New York, 1975.

34. Kauffman, A., *Introduction a la Theorie des Sousensambles,* Vol 1. Flous, Masson et Cie, Paris, France, 1977.

35. Kim, K.H. and Roush, F.W., Fuzzy flows on networks, *Fuzzy Sets and Systems*, 8 (1982) 35-38.

36. Klein, C. M., Fuzzy shortest paths, *Fuzzy Sets and Systems*, 39 (1991) 27-41.

37. Klein, C., A model for transportation of hazardous waste, Department of Industrial Engineering, University of Missouri-Columbia, working paper series #1210388 (1988).

38. Klein, C. M. and Kincaid, Fuzzy location problems, in G. Evans, W. Karwowski, M. Wilhelm, Eds., Applications of Fuzzy Set Methodologies in Industrial and Systems Engineering, Elsevier, Amsterdam, 1989, 59-72.

39. Klein, C.M. and Ventura, J.A., Submodular function and convex related results, *AMSE Rev.*, 8 (1983) 43-64.

40. Klein, C.M. and Yeralan, S., Network flow models with fuzzy auxilary edge and vertex attributes, in J.L. Chameau and J. Yao, Eds, Proc. of Nafips Workship, Purdue University, West Lafayette, IN (May 5-7, 1987) 431-444.

41. Koczy L.T., Fuzzy trees in distributed systems design, preprints of International Workshop on Fuzzy Systems Applications, Fukuoka, Japan, 1988.

42. Lawler, E., *Combinatorial Optimization; Networks and Matroids*, Holt Reinhart and Winston, New York 1976.

43. Lovász, D., L., Submodular functions and convexity in: A. Bachem, M. Grötschel and B. Korte, Eds., Mathematical Programming the State of the Art-Bonn 1982 (Springer-Verlag, New York, 1983) 235-257.

44. Mares, M. and Horak, J., Fuzzy quantities in networks, *Fuzzy Sets and Systems,* 10 (1988) 123-134.

45. McAllister, M., Fuzzy matrices and diagraphs, Communications of First I.F.S.A Congress, Mallorca, Spain, 1985.

46. Morioka, M. and Yamasita, H., Extraction method of difference between fuzzy graphs, preprints of IFAC Symposium on Fuzzy Information, Knowledge Representation and Decision Analysis, (Marseille, France) E. Sanchez and M.M. Gupta (eds.), Pergamon Press, New York, 1983.

47. Nair, P.S., Extension principle revisited, JCIS'98, Vol. II, (1998) 11-14.

48. Rosenfeld, A., Fuzzy graphs, In: Zadeh, L.A., Fu, K.S., Tanaka, K., and Shimura, M., eds, Fuzzy Sets and Their Applications to Cognitive and Decision Processes, Academic Press, New York, 1975, 75-95.

49. Takeya M., A clustering method for a fuzzy diagraph based on connectedness and its applications to instructional evaluation, preprints of IFAC Symposium on Fuzzy Information, Knowledge Repsentaion and Decision Analysis, (Marseille, France) E. Sanchez and M. M. Gupta (eds.), Pergamon Press, New York, 1983.

50. Ventura, J.A. and Klein, C. M., An algorithm for certain convex network problems, *AMSE Rev.,* 8 (1988) 31-42.

51. Yager R., Paths of least resistence in possibilistic production systems, *Fuzzy Sets and Systems,* 19 (1986) 121-132.

52. Yeh R.T. and Bang S.Y., Fuzzy relations, fuzzy graphs, and their applications to clustering analysis, In: Zadeh, L.A., Fu, K.S., Tanaka, K., and Shimura, M., eds., Fuzzy Sets and Their Applications to Cognitive Decision Processes, Academic Press, London, 1975, 125-150.

5

Fuzzy Machines, Languages, and Grammars

5.1 Max-Product Machines

In this and the next section, we present the work of Santos [18].

Max-product machines may be considered as models of fuzzy systems [30]. In this section, we consider the minimization problem of max-product machines. We introduce various types of equivalence relations and minimal forms. Among the minimal forms considered are those which are similar to the reduced [5] and minimal state [1] forms of stochastic machines.

In order to investigate the properties of max-product machines, a type of algebra, called the **max-product algebra**, is developed in this section. The role played by max-product algebra in the theory of max-product machines is the same as that played by (convex) linear algebra in the theory of stochastic machines.

By the use of max-product algebra, complete solutions for the minimization problem of max-product machines are obtained for both the equivalence relations and minimal forms.

Definition 5.1.1 *Let $A = (a_{ij})$ be $n \times p$ and $B = (b_{ij})$ be $p \times m$ matrices of nonnegative real numbers. Let $A \odot B$ be the $n \times m$ matrix (c_{ij}), where*

$$c_{ij} = \vee\{a_{ik}b_{kj} \mid k = 1, 2, \ldots, p\}.$$

Definition 5.1.1 is applicable even if n or m is infinite.

Proposition 5.1.2 *Let A_1 be $n \times p$, A_2 be $p \times m$, and A_3 be $m \times t$ matrices of nonnegative real numbers. Then*

$$(A_1 \odot A_2) \odot A_3 = A_1 \odot (A_2 \odot A_3),$$

that is, the operation $\odot$ is associative.

We introduce the following notation: a, b, and c (with or without subscripts) will denote nonnegative real numbers; x and y (with or without subscripts) will denote (finite or infinite) sequences of nonnegative real numbers. Superscripts will be used to denote the particular term of the sequence, e.g., x^k will denote the k-th term of the sequence x. X and Y (with or without subscripts) will denote collections of (finite or infinite) sequences of nonnegative real numbers. That is, if S denotes the set of all sequences of nonnegative real numbers, then $X, Y \in \mathcal{P}(S)$.

(1) Let ax denotes the sequence whose k-th term is ax^k.

(2) Let $x_1 \vee x_2 \vee \ldots \vee x_n$ or $\vee\{x_i \mid i = 1, \ldots, n\}$ denote the sequence whose k-th term is $x_1^k \vee x_2^k \vee \ldots \vee x_n^k$.

(3) MP stands for the max-product.

Definition 5.1.3 *Let $X = \{x_1, x_2, \ldots, x_n\}$. An **MP-combination** of X is an expression of the form*

$$\vee\{a_i x_i \mid i = 1, \ldots, n\}, \tag{5.1}$$

where a_i is a nonnegative real number, $i = 1, 2, \ldots, n$.

*If $0 \leq a_i \leq 1$ for $i = 1, 2, \ldots, n$, then (5.1) is called a **convex MP-combination** of X.*

Definition 5.1.4 *The (**convex**) **MP-span** of X is the collection of all (convex) MP-combinations of finite subsets of X.*

$C(X)$ will denote the convex MP-span of X.

Proposition 5.1.5 *(i) $X \subseteq C(X)$;*
(ii) $C(C(X)) = C(X)$;
(iii) If $X_1 \subseteq X_2$, then $C(X_1) \subseteq C(X_2)$. ∎

Hence we can think of C as a function of $\mathcal{P}(S)$ into $\mathcal{P}(S)$.

Definition 5.1.6 *Y is called a convex **MP-set** if for every $y_1, y_2 \in Y$ all convex MP-combinations of $\{y_1, y_2\}$ are also in Y.*

Proposition 5.1.7 *Y is a convex MP-set if and only if $Y = C(Y)$.* ∎

Proposition 5.1.8 *For every X, $C(X)$ is a convex MP-set.*

Proof. The proof follows from Propositions 5.1.7 and 5.1.5(ii). ∎

In the rest of this section, Y will always denote a convex MP-set.

Definition 5.1.9 *Let $X \subseteq Y$. X is called a* **set of generators** *of Y if $Y = C(X)$. If X does not contain any proper subset which is itself a set of generators of Y, then X is called a* **set of vertices** *of Y.*

Proposition 5.1.10 *Let $X \subseteq Y$. Then X is a set of vertices of Y if and only if*
(i) $Y = C(X)$ and
(ii) if $x \in X$, then $x \notin C(X\backslash\{x\})$. ∎

Proposition 5.1.11 *Let X be a set of generators of Y such that X is finite. Then there exists $X' \subseteq X$ such that X' is a set of vertices of Y.*

Proof. The proof follows from Propositions 5.1.10 and the fact that X is finite. ∎

We write $x_1 \geq x_2$ if $x_1^k \geq x_2^k$ for all k.

Proposition 5.1.12 *Let X be a set of vertices of Y and X' be a set of generators of Y. Then $X \subseteq X'$.*

Proof. Let $x \in X$. By Proposition 5.1.10(i),

$$x = \vee\{a_i x_i' \mid i = 1, \ldots, n\}, \tag{5.2}$$

where for $i = 1, 2, \ldots, n$, $x_i' \in X'$ and $0 \leq a_i \leq 1$. Moreover

$$x_i' = \vee\{b_{ij} x_j \mid j = 0, \ldots, m\}, \tag{5.3}$$

where for $j = 0, 1, 2, \ldots, m$, $x_j \in X$ and $0 \leq b_{ij} \leq 1$. Thus

$$x = \vee\{c_j x_j \mid j = 0, \ldots, m\},$$

where

$$c_j = \vee\{a_i b_{ij} \mid i = 1, \ldots, n\}. \tag{5.4}$$

By Proposition 5.1.10(ii), x must occur among the x_j, say $x = x_0$, and $c_0 = 1$. Therefore, by (5.4), there exists k such that $a_k = 1$ and $b_{k0} = 1$. It follows from (5.2) and (5.3), that

$$x \geq x_k' \qquad \text{and} \qquad x_k' \geq x_0 = x.$$

Thus $x = x_k' \in X'$. Consequently, $X \subseteq X'$. ∎

Theorem 5.1.13 *The set of vertices of Y is unique if it exists.*

Proof. The proof follows from Proposition 5.1.12. ■

Proposition 5.1.14 *Let X be the set of vertices of Y. If X_1 and X_2 are sets of generators of Y such that $|X_1| > |X_2|$, then there exists $x \in X_1$ such that $x \in C(X_1 \backslash \{x\})$.*

Proof. From Proposition 5.1.12, $X \subseteq X_1$ and $X \subseteq X_2$. Therefore $|X| \leq |X_2| < |X_1|$. Thus X is a proper subset of X_1. The conclusion follows from Proposition 5.1.10. ■

Definition 5.1.15 *Let X_1, $X_2 \subseteq Y$. X_1 is called a* ***basis*** *of X_2 if every $x \in X_2$ can be expressed uniquely as a convex MP-combination of a finite subset of X_1.*

Proposition 5.1.16 *Let X be a basis of Y. Then X is a set of vertices of Y.*

Proof. The proof follows from Proposition 5.1.10. ■

Equivalences

Definition 5.1.17 *A* ***max-product sequential-like machine (MPSM)*** *is a quadruple (U, S, V, μ), where U, S, V are finite nonempty sets and μ is a function from $S \times U \times V \times S$ into $[0, 1]$.*

The sets U and V are the input and output alphabets, respectively. S is the set of internal states and μ is the transition function. We can regard $\mu(s, u, v, s')$ as the grade of membership that the MPSM will enter state s' and produce output v given that the present state is s and the input is u.

The readers should note the differences between the above definition and the one given in [17].

We call finite sequences of elements of U **input tapes** or **strings**. Similarly, we call finite sequences of elements of V **output tapes** or **strings**. The collection of all input (output) tapes will be denoted by $U^*(V^*)$. Let Λ denote the empty tape. Then Λ has the property $x\Lambda = \Lambda = \Lambda x$ for all tapes x. The **length** of the tape x, denoted by $len(x)$, is the number of elements in the sequence x. Clearly, $len(\Lambda) = 0$. Moreover,

$$(U \times V)^* = \{(x, y) \mid x \in U^*, y \in V^*,\ len(x) = len(y)\}.$$

Unless stated otherwise, the symbol M, with or without subscripts, will denote an MPSM. All MPSMs will be assumed to have the same input and output sets. Since all MPSMs are assumed to have the same input set U and the same output set V, for the ease of notation, we represent an MPSM by (S, μ). Furthermore, we assume that $(U \times V)^*$ is ordered in such a way that $len(x_1) < len(x_2)$ implies $(x_1,y_1) < (x_2,y_2)$. The order of $(U \times V)^*$ will be kept fixed and assumed to be the same for all MPSMs under consideration.

Definition 5.1.18 *Let $M = (S, \mu)$ be an MPSM. The* ***extended transition function*** *μ^* of M is a function from $S \times (U \times V)^* \times S$ into $[0,1]$ defined inductively on $len(x)$, $x \in U^*$, as follows:*

$$\mu^*(s', \Lambda, \Lambda, s'') = \begin{cases} 1 & \text{if } s' = s'' \\ 0 & \text{if } s' \neq s''; \end{cases}$$

$$\mu^*(s', ux, vy, s'') = \vee_{s \in S}\{\mu(s', u, v, s) \times \mu^*(s, x, y, s'')\},$$

where $u \in U$, $v \in V$, and $y \in V^*$, $len(x) = len(y)$. Moreover, the output function q^M of M is a function from $S \times (U \times V)^*$ into $[0,1]$ defined as follows:

$$q^M(s, x, y) = \vee_{s' \in S}\mu^*(s, x, y, s').$$

If no confusion arises, we will write q for q^M.

For ease of notation, we assume that $S = \{s_1, s_2, \ldots, s_n\}$. Furthermore,

(1) $P^M(x, y) = (p_{ij})$, is a matrix, where $p_{ij} = \mu^*(s_i, x, y, s_j)$;

(2) $Q^M(x, y) = (q_i)$ is the column matrix, where the i-th row $q_i = q^M(s_i, x, y)$;

(3) E is the column matrix with all entries 1.

Proposition 5.1.19 *Let $M = (S, \mu)$ and $(x_1, y_1), (x_2, y_2) \in (U \times V)^*$. Then*

(i) $P^M(x_1x_2, y_1y_2) = P^M(x_1, y_1) \odot P^M(x_2, y_2)$,

(ii) $Q^M(x_1x_2, y_1y_2) = P^M(x_1, y_1) \odot Q^M(x_2, y_2) = P^M(x_1x_2, y_1y_2) \odot E$.

Proof. The proof follows from Proposition 5.1.2. ■

Definition 5.1.20 *Let $M = (S, \mu)$.*

(i) An ***initial distribution (i.d.)*** *of M is a function ι from S into $[0,1]$.*

(ii) ι is said to be ***concentrated*** *at $s \in S$ if $\iota(s) = 1$ and 0 elsewhere.*

We will also use the symbol ι to denote the row matrix whose i-th row is $\iota(s_i)$.

Definition 5.1.21 *Let M be an MPSM and ι be an i.d. of M. Then the ordered pair (M, ι) is called an* ***initialized max-product sequential-like machine (IMPSM)****.*

If ι is concentrated at s, we shall also write (M, s) for (M, ι).

Definition 5.1.22 *Let $I = (M, \iota)$ be an IMPSM. The* ***response function*** *r^I of I is a function from $(U \times V)^*$ into $[0,1]$ defined as follows:*

$$r^I(x,y) = \vee_{s \in S}\{\iota(s) \cdot q^M(s,x,y)\}.$$

A^M will denote the semi-infinite matrix whose columns are $Q^M(x,y)$ arranged in the order of $(U \times V)^*$.

Let A be a matrix. $\rho(A)$ will denote the collection of all rows of A.

If $R^I = \iota \odot A^M$, then R^I is a row matrix whose entries are $r^I(x,y)$. Moreover, R^I is a convex MP-combination of $\rho(A^M)$.

Definition 5.1.23 *Let I_1 and I_2 be IMPSMs. The I_1 and I_2 are called* ***equivalent****, denoted by $I_1 \sim I_2$, if $r^{I_1} = r^{I_2}$.*

Definition 5.1.24 *Let ι_1 and ι_2 be i.d.s of an MPSM M. Then ι_1 and ι_2 are called* ***M-equivalent****, denoted by $\iota_1 \sim \iota_2$, if $(M, \iota_1) \sim (M, \iota_2)$.*

If ι_1 is concentrated at s, then we shall also write $s \sim \iota_2$ for $\iota_1 \sim \iota_2$.

Proposition 5.1.25 *Let $\iota_1$and ι_2 be i.d.s of an MPSM M. Then $\iota_1 \sim \iota_2$ if and only if*

$$\iota_1 \odot A^M = \iota_2 \odot A^M. \blacksquare$$

Definition 5.1.26 *Let $M_1 = (S_1, \mu_1)$ and $M_2 = (S_2, \mu_2)$:*

(i) M_1 and M_2 are called ***statewise equivalent,*** *written $M_1 \sim M_2$, if for every $s' \in S_1$, there exists $s'' \in S_2$ such that $(M_1, s') \sim (M_2, s'')$ and vice versa.*

(ii) M_1 and M_2 are called ***compositewise equivalent****, written $M_1 \simeq M_2$, if for every $s \in S_1$, there exists i.d. ι of M_2 such that (M_2, ι) and vice versa.*

(iii) M_1 and M_2 are called ***distributionwise equivalent,*** *written $M_1 \approx M_2$, if for every i.d. ι_1 of M_1, there exists i.d. ι_2 of M_2 such that $(M_1, \iota_1) \sim (M_2, \iota_2)$ and vice versa.*

Theorem 5.1.27 *Let M_1 and M_2 be MPSMs. Then the following assertions hold.*

(i) $M_1 \sim M_2$ if and only if $\rho(A^{M_1}) = \rho(A^{M_2})$;

(ii) $M_1 \simeq M_2$ if and only if $\rho(A^{M_1}) \subseteq C[\rho(A^{M_2})]$ and $\rho(A^{M_2}) \subseteq C[\rho(A^{M_1})]$;

(iii) $M_1 \approx M_2$ if and only if $C[\rho(A^{M_1})] \subseteq C[\rho(A^{M_2})]$. $\blacksquare$

Proposition 5.1.28 *Let M_1 and M_2 be MPSMs. If $M_1 \sim M_2$, then $M_1 \simeq M_2$.*

Proof. The proof follows from Proposition 5.1.5(i) and Theorem 5.1.27. $\blacksquare$

Proposition 5.1.29 *Let M_1 and M_2 be MPSMs. Then $M_1 \simeq M_2$ if and only if $M_1 \approx M_2$.*

Proof. The proof follows from Proposition 5.1.5 and Theorem 5.1.27. $\blacksquare$

5.2 Irreducibility and Minimality

In this section, we study the irreducibility and minimality of an MPSM.

Definition 5.2.1 *Let $M = (S, \mu)$.*

(i) M is called ***statewise irreducible*** *if for every s', $s'' \in S$, $s' \sim s''$ implies $s' = s''$.*

(ii) M is called ***compositewise irreducible*** *if for every $s \in S$ and i.d. ι of M, $s \sim \iota$ implies $\iota(s) > 0$.*

(iii) M is called ***distributionwise irreducible*** *if for every i.d. ι_1 and ι_2 of M, $\iota_1 \sim \iota_2$ implies $\iota_1 = \iota_2$.*

Theorem 5.2.2 *Let M be an MPSM.*

(i) M is statewise irreducible if and only if no two rows of A^M are identical.

(ii) M is compositewise irreducible if and only if $\rho(A^M)$ is a set of vertices of $C[\rho(A^M)]$, i.e., no row of A^M is a convex MP-combination of the other rows of A^M.

(iii) M is distributionwise irreducible if and only if $\rho(A^M)$ is a basis of $C[\rho(A^M)]$.

Proof. (i) follows from Proposition 5.1.25, (ii) follows from Propositions 5.1.25 and 5.1.10, and (iii) follows from Proposition 5.1.25 and the definition of basis. ■

Proposition 5.2.3 *Let M be an MPSM. If M is distributionwise irreducible, then M is compositewise irreducible.* ■

Proposition 5.2.4 *Let M be an MPSM. If M is compositewise irreducible, then M is statewise irreducible.* ■

Definition 5.2.5 *Let M be an MPSM. Then M is called* ***statewise (compositewise) minimal*** *if M is not statewise (compositewise) equivalent to any MPSM with a fewer number of states.*

Let $M = (S, \mu)$. Then the cardinality of M, is defined by $|M| = |S|$.

Theorem 5.2.6 *Let M be an MPSM. Then M is statewise minimal if and only if M is statewise irreducible.*

Proof. Let $M = (S, \mu)$. Suppose that M is not statewise minimal. Then there exists MPSM M' such that $|M'| < |M|$ and $M' \sim M$. By Theorems 5.1.27(i) and 5.2.2(i), M is not statewise irreducible. Conversely, suppose that M is not statewise irreducible. Then there exist $s', s'' \in S$ such that $s' \sim s''$ and $s' \neq s''$. By renumbering the elements of S, if necessary, we may assume that $s' = s_{n-1}$ and $s'' = s_n$, where $n = |S|$. Let $M' = (S', \mu')$, where $S' = S \backslash \{s_n\}$ and for $i = 1, 2, \ldots, n-1$,

$$\mu'(s_i, u, v, s_j) = \begin{cases} \mu(s_i, u, v, s_j) & \text{if } j = 1, 2, \ldots, n-2 \\ \mu(s_i, u, v, s_{n-1}) \vee \mu(s_i, u, v, s_n) & \text{if } j = n-1. \end{cases}$$

We can show that $\rho(A^M) = \rho(A^{M'})$. By Theorem 5.1.27(a), $M \sim M'$. Thus M is not statewise minimal. ■

Theorem 5.2.7 *Let M be an MPSM. Then M is compositewise minimal if and only if M is compositewise irreducible.*

Proof. Let $M = (S, \mu)$. Suppose that M is not compositewise minimal. Then there exists an MPSM M' such that $|M'| < |M|$ and $M' \simeq M$. By Theorems 5.1.27 and 5.2.2 and Proposition 5.1.14, M is not compositewise irreducible. Conversely, suppose that M is not compositewise irreducible. Then there exists $s' \in S$ and i.d. ι of M such that $s' \sim \iota$ and $\iota(s') = 0$. By renumbering the elements of S, if necessary, we may assume that $s' = s_n$, where $n = |S|$. Since $s_n \sim \iota$ and $\iota(s_n) = 0$,

$$q^M(s_n, x, y) = \vee \{\iota(s_i) \cdot q^M(s_i, x, y) \mid i = 1, 2, \ldots, n\}$$

for all $(x, y) \in (U \times V)^*$. Let $M' = (S', \mu')$, where $S' = S \backslash \{s_n\}$ and for $i, j = 1, 2, \ldots, n-1$,

$$\mu'(s_i, u, v, s_j) = \mu(s_i, u, v, s_j) \vee \iota(s_j) \cdot \mu(s_i, u, v, s_n).$$

Since for every $(x, y) \in (U \times V)^*$ and $i = 1, 2, \ldots, n-1$, $q^M(s_n, x, y) = q^M(s_i, x, y)$, we have $C[\rho(A^M)] = C[\rho(A^{M'})]$. By Theorem 5.1.27, $M' \simeq M$. Thus M is not compositewise minimal. ■

Proposition 5.2.8 *Let M_1 and M_2 be MPSMs. Suppose that M_1 is statewise minimal and $M_1 \sim M_2$. Then M_2 is statewise minimal if and only if $|M_1| \leq |M_2|$.* ■

Proposition 5.2.9 *Let M_1 and M_2 be MPSMs. Suppse that $M_1$is statewise minimal and $M_1 \simeq M_2$. Then M_2 is compositewise minimal if and only if $|M_1| \leq |M_2|$.* ■

Proposition 5.2.10 *Let M_1 and M_2 be MPSMs which are compositewise minimal. Then $M_1 \sim M_2$ if and only if $M_1 \simeq M_2$.*

Proof. The proof follows from Proposition 5.1.28 and Theorems 5.1.13, 5.2.2, and 5.2.7 ■

Proposition 5.2.11 *Let M_1 and M_2 be MPSMs. Suppose that M_1 and M_2 are statewise minimal and $M_1 \sim M_2$. Then for every $(x, y) \in (U, V)^*$,*

$$P^{M_1}(x, y) \odot A^{M_1} = P^{M_2}(x, y) \odot A^{M_1}$$

after an appropriate rearrangement of states.

Proof. From Theorems 5.1.27(i), 5.2.2(i), and 5.2.6, $A^{M_1} = A^{M_2}$ after an appropriate rearrangement of states. ∎

Definition 5.2.12 *Let M_1 and M_2 be MPSMs. M_1 and M_2 are called* ***isomorphic*** *($\equiv$) if they are equal up to a permutation of states.*

Definition 5.2.13 *Let M be an MPSM. M is called* ***simple*** *if there exists no MPSM M' such that $|M'| \leq |M|$, $M' \sim M$, but $M' \equiv M$.*

Let M be an MPSM. Then we let

$$\rho(M) = \cup_{u \in U} \cup_{v \in V} \rho[P^M(u,v)]$$

and

$$\overline{\rho}\,(M) = \{\iota \odot A^M \mid \iota \in \rho(M)\}.$$

Theorem 5.2.14 *Let M be an MPSM such that M is statewise minimal. Then M is simple if and only if $\rho(A^M)$ is a basis of $\overline{\rho}\,(M)$.*

Proof. Suppose that $\rho(A^M)$ is a basis of $\overline{\rho}\,(M)$. Let M' be an MPSM such that $|M'| \leq |M|$ and $M' \sim M$. By Proposition 5.2.8, M' is statewise minimal. It follows from Proposition 5.2.11 that $P^M(x,y) = P^{M'}(x,y)$ for all $(x,y) \in (U \times V)^*$ after an appropriate rearrangement of states. Thus $M \equiv M'$. Hence M is simple. Conversely, suppose that $\rho(A^M)$ is not a basis of $\overline{\rho}\,(M)$. There exists $j \in \rho(M)$ such that

$$j \odot A^M = j' \odot A^M \tag{5.5}$$

for some $j' \neq j$. Let j be the i-th row of the matrix $P^M(u,v)$. Construct M' from M by replacing the i-th row of the matrix $P^M(u,v)$ by j' and leaving the rest unaltered. By (5.5), $A^M = A^{M'}$. Thus $|M'| = |M|$ and $M' \sim M$. However, $M' \not\equiv M$. Hence M is not simple. ∎

Proposition 5.2.15 *Let M be an MPSM such that M be statewise (compositewise) minimal. Then there exists no statewise (compositewise) minimal MPSM which is statewise (compositewise) equivalent, but not isomorphic to M if and only if $\rho(A^M)$ is a basis of $\overline{\rho}\,(M)$.*

Proof. If M is statewise minimal, the desired result follows immediately from Theorem 5.2.14. The case when M is compositewise minimal follows from Theorem 5.2.14 and Propositions 5.2.9 and 5.2.10. ∎

Proposition 5.2.16 *Let M be an MPSM such that M is distributionwise irreducible. Then there exists no statewise (compositewise) minimal MPSM which is statewise (compositewise) equivalent, but not isomorphic to M.*

Proof. The proof follows from Theorem 5.2.2(iii) and Proposition 5.2.15.
∎

5.3 On Reductions of Maximin Machines

The material in this section is from [19].

The reduction problem of maximin sequential machines and maximin automata was studied by Santos in [15,16]. However, the solutions provided there were not complete. In the present section, the same problem is reexamined. Hence a maximin sequential-like machine [17] is used. Most of the results are applicable to the other two variations of maximin machines.

Three types of equivalence relations are considered, namely, statewise, compositewise, and distributionwise equivalence. We show that the last two are equivalent. From the first two equivalence relations, two minimal forms are defined. The counterparts of these two minimal forms in the theory of stochastic sequential machines are the reduced [5] and minimal state forms [1].

For stochastic sequential machines, (convex) linear algebra is a useful theory. Unfortunately, no similar theory is available for maximin machines. For this reason, the next section is devoted to the development of a type of algebra, called the **maximin algebra**. The maximin algebra is a very useful tool for dealing with maximin machines. Although maximin algebra resembles linear algebra and the max-product algebra of Section 5.1 in certain respects, they are almost completely unrelated.

Most of the results have counterparts in the theory of stochastic sequential machines and max-product machines. These counterparts are either known to be true [1,5,13,18] or can be shown to be true. The section concludes with a look at nondeterministic and deterministic machines which are special cases of maximin machines. Many of the results for maximin machines are strengthened for these particular cases.

Definition 5.3.1 *Let $A = (a_{ij})$ be $n \times p$ and $B = (b_{ij})$ be $p \times m$ matrices of nonnegative real numbers. Let $A \otimes B$ be the $n \times m$ matrix (c_{ij}), where*

$$c_{ij} = \vee\{a_{ik} \wedge b_{kj} \mid k = 1, \ldots, p\}.$$

Remark 5.3.2 *Definition 5.3.1 is applicable even if n or m is infinite.*

Clearly, the operation $\otimes$ is associative.

As in the previous two sections, a, b, and c (with or without subscripts) denote real numbers. x and y (with or without subscripts) denote (finite or infinite) sequences of real numbers. A superscript is used to denote the particular term of the sequence, e.g., x^k denotes the k-th term of the sequence x. X and Y (with or without subscripts) denote collections of (finite or infinite) sequences of real numbers. Recall that $a \wedge x$ denotes the sequence whose k-th term is $a \wedge x^k$.

Definition 5.3.3 *Let $X = \{x_1, x_2, \dots, x_n\}$. A* ***maximin combination*** *of X is an expression of the form*

$$\vee\{a_i \wedge x_i \mid i = 1, \dots, n\}, \tag{5.6}$$

where a_i is a nonnegative real number, $i = 1, 2, \dots, n$.

If $0 \leq a_i \leq 1$ for $i = 1, 2, \dots, n$, then (5.6) is a convex maximin combination of X.

Definition 5.3.4 *Let T_x be the set of all distinct terms of x. x is called* ***admissible*** *if T_x is finite and T_x can be effectively constructed from x. X is called* ***admissible*** *if every x in X is admissible.*

Proposition 5.3.5 *Let $x = \vee\{a_i \wedge x_i \mid i = 1, \dots, n\}$. If T_x is finite, then*

$$x = \vee\{b_i \wedge x_i \mid i = 1, \dots, n\},$$

where for $i = 1, 2, \dots, n$, $b_i \in T_x$.

Proof. Choose b_i to be the largest number in T_x such that $b_i \leq a_i$. ∎

Proposition 5.3.6 *Suppose that T_x is admissible and X is finite. Then it is decidable whether or not x is a (convex) maximin combination of X.*

Proof. The proof follows from Proposition 5.3.5. ∎

Definition 5.3.7 *The* ***(convex) maximin span*** *of X is the collection of all (convex) maximin combinations of finite subsets of X.*

Let $C(X)$ denote the convex maximin span of X.

Proposition 5.3.8 *(i) $X \subseteq C(X)$;*
(ii) $C(C(X)) = C(X)$;
(iii) If $X_1 \subseteq X_2$, then $C(X_1) \subseteq C(X_2)$. ∎

Definition 5.3.9 *Y is called a* ***convex maximin set*** *if for every y_1, $y_2 \in Y$, all convex maximin combinations of $\{y_1, y_2\}$ are also in Y.*

Proposition 5.3.10 *Y is a convex maximin set if and only if $Y = C(Y)$.* ∎

Proposition 5.3.11 *For every X, $C(X)$ is a convex maximin set.*

Proof. The proof follows from Propositions 5.3.10 and 5.3.8(ii). ∎

In the rest of this section, Y always denotes a convex maximin set.

Definition 5.3.12 *Let $X \subseteq Y$.*
(i) X is called a ***set of generators*** *of Y if $Y = C(X)$.*
(ii) If X does not contain any proper subset which is itself a set of generators of Y, then X is called a ***set of vertices*** *of Y.*

Proposition 5.3.13 *X is a set of generators of $C(X)$.* ∎

Proposition 5.3.14 *Let $X \subseteq Y$. Then X is a set of vertices of Y if and only if*
(i) $Y = C(X)$ and
(ii) if $x \in X$, then $x \notin C(X \backslash \{x\})$. ∎

Proposition 5.3.15 *Suppose that X is a finite set of generators of Y. Then there exists $X' \subseteq X$ such that X' is a set of vertices of Y. Moreover, X' can be effectively constructed provided X is admissible.*

Proof. The proof follows from Propositions 5.3.14, 5.3.6, and the fact that X is finite. ∎

Definition 5.3.16 *Y is called **finitary** if it contains a set of generators which is finite.*

Proposition 5.3.17 *Every finitary convex maximin set has at least one set of vertices.*

Proof. The proof follows from Proposition 5.3.15. ∎

Remark 5.3.18 *Proposition 5.3.17 is not true in general for arbitrary convex maximin sets.*

Proposition 5.3.19 *Suppose that Y is finitary. Then every set of generators of Y contains a finite subset which is a set of vertices of Y.*

Proof. By Proposition 5.3.17, there exists a set X of vertices of Y which is finite. Let X' be a set of generators of Y. Now $\forall x \in X$, x is a convex maximin combination of a finite subset of X'. Let X'' be the set of all elements of X' which appear in one or more of these convex maximin combinations. Clearly, X'' is finite and is itself a set of generators of Y. The result now follows from Proposition 5.3.15. ∎

Proposition 5.3.20 *Suppose that Y is finitary. Then every set of vertices of Y is finite.*

Proof. The proof follows from Proposition 5.3.19. ∎

Definition 5.3.21 *Let X be a set of vertices of Y. X is called **fundamental** if for every $x \in X$, there exists no $y \in Y$ such that $x \neq y = a \wedge x$ for some $0 \leq a \leq 1$ and $(X \backslash \{x\}) \cup \{y\}$ is a set of vertices of Y.*

We write $x \leq a$ if $x^k \leq a$ for all k, $x_1 \leq x_2$ means $x_1^k \leq x_2^k$ for all k, and $x = a$ means $x^k = a$ for all k.

Proposition 5.3.22 *Let X be a set of vertices of Y. Then X is fundamental if and only if for every $x \in X$,*

$$x = (a \wedge x) \vee y,$$

where $y \in C(X\backslash\{x\})$ *and* $0 \leq a \leq 1$, *implies* $x = a \wedge x$.

Proof. Let $x \in X$ be such that $x = (a \wedge x) \vee y$, where $y \in C(X\backslash\{x\})$ and $0 \leq a \leq 1$, but $x \neq a \wedge x$. Let $y' = a \wedge x$ and $X' = (X\backslash\{x\}) \cup \{y'\}$. Then X' is a set of vertices of Y. Thus X is not fundamental. Conversely, suppose that X is not fundamental. Then there exists $x \in X$, $y' \in Y$ such that $x \neq y' = a \wedge x$ for some $0 \leq a \leq 1$ and $(X\backslash\{x\}) \cup \{y'\}$ is a set of vertices of Y. Thus x is a convex maximin combination of a finite subset X' of $(X\backslash\{x\})\cup\{y'\}$. By Proposition 5.3.14, $x \in C(X')$, i.e., $x = (b \wedge y') \vee y$, where $y \in X'\backslash\{y'\}$ and $0 \leq b \leq 1$. Let $c = a \wedge b$. Then $x = (c \wedge x) \vee y$. Since $c \wedge x = b \wedge y' \leq y' = a \wedge x \leq x$, $x \neq c \wedge x$. ■

Proposition 5.3.23 *Suppose that* X *is a fundamental set of vertices of* $C(X)$ *and* $x \in X$. *Then* $X\backslash\{x\}$ *is a fundamental set of vertices of* $C(X\backslash\{x\})$.

Proof. The proof follows from Propositions 5.3.14 and 5.3.22. ■

Proposition 5.3.24 *Let* X' *and* X'' *be fundamental sets of vertices of* Y *and* $\bar{x} \in X' \cap X''$. *Then* $C(X'\backslash\{\bar{x}\}) = C(X''\backslash\{\bar{x}\})$.

Proof. Let $x \in X'\backslash\{\bar{x}\}$. Then

$$x = \vee\{a_i \wedge x_i'' \mid i = 0, 1, \ldots, n\}, \tag{5.7}$$

where for $i = 0, 1, 2, \ldots, n$, $x_i'' \in X''$ and $0 \leq a_i \leq 1$. If $\bar{x}$ does not occur among the x_i'', then $x \in C(X\backslash\{\bar{x}\})$. Therefore, let $x_0'' = \bar{x}$. Now for $i = 1, 2, \ldots, n$,

$$x_i'' = \vee\{b_{ij} \wedge x_j' \mid j = 0, 1, \ldots m\}, \tag{5.8}$$

where for $j = 0, 1, 2, \ldots m$, $x_j' \in X''$ and $0 \leq b_{ij} \leq 1$. Thus

$$x = \vee\{(a_0 \wedge \bar{x}),\ \vee\{c_j \wedge x_j' \mid j = 0, 1, \ldots m\}\},$$

where

$$c_i = \vee\{a_i \wedge b_{ij} \mid i = 1, 2, \ldots, n\}.$$

Since X' is a set of vertices of Y, x must appear among the x_j', say x_0'. Also, since X' is fundamental, by Proposition 5.3.22,

$$x = c_0 \wedge x.$$

Therefore,

$$x \leq c_0 = \vee\{a_i \wedge b_{i0} \mid i = 1, 2, \ldots, n\} = a_k \wedge b_{k0}$$

for some $1 \leq k \leq n$. Thus $x \leq a_k$ and $x \leq b_{k_0}$. By (5.7) and (5.8),

$$x \geq a_k \wedge x''_k \geq a_k \wedge b_{k_0} \wedge x'_0 = x.$$

Hence $x = a_k \wedge x''_k$ or $x \in C(X''\backslash\{\bar{x}\})$. Thus

$$C(X'\backslash\{\bar{x}\}) \subseteq C(X''\backslash\{\bar{x}\}).$$

Similarly,

$$C(X''\backslash\{\bar{x}\}) \subseteq C(X'\backslash\{\bar{x}\}).$$

Consequently,

$$C(X'\backslash\{\bar{x}\}) = C(X''\backslash\{\bar{x}\}). \blacksquare$$

Theorem 5.3.25 *If Y is finitery, then the fundamental set of vertices of Y is unique.*

Proof. Let X' and X'' be fundamental sets of vertices of Y. By Proposition 5.3.20, both X' and X'' are finite. Let $X' = \{x'_1, x'_2, \ldots, x'_n\}$ and $X'' = \{x''_1, x''_2, \ldots, x''_m\}$, where $n \leq m$. We prove the result by induction on m. If $m = 1$, the result is trivial. Suppose that the theorem is true for all $m < p$. We show that it is also true for $m = p$. Let

$$x'_i = \vee\{a_{ij} \wedge x''_j \mid j = 1, 2, \ldots, m\},$$

and

$$x''_j = \vee\{b_{jk} \wedge x'_k \mid k = 1, 2, \ldots, n\}.$$

Then

$$x'_i = \vee\{c_{ik} \wedge x'_k \mid k = 1, 2, \ldots, n\},$$

where

$$c_{ik} = \vee\{a_{ij} \wedge b_{jk} \mid j = 1, 2, \ldots, m\}.$$

Since X' is fundamental, by Proposition 5.3.22, $x_1' = c_{11} \wedge x_1'$. Hence

$$x_1' \leq c_{11} = \vee\{a_{1j} \wedge b_{j1} \mid j = 1, 2, \ldots, m\} = a_{1j_1} \wedge b_{j_1 1}$$

for some j_1. This implies that $x_1' \leq a_{1_{j_1}}$ and $x_1' \leq b_{j_1 1}$. Thus

$$x_{j_1}'' \geq b_{j_1 1} \wedge x_1' = x_1'.$$

Similarly, there exists j_2 such that $x_{j_2}'' \geq x_{j_1}''$. By repeating this process, we obtain a sequence $j_1, j_2, \ldots, j_n, \ldots$, where

$$x_{j_{2k-1}}'' \leq x_{j_{2k}}' \leq x_{j_{2k+1}} \leq x_{j_{2k+2}}.$$

Since X'' is finite, we must have $x_{j_{2k-1}}'' = x_{j_{2l-1}}''$, for some k and l, where $l > k$. Thus $x_{j_{2k-1}}'' = x_{j_{2k}}' = \bar{x}$. Now consider $X' \backslash \{\bar{x}\}$ and $X'' \backslash \{\bar{x}\}$. By Propositions 5.3.23 and 5.3.24 and induction hypothesis, $X' \backslash \{\bar{x}\} = X'' \backslash \{\bar{x}\}$. Hence $X' = X''$. ■

Proposition 5.3.26 *Let Y be admissible and $X = \{x_1, x_2, \ldots, x_n\}$ be a set of vertices of Y. Then there exists a fundamental set X' of vertices of Y such that $|X'| = |X|$. Furthermore, X' can be effectively constructed from X.*

Proof. We define x_i', $i = 1, 2, \ldots n$, inductively as follows: Let

$$\begin{aligned} A_1 &= \{a \mid x_1 = \vee\{a \wedge x_1, \vee\{a_i \wedge x_i \mid i > 1\}\} \\ &\qquad \text{for some } 0 \leq a_i \leq 1,\ i \neq 1\}, \end{aligned}$$

and $a_1' = \wedge A_1$. Define $x_1' = a_1' \vee x_1$. Suppose that x_{k-1}' has been defined. Let

$$\begin{aligned} A_k &= \{a \mid x_k = \vee\{a \wedge x_k, \vee\{a_i \wedge x_i' \mid i < k\}, \vee\{a_i \wedge x_i\ i > k\}\} \\ &\qquad \text{for some } 0 \leq a_i \leq 1,\ i \neq k\}, \end{aligned}$$

$a_k' = \wedge A_k$ and $x_k' = a_k' \wedge x_k$. Let $X' = \{x_1', x_2', \ldots, x_n'\}$. Clearly $|X'| = |X|$. By Proposition 5.3.6, X' can be effectively constructed from X. Moreover, $a_k' \in A_k$ for all k. Thus X' is a set of vertices of Y. We now show that X' is fundamental. Let

$$x_k' = (a_k \wedge x_k') \vee y,$$

where $y \in C(X' \backslash \{\bar{x}_k'\})$. Let $X_k = \{x_1', x_2', \ldots, x_{k-1}', x_{k+1}, \ldots, x_n\}$. Since $x_k = x_k' \vee y'$, where $y' \in C(X_k)$, it follows that $x_k = (a_k \wedge a_k' \wedge x_k) \vee\ y''$, where $y'' \in C(X_k)$. By the definition of a_k', $a_k' \leq a_k \wedge a_k'$ or $a_k \geq a_k'$. Thus

$$x_k' \geq a_k \wedge x_k' = a_k \wedge a_k' \wedge x_k = a_k' \wedge x_k = x_k'.$$

Hence $x_k' = a_k \wedge x_k'$. By Proposition 5.3.22, X' is fundamental. ■

Theorem 5.3.27 *Every set of vertices of Y has the same number of elements provided Y is finitary and admissible.*

Proof. The proof follows from Propositions 5.3.20, 5.3.26, and Theorem 5.3.25. ■

Proposition 5.3.28 *Let Y be finitary and admissible and let X_1, X_2 be sets of generators of Y. If $|X_1| > |X_2|$, then there exists $x \in X_1$ such that $x \in C(X_1 \backslash \{x\})$.*

Proof. By Proposition 5.3.19, there exist $X_1' \subseteq X_1$ and $X_2' \subseteq X_2$, such that X_1' and X_2' are sets of vertices of Y. By Theorem 5.3.27, $|X_1'| = |X_2'|$. Thus $|X_1'| = |X_2'| \leq |X_2| < |X_1|$. This implies that X_1' is a proper subset of X_1. The result now follows from Proposition 5.3.14. ■

Proposition 5.3.29 *Let X be a collection of sequences consisting of 0 and 1. If X is a set of vertices of $C(X)$, then X is fundamental.*

Proof. Let $x \in X$ and $x = (a \wedge x) \vee y$, where $0 \leq a \leq 1$ and $y \in C(X \backslash \{x\})$. Suppose that $a < 1$ and let $x_0 = a \wedge x$. If $x^k = 0$, then $y^k = 0$. If $x^k = 1$, then since $x_0^k = a < 1$, it follows that $y^k = 1$. Thus $x = y$, which is a contradiction. Hence $a = 1$ and $x = x_0$. By Proposition 5.3.22, X is fundamental. ■

Definition 5.3.30 *Y is called* ***fundamental*** *if every set of vertices of Y is fundamental.*

Remark 5.3.31 *Not every convex maximin set is fundamental.*

Theorem 5.3.32 *Let Y be finitary. Then Y is fundamental if and only if Y contains only one set of vertices.*

Proof. The proof follows from Proposition 5.3.17 and Theorem 5.3.25. ■

Definition 5.3.33 *Let X_1, $X_2 \subseteq Y$. Then X_1 is called a* ***basis*** *of X_2 if every $x \in X_2$ can be expressed uniquely as a convex maximin combination of a unique finite subset of X_1.*

Proposition 5.3.34 *Let X be a basis of Y. Then X is a fundamental set of vertices of Y.*

Proof. The proof follows from Propositions 5.3.14 and 5.3.22. ■

Proposition 5.3.35 *Let Y be finitary and have a basis. Then Y is fundamental.*

Proof. Let $X = \{x_1, x_2, \ldots, x_n\}$ be a basis of Y. By Proposition 5.3.34, X is a fundamental set of vertices of Y. Suppose that Y is not fundamental.

Then by Theorem 5.3.32, Y has a set X' of vertices such that $X' \neq X$. By Proposition 5.3.26, we may assume, without loss of generality, that

$$X' = \{x_1, x_2, \dots, x_{n-1}, x'_n\},$$

where $x'_n \neq x_n$. Furthermore,

$$x_n = a \wedge x'_n,$$

for some $0 \leq a \leq 1$ and

$$x'_n = x_n \vee y,$$

where $y \in C(X \backslash \{x_n\})$. Thus,

$$x_n = (a \wedge x_n) \vee (a \wedge y).$$

Since X is a basis of Y, $a = 1$. Hence $x'_n = x_n$, which is a contradiction. Therefore, Y is fundamental. ■

Equivalences

Definition 5.3.36 ***A maximin sequential-like machine (MSLM)*** *is a quadruple* (U, S, V, μ), *where* U, S, V *are finite nonempty sets and* μ *is a function from* $S \times U \times V \times S$ *into* $[0, 1]$.

The set U is the input alphabet. The set V is the output alphabet. S is the set of internal states and μ is the transition function. We may interprete $\mu(s, u, v, s')$ as the grade of membership that the MSLM will enter state s' and produce output v given that the present state is s and the input is u.

The readers should note the difference between the above definition and the one given previously.

As before finite sequences of elements of $U(V)$ will be called input (output) tapes. The collection of all input (output) tapes will be denoted by $U^*(V^*)$. Moreover,

$$(U \times V)^* = \{(x, y) \mid x \in U^*,\ y \in V^*,\ len(x) = len(y)\}.$$

In what follows, the symbol M, with or without subscript, will always denote an MSLM.

As before, we assume that all MSLMs have the same input set U and output set V. Therefore, by suppressing U and V, an MSLM may be represented by (S, μ). Also, $(U \times V)^*$ is ordered in such a way that

$len(x_1) < len(x_2)$ implies $(x_1, y_1) < (x_2, y_2)$. The order of $(U \times V)^*$ is the same for all MSLMs and will be kept fixed throughout the entire section.

Definition 5.3.37 *Let $M = (S, \mu)$ be a MSLM. The* ***extended transition function*** *μ^* of M is a function from $S \times (U \times V)^* \times S$ into $[0,1]$ defined inductively on $len(x)$, $x \in U^*$, as follows:*

$$\mu^*(s', \Lambda, \Lambda, s'') = \begin{cases} 1 & if \quad s' = s'' \\ 0 & if \quad s' \neq s'', \end{cases}$$

$$\mu^*(s', ux, vy, s'') = \vee_{s \in S}\{\mu(s', u, v, s) \wedge \mu^*(s, x, y, s'')\},$$

where $u \in U$, $v \in V$, and $y \in V^*$ with $len(x) = len(y)$. Furthermore, the overall transition function q^M of M is a function from $S \times (U \times V)^*$ into $[0,1]$ defined as follows:

$$q^M(s, x, y) = \vee_{s' \in S}\mu^*(s, x, y, s').$$

If no confusion arises, we will write q for q^M.

For ease of notation, we shall assume that $S = \{s_1, s_2, \ldots, s_n\}$. Furthermore,

(1) $P^M(x, y) = (p_{ij})$, is a matrix, where $p_{ij} = \mu^*(s_i, x, y, s_j)$;

(2) $Q^M(x, y) = (q_i)$ is the column matrix, where the i-th row $q_i = q^M(s_i, x, y)$;

(3) E is the column matrix with all entries 1.

Proposition 5.3.38 *Let $M = (S, \mu)$ be an MSLM. Then for every (x_1, y_1), $(x_2, y_2) \in (U \times V)^*$,*

(i) $P^M(x_1x_2, y_1y_2) = P^M(x_1, y_1) \otimes P^M(x_2, y_2)$, and

(ii) $Q^M(x_1x_2, y_1y_2) = P^M(x_1, y_1) \otimes Q^M(x_2, y_2) = P^M(x_1x_2, y_1y_2) \otimes E$.

Proof. The proof follows immediately since $\otimes$ is associative. ∎

(1) Let A^M denote the matrix whose columns are $Q^M(x, y)$ arranged in the order of $(U \times V)^*$.

(2) For any nonnegative integer n, A_n^M denotes the submatrix of A^M consisting of only those columns corresponding to $Q^M(x, y)$ with $len(x) \leq n$.

(3) Let B^M and B_n^M denote, respectively, the matrix obtained from A^M and A_n^M by omitting all columns which are convex maximin combination of previous columns.

(4) Let A be a (finite or infinite) matrix. $\mid A \mid$ denotes the number of columns of A and $\rho(A)$ denotes the set of distinct rows of A.

Proposition 5.3.39 *If $B_{n-1}^M = B_n^M$ for some n, then $B_m^M = B_{n-1}^M$ for all $m \geq n - 1$.*

Proof. It suffices to show that $B_{n+1}^M = B_n^M$. Let $(x, y) \in (U \times V)^*$, where $len(x) = n$, then $Q^M(x, y)$ is a convex maximin combination of columns of B_{n-1}^M. Since

$$Q^M(ux, vy) = P^M(u, v) \otimes Q^M(x, y),$$

the result now follows from Proposition 5.3.38 and the fact that $\otimes$ is associative. ■

Theorem 5.3.40 *Let $M = (S, \mu)$ be an MSLM. Then $B^M = B_m^M$, where $m \le a^{|S|} - 1$ and a is the number of distinct entries in A^M.*

Proof. Clearly, $| B_n^M | \le | B_{n+1}^M | \le | B^M | \le a^{|S|}$ for every n. Thus $B_{m-1}^M = B_m^M$, where $m = a^{|S|} - 1$. The result now follows from Proposition 5.3.39. ■

Proposition 5.3.41 *The matrix B^M can be constructed effectively from M.* ■

Proposition 5.3.42 *$C[\rho(A^M)]$ is admissible and finitary.* ■

Definition 5.3.43 *Let $M = (S, \mu)$ be a MSLM. A* ***state distribution*** **(sd)** *of M is a function η from S into $[0, 1]$. η is said to be* **concentrated** *at $s \in S$ if $\eta(s) = 1$ and 0 elsewhere.*

We will also use the symbol η to denote the row matrix whose i-th row is $\eta(s_i)$.

Definition 5.3.44 *An* ***initialized maximin sequential-like machine (IMSLM)*** *is an ordered pair (M, η), where M is an MSLM and η is an sd of M.*

If η is concentrated at s, we write (M, s) for (M, η).

Definition 5.3.45 *Let $I = (M, \eta)$ be an IMSLM. The* ***response function*** *r^I or I is a function from $(U \times V)^*$ into $[0, 1]$ where*

$$r^I(x, y) = \vee_{s \in S}\{\eta(s) \wedge q^M(s, x, y)\}.$$

Let $R^I = \eta \otimes A^M$. Then R^I is a row matrix whose entries are $r^I(x, y)$. Furthermore, R^I is a convex maximin combination of $\rho(A^M)$.

Definition 5.3.46 *Let I_1 and I_2 be IMSLMs. Then I_1 and I_2 are called* ***equivalent*** *($\sim$) if $r^{I_1} = r^{I_2}$.*

Definition 5.3.47 *Let η_1 and η_2 be state distributions of M. Then η_1 and η_2 are called M-**equivalent** ($\overset{M}{\sim}$) if (M, η_2).*

If no confusion arises, we will use the symbol $\sim$ for $\overset{M}{\sim}$.

If η_1 is concentrated at s, then we write $s \sim \eta_2$ for $\eta_1 \sim \eta_2$.

Proposition 5.3.48 *Let η_1 and η_2 be state distributions of M. Then $\eta_1 \sim \eta_2$ if and only if*

$$\eta_1 \otimes B^M = \eta_2 \otimes B^M. \blacksquare$$

Definition 5.3.49 *Let $M_1 = (S_1, \mu_1)$ and $M_2 = (S_2, \mu_2)$ be MSLMs.*

(i) M_1 and M_2 are called ***statewise equivalent*** *($\sim$) if for every $s' \in S_1$, there exists $s'' \in S_2$ such that $(M_1, s') \sim (M_2, s'')$ and vice versa.*

(ii) M_1 and M_2 are called ***compositewise equivalent*** *($\simeq$) if for every $s \in S_1$, there exists an sd η of M_2 such that $(M_1, s) \sim (M_2, \eta)$ and vice versa.*

(iii) M_1 and M_2 are called ***distributionwise equivalent*** *($\approx$) if for every sd η_1 of M_1, there exists an sd η_2 of M_2 such that $(M_1, \eta_1) \sim (M_2, \eta_2)$ and vice versa.*

Theorem 5.3.50 *Let M_1 and M_2 be MSLMs.*

(i) $M_1 \sim M_2$ if and only if $\rho(A^{M_1}) = \rho(A^{M_2})$.

(ii) $M_1 \simeq M_2$ if and only if $\rho(A^{M_1}) \subseteq C[\rho(A^{M_2})]$ *and* $\rho(A^{M_2}) \subseteq C[\rho(A^{M_1})]$.

(iii) $M_1 \approx M_2$ if and only if $C[\rho(A^{M_1})] = C[\rho(A^{M_2})]$. ■

Proposition 5.3.51 *Let M_1 and M_2 be MSLMs such that $M_1 \sim M_2$. Then $M_1 \simeq M_2$.*

Proof. The proof follows from Propositions 5.3.8(i) and Theorem 5.3.50. ■

Proposition 5.3.52 *Let M_1 and M_2 be MSLMs.Then $M_1 \simeq M_2$ if and only if $M_1 \approx M_2$.*

Proof. The proof follows from Theorem 5.3.50(ii) and (iii), it follows that $M_1 \approx M_2$ implies $M_1 \simeq M_2$. The converse follows from Propositions 5.3.8(ii) and (iii) and Theorem 5.3.50. ■

Irreducibility and Minimality

Definition 5.3.53 *Let $M = (S, \mu)$ be an MSLM.*

(i) M is called ***statewise irreducible*** *if for every s', $s'' \in S$, $s' \sim s''$ implies $s' = s''$.*

(ii) M is called ***compositewise irreducible*** *if for every $s \in S$ and sd η of M, $s \sim \eta$ implies $\eta(s) > 0$.*

(iii) M is called ***distributionwise irreducible*** *if for every sd η_1 and η_2 of M, $\eta_1 \sim \eta_2$ implies $\eta_1 = \eta_2$.*

Theorem 5.3.54 *Let M be an MSLM.*

(i) M is statewise irreducible if and only if no two rows of B^M are identical.

(ii) M is compositewise irreducible if and only if $\rho(B^M)$ is a set of vertices of $C[\rho(B^M)]$, i.e., no row of B^M is a convex maximin combination of the other rows of B^M.

(iii) M is distributionwise irreducible if and only if $\rho(B^M)$ is a basis of $C[\rho(B^M)]$.

Proof. (i) The proof follows from Propositions 5.3.48.

(ii) The proof follows from Propositions 5.3.48 and 5.3.14.

(iii) The proof follows from Proposition 5.3.48 and the definition of basis.
■

Remark 5.3.55 *All assertions of Theorem* 5.3.54 *are also valid if B^M is replaced by A^M.*

Proposition 5.3.56 *Let M be an MSLM. If M is distributionwise irreducible, then M is compositewise irreducible.* ■

Proposition 5.3.57 *Let M be an MSLM. If M is compositewise irreducible, then M is statewise irreducible.* ■

Definition 5.3.58 *Let M be an MSLM. M is called* ***statewise (compositewise) minimal*** *if M is not statewise (compositewise) equivalent to an MSLM with a fewer number of states.*

Let $M = (S, p)$. The cardinality of M, is defined by $|M| = |S|$.

Theorem 5.3.59 *Let M be an MSLM. Then M is statewise minimal if and only if M is statewise irreducible.*

Proof. Let $M = (S, \mu)$. Suppose that M is not statewise minimal. Then for some MSLM M' with $|M'| < |M|$, $M' \sim M$. By Theorems 5.3.50(i) and 5.3.54(i), M is not statewise irreducible. Conversely, suppose that M is not statewise irreducible. There exist $s', s'' \in S$ such that $s' \sim s''$ but $s' \neq s''$. By renumbering the elements of S, if necessary, we may assume that $s' = s_{n-1}$ and $s'' = s_n$, where $n = |S|$. Let $S' = S \backslash \{s_n\}$ and $M' = (S', \mu')$, where for $i = 1, 2, \ldots, n-1$,

$$\mu'(s_i, u, v, s_j) = \begin{cases} \mu(s_i, u, v, s_j) & \text{if } j = 1, 2, \ldots, n-2 \\ \mu(s_i, u, v, s_{n-1}) \vee \mu(s_i, u, v, s_n) & \text{if } j = n-1. \end{cases}$$

Now for every $(x, y) \in (U \times V)^*$,

$$q^M(s_i, x, y) = \begin{cases} q^{M'}(s_i, x, y) & \text{if } i = 1, 2, \ldots, n-1 \\ q^{M'}(s_{n-1}, x, y) & \text{if } i = n. \end{cases}$$

Therefore, $\rho(A^M) = \rho(A^{M'})$. By Theorem 5.3.50(i), $M \sim M'$. Thus M is not statewise minimal. ■

Theorem 5.3.60 *Let M be an MSLM. Then there exists an effective procedure for constructing a statewise minimal MSLM which is statewise equivalent to M.*

Proof. Consider the following procedure:

Step 1. Construct B^M.

Step 2. If there exist two rows of B^M which are identical, then proceed to Step 3; otherwise stop.

Step 3. Construct M' as given in the proof of Theorem 5.3.59. Return to Step 1 with M' replacing M.

Since $|M|$ is finite, the procedure must terminate in a finite number of steps. By Proposition 5.3.41, B^M can be effectively constructed. Thus the procedure is effective. By Theorem 5.3.59, the resulting MSLM is the desired MSLM. ■

Theorem 5.3.61 *Let M be an MSLM. Then M is compositewise minimal if and only if M is compositewise irreducible.*

Proof. Let $M = (S, \mu)$. Suppose that M is not compositewise minimal. Then there exists an MSLM M' with $|M'| < |M|$ such that $M' \simeq M$. By Theorems 5.3.50(iii), 5.3.54(ii), and Propositions 5.3.15, 5.3.28, and 5.3.42, M is not compositewise irreducible. Conversely, suppose that M is not compositewise irreducible. Then there exist $s' \in S$ and sd η of M such that $s' \sim \eta$, but $\eta(s') = 0$. By renumbering the elements of S, if necessary, we may assume that $s' = s_n$, where $n = |\ S\ |$. Since $s_n \sim \eta$ and $\eta(s_n) = 0$,

$$q^M(s_n, x, y) = \vee\{\eta(s_i) \wedge q^M(s_i, x, y) \mid i = 1, 2, \ldots, n\},$$

for all $(x, y) \in (U \times V)^*$. Let $M' = (S', \mu')$, where $S' = S - \{s_n\}$ and for $i, j = 1, 2, \ldots, n-1$,

$$\mu'(s_i, u, v, s_j) = \mu(s_i, u, v, s_j) \vee (\eta(s_j) \wedge \mu(s_i, u, v, s_n)).$$

Now for every $(x, y) \in (U \times V)^*$,

$$q^M(s_i, x, y) = \begin{cases} q^{M'}(s_i, x, y) & \text{if } i = 1, \ldots, n-1 \\ \vee\ \{\eta(s_j) \wedge q^{M'}(s_j, x, y) \mid j = 1, 2, \ldots, n\} & \text{if } i = n. \end{cases}$$

Therefore $C[\rho(A^M)] = C[\rho(A^{M'})]$. By Theorem 5.3.50(iii) and Proposition 5.3.52, $M' \simeq M$. Thus M is not compositewise minimal. ■

Theorem 5.3.62 *Let M be an MSLM. There exists an effective procedure for constructing a compositewise minimal MSLM which is compositewise equivalent to M.*

Proof. Consider the following procedure:

Step 1. Construct B^M.

Step 2. If there exists a row of B^M which is a convex maximin combination of the other rows of B^M, then proceed to Step 3; otherwise stop.

Step 3. Construct M' as given in the proof of Theorem 5.3.61. Return to Step 1 with M' replacing M.

Since $|M|$ is finite, the procedure must terminate in a finite number of steps. By Proposition 5.3.52, B^M can be constructed effectively. By Propositions 5.3.6 and 5.3.41 and the fact that $|M|$ is finite, Step 2 can be carried out effectively. Thus the procedure is effective. By Theorem 5.3.61, the resulting MSLM is the desired MSLM. ■

Proposition 5.3.63 *Let M_1 and M_2 be MSLMs. Let $\rho(B^{M_1})$ and $\rho(B^{M_2})$ be, respectively, fundamental set of vertices of $C[\rho(B^{M_1})]$ and $C[\rho(B^{M_2})]$. $M_1 \simeq M_2$ if and only if $M_1 \sim M_2$.*

Proof. The proof follows from Theorems 5.3.50(iii), 5.3.25 and Proposition 5.3.52, Theorem 5.3.50(i) and (ii). ■

Definition 5.3.64 *Let M be an MSLM. M is **fundamental** if $C[\rho(B^M)]$ is fundamental.*

Proposition 5.3.65 *Let M_1 and M_2 be MSLMs. If M_1, M_2 are compositewise minimal, M_1 fundamental and $M_1 \simeq M_2$, then M_2 is fundamental and $M_1 \sim M_2$.*

Proof. The proof follows from Theorems 5.3.25, 5.3.54(ii), 5.3.61, and Proposition 5.3.63. ■

Proposition 5.3.66 *Let M_1 and M_2 be MSLMs. If M_1 is distributionwise irreducible, M_2 is compositewise minimal and $M_1 \simeq M_2$, then M_2 is fundamental and $M_1 \sim M_2$.*

Proof. The proof follows from Propositions 5.3.35 and 5.3.65. ■

Proposition 5.3.67 *Let M_1 and M_2 be MSLMs. If M_1, M_2 are statewise minimal and $M_1 \sim M_2$, then for every $(x, y) \in (U \times V)^*$,*

$$P^{M_1}(x,y) \otimes B^{M_1} = P^{M_2}(x,y) \otimes B^{M_1},$$

after an appropriate rearrangement of states.

Proof. From Theorems 5.3.50(i), 5.3.54(i), and 5.3.59, $A^{M_1} = A^{M_2}$ after an appropriate rearrangement of states. ■

Definition 5.3.68 *Let M_1 and M_2 be MSLMs. M_1 and M_2 are called* ***isomorphic*** *($\equiv$) if they are equal up to a permutation of states.*

Definition 5.3.69 *Let M be statewise (compositewise) minimal. M is called* ***statewise (compositewise) simple*** *if there exist no statewise (compositewise) minimal MSLM which is statewise (compositewise) equivalent to M but not isomorphic to M.*

Let M be an MSLM. Then
(1) $\rho(M) = \cup_{u \in U} \cup_{v \in V} \rho[P^M(u, v)]$,
(2) $\bar{\rho}(M) = \{j \otimes B^M \mid j \in \rho(M)\}$.

Theorem 5.3.70 *Let M be statewise minimal. Then M is statewise simple if and only if $\rho(B^M)$ is a basis of $\bar{\rho}(M)$.*

Proof. Let $\rho(B^M)$ be a basis of $\bar{\rho}(M)$ and M' be a statewise minimal MSLM such that $M' \sim M$. By Proposition 5.3.67, for every $u \in U$, $v \in V$,

$$P^M(u, v) = P^M(u, v),$$

after an appropriate rearrangement of states. Hence $M' \equiv M$. Thus M is statewise simple. Conversely, suppose $\rho(B^M)$ is not a basis of $\bar{\rho}(M)$. There exist $j \in \rho(M)$ such that

$$j \otimes B^M = j' \otimes B^M, \tag{5.9}$$

for some $j' \neq j$. Let j be the i-th row of the matrix $P^M(u, v)$. Construct M' from M by replacing the i-th row of the matrix $P^M(u, v)$ by j' and leaving the rest unchanged. By (5.9), we have $A^M = A^{M'}$. By Theorems 5.3.50(i), 5.3.54(i), and 5.3.59, M' is statewise minimal and $M' \sim M$. However, $M' \not\equiv M$. Thus M is not statewise simple. ■

Proposition 5.3.71 *Let M be statewise minimal. If M is distributionwise irreducible, then M is statewise simple.*

Proof. The proof follows from Theorem 5.3.54(iii) and 5.3.70. ■

Theorem 5.3.72 *Let M be compositewise minimal and fundamental. Then M is compositewise simple if and only if $\rho(B^M)$ is a basis of $\bar{\rho}(M)$.*

Proof. Let $\rho(B^M)$ be a basis of $\bar{\rho}(M)$ and M' be a compositewise minimal MSLM such that $M' \simeq M$. By Proposition 5.3.57 and Theorems 5.3.59 and 5.3.61, both M and M' are statewise minimal. Moreover, by Proposition 5.3.65, $M' \sim M$. The remainder of the proof is similar to Theorem 5.3.70. ■

Proposition 5.3.73 *Let M be compositewise minimal. If M is distributionwise irreducible, then M is compositewise simple.*

Proof. The proof follows from Theorems 5.3.54(iii), 5.3.72, and Proposition 5.3.35. ■

Nondeterministic and Deterministic Case

Definition 5.3.74 *A **nondeterministic sequential-like machine (NSLM)** is an MSLM (S, μ) such that $Im(\mu) \subseteq \{0, 1\}$. If for every $s \in S$, $u \in U$, there exist unique $v \in V$ and $s' \in S$ such that $\mu(s, u, v, s') = 1$, then it is called a **deterministic sequential-like machine (DSLM)**.*

Proposition 5.3.75 *Let M be a DSLM. Then for every $s \in S$, $x \in U^*$, there exists unique $y \in V^*$ with $len(x) = len(y)$ such that $q^M(s, x, y) = 1$.* ■

Due to the special characters of NSLMs and DSLMs, several of the above results can be strengthened.

Proposition 5.3.76 *Let M be a NSLM. Then M is statewise minimal if and only if M is not statewise equivalent to any NSLM with a fewer number of states.*

Proof. The proof follows from the proof of Theorem 5.3.59. ■

Proposition 5.3.77 *Let M be a NSLM. Then there exists an effective procedure for constructing a statewise minimal NSLM which is statewise equivalent to M.* ■

Proposition 5.3.78 *Let M be a NSLM. Then M is compositewise minimal if and only if M is not compositewise equivalent to any NSLM with a fewer number of states.*

Proof. The proof follows from Proposition 5.3.5 and the proof of Theorem 5.3.61. ■

Proposition 5.3.79 *Let M be a NSLM. Then there exists an effective producer for constructing a compositewise minimal NSLM which is compositewise equivalent to M.* ■

Proposition 5.3.80 *Let M be a DSLM. Then M is statewise minimal if and only if M is not statewise equivalent to any DSLM with a fewer number of states.*

Proof. The proof follows from the proof of Theorem 5.3.59. ■

Proposition 5.3.81 *Let M be a DSLM. Then there exists an effective procedure for constructing a statewise minimal DSLM which is statewise equivalent to M.* ■

Theorem 5.3.82 *Let M be a DSLM. Then M is compositewise minimal if and only if M is statewise minimal.*

Proof. Suppose that M is not compositewise minimal. Then there exists a row of B^M, say the n-th row where $n = |M|$, which is a convex maximin combination of the other rows of B^M. By Proposition 5.3.5, we may assume, without loss of generality, that

$$x_n = \vee\{x_i \mid i = 1, 2, \ldots, m\},$$

where $m < n$, after an appropriate rearrangement of states. Here, x_i denotes the i-th row of B^M. Since $x_n^k = 0$ implies $x_i^k = 0$ for $i = 1, 2, \ldots, m$, we have by Proposition 5.3.75 that $x_n = x_i$ for $i = 1, 2, \ldots, m$. Thus M is not statewise minimal. The converse is straightforward. ■

Proposition 5.3.83 *Let M_1 and M_2 be compositewise minimal NSLM. Then $M_1 \simeq M_2$ if and only if $M_1 \sim M_2$.*

Proof. The proof follows from Propositions 5.3.29 and 5.3.63. ■

Definition 5.3.84 *Let M be statewise (compositewise) minimal NSLM. M is called* ***statewise (compositewise) ND-simple*** *if there exists no statewise (compositewise) minimal NSLM which is statewise (compositewise) equivalent to M but not isomorphic to M.*

Definition 5.3.85 *Let M be a statewise minimal DSLM. M is called* ***statewise D-simple*** *if there exists no statewise minimal DSLM which is statewise equivalent to M, but not isomorphic to M.*

Theorem 5.3.86 *Let M be a statewise (compositewise) minimal NSLM. Then M is statewise (compositewise) ND-simple if and only if for every $x_0 \in \bar{\rho}(M)$, there exists uniquely $L \subseteq \rho(B^M)$ such that*

$$x_0 = \vee\{x \mid x \in L\}.$$

Proof. The proof is similar to the proofs of Theorems 5.3.70 and 5.3.72. For the compositewise case, Proposition 5.3.83 is needed. ■

Theorem 5.3.87 *Let M be statewise (compositewise) minimal DSLM. Then M is statewise (compositewise) ND-simple.*

Proof. The proof follows from Theorem 5.3.86 and the fact that every row of any matrix $P^M(u, v)$ has at most one nonzero entry and this nonzero entry is a 1. ■

Corollary 5.3.88 *Let M be a statewise minimal DSLM. Then M is statewise D-simple.* ■

5.4 Context-Free Max-Product Grammars

In [22], two different formulations of context-free max-product grammars were introduced. The first formulation corresponds to the maximal interpretation [14] of probabilistic grammars [20] and the second is an attempt to provide a grammar to generate fuzzy languages [21]. The second formulation is simpler. Consequently, it is adopted here and the next section. Further details can be found in [22].

Let C be a nonempty set. Then C^* is the free semigroup with identity Λ generated by C and $C^+ = C^* \backslash \{\Lambda\}$.

Let $\mathbb{R}^{\geq 0}$ be the set of all nonnegative real numbers and $\mathbb{R}^{\geq 0}_{\infty} = \mathbb{R}^{\geq 0} \cup \{\infty\}$.

Definition 5.4.1 *A* ***context-free max-product grammar*** *(CMG) is a quadruple $G = (T, N, P, \eta)$ such that*

(i) T and N are disjoint finite nonempty sets;

(ii) P is a finite collection of fuzzy productions each of which is of the form $A \to^p x$, where $A \in N$, $x \in (T \cup N)^*$, and $p \in \mathbb{R}^{\geq 0}$; and

(iii) η is a function from N into $\mathbb{R}^{\geq 0}$. If, in addition, for every $(A \to^p x) \in P$, $p \in [0, 1]$, and η is a function from N into $[0, 1]$, then G is a strict CMG.

In Definition 5.4.1,

(1) elements of T are called terminals,

(2) elements of N are called nonterminals,

(3) $\eta(A)$ is the grade of membership [31] that A is the start symbol of G.

(4) $A \to^p x$ means the grade of membership is p that A will be replaced by x.

Let $G = (T, N, P, \eta)$ be a CMG. Then we write

$$x \rightrightarrows^p y (\text{mod } w),$$

where $w = (\rho_1, k_1)(\rho_2, k_2) \cdots (\rho_n, k_n)$, $x, y \in (T \cup N)^*$, $\rho_i = (A_i \to^{p_i} x_i) \in P$, $k_i \in \mathbb{N}$, $i = 1, 2, \ldots, n$, and $p = p_1 p_2 \cdots p_n$ if there exist $z_i \in (T \cup N)^*$, $i = 0, 1, 2, \ldots, n$, such that $z_0 = x$, $z_n = y$, and for each $i = 1, 2, \ldots, n$, z_i is obtained from z_{i-1} by replacing the k_ith occurrence of A_i in z_{i-1} by x. Otherwise we write

$$x \rightrightarrows^0 y (\text{mod } w).$$

If, in addition, for every $i = 1, 2, \ldots, n$, $k_i = 1$ and $z_{i-1} = u A_i v$, where $u \in T^*$, then we write

$$x \rightrightarrows^p_L y (\text{mod } w).$$

Otherwise we write

$$x \rightrightarrows^0_L y ((\text{mod } w).$$

Clearly in the above notation, p is uniquely determined by $x, y \in (T \cup N)^*$ and $w \in (P \times \mathbb{N})^*$. This allows us to define, for each $w \in (P \times \mathbb{N})^*$, functions f_w and f_w^L, both from $(T \cup N)^* \times (T \cup N)^*$ into $\mathbb{R}^{\geq 0}$, where $f_w(y \mid x) = p$ if and only if $x \rightrightarrows^p y \pmod w$, and $f_w^L(y \mid x) = p$ if and only if $x \rightrightarrows_L^p y \pmod w$.

Let $G = (T, N, P, \eta)$ be a CMG. Let $W = P \times \mathbb{N}$. Define $f_G : T^* \to \mathbb{R}^{\geq 0}$and $f_G^L : T^* \to \mathbb{R}^{\geq 0}$ by $\forall x \in T^*$,

$$f_G(x) = \wedge_{w \in W^*} \vee_{A \in N} \eta(A) f_w(x \mid A)$$

and

$$f_G^L(x) = \wedge_{w \in W^*} \vee_{A \in N} \eta(A) f_w^L(x \mid A).$$

In the above definition, as well as in the rest of this section, we assume that the usual arithmetic of the extended real number system including the fact that $0 \cdot \infty = 0$. Furthermore, the concept of least upper bound is extended in a natural manner to include ∞.

Definition 5.4.2 *A* ***fuzzy language*** *f over C is a function from C^* into $\mathbb{R}_\infty^{\geq 0}$. A* ***finitary fuzzy language*** *over C is a function from C^* into $\mathbb{R}^{\geq 0}$.*

Remark 5.4.3 *Let f be a fuzzy langauge over C. Then $f(x)$ is the grade of membership that x is a member of the language.*

Remark 5.4.4 *Let G be a CMG. Then*

(i) f_G is the fuzzy language generated by G,

(ii) f_G^L is the fuzzy language generated by G using left most derivations only.

Theorem 5.4.5 *Let G be a CMG. Then $f_G = f_G^L$.*

Proof. Let $G = (T, N, P, \eta)$. It suffices to show that for every $A \in N$, $x \in T^*$ and $p \in \mathbb{R}^{\geq 0}$, if $A \rightrightarrows^p x \pmod w$ for some $w \in (P \times \mathbb{N})^*$, then

$$A \rightrightarrows_L^p x (\text{mod } w')$$

for some $w' \in (P \times \{1\})^*$. This can be accomplished by induction (see [9]) on $len(x)$ and so we leave the details for the reader. ∎

Definition 5.4.6 *Let f be a (finitary) fuzzy language over T. f is called a* ***(finitary) context-free fuzzy language (CFFL)*** *if $f = f_G$ for some CMG G.*

Proposition 5.4.7 *Let f be a CFFL over T. Then $f = f_G$ for some CMG $G = (T, N, P, \eta)$ such that the following properties hold:*

(i) There exists a distinguished symbol $A_0 \in N$ such that $\eta(A_0) = 1$, and $(A \to^p x) \in P$ implies A_0 does not occur in x.

(ii) For every $A \in N$ and $x \in (T \cup N)^*$, $(A \to^{p_1} x) \in P$ and $(A \to^{p_2} x) \in P$ implies $p_1 = p_2 > 0$.

(iii) For every $A \in N$, there exist $u, v \in T^*$ and $w \in (P \times \mathbb{N})^*$ such that $f_w(uAv \mid A_0) > 0$.

(iv) For every $A \in N$, there exists $x \in T^*$ and $w \in (P \times \mathbb{N})^*$ such that $f_w(x \mid A) > 0$.

A CMG satisfying conditions (i) to (iv) of Proposition 5.4.7 is said to be **reduced**. In this case, we shall also write (T, N, P, A_0) for (T, N, P, η).

Proposition 5.4.8 *Let $G = (T, N, P, A_0)$ be a reduced CMG. Then f_G is finitary if and only if for every $A \in N$ and $w \in (P \times \mathbb{N})^*$, $f_w(A \mid A) \leq 1$.*

Lemma 5.4.9 *Let f be a finitary CFFL. Then $f = f_G$ for some reduced CMG $G = (T, N, P, A)$, where P does not contain any fuzzy productions of the form $A \to^p \Lambda$ and where $A \in N \backslash \{A_0\}$.*

Proof. Let $f = f_{G_0}$, where $G_0 = (T, N, P_0, A_0)$ is a CMG. For every fuzzy production $A \to^p x_0A_1x_1A_2x_2 \cdots A_nx_n$ in P_0, where $n \geq 0$, each $A_i \in N$ and $x = x_0x_1 \cdots x_n \in (T \cup N)^+$, let $A \to^{p'} x$ be a fuzzy production such that $p' = p\prod_{i=1}^n q(A_i) > 0$. Then for each $B \in N$, $q(B) = \wedge_{w \in W^*} f_w(\Lambda \mid B)$, where $W = P_0 \times \mathbb{N}$. Let $G = (T, N, P, A_0)$ be the CMG, where P is the collection of all fuzzy productions obtained in the manner described above plus the fuzzy production $A_0 \to^p \Lambda$, where $p = f_G(\Lambda)$. Clearly, $(A \to^p \Lambda) \in P$ implies $A = A_0$. Furthermore, it can be shown that $f = f_G$. If G is not a reduced CMG, it can be easily placed in reduced form and still keep the desired property. ∎

Theorem 5.4.10 *f is a finitary CFFL if and only if $f = f_G$ for some reduced CMG $G = (T, N, P, A_0)$, where P does not contain any fuzzy productions of the forms $A \to^p B$, $A, B \in N$, and $A \to^p \Lambda$, $A \in N \backslash \{A_0\}$.*

Proof. Suppose that f is a finitary CFFL. By Lemma 5.4.9, $f = f_G$ for some reduced CMG $G_0 = (T, N, P_0, A_0)$, where $(A \to^p \Lambda) \in P_0$ implies $A = A_0$. Let ξ be the function from $(T \cup N)^* \times (T \cup N)^*$ into $\mathbb{R}^{\geq 0}$ such that for all $s, t \in (T \cup N)^*$,

$$\xi(t \mid s) = \begin{cases} p & \text{if } (s \to^p t) \in P \\ 0 & \text{otherwise.} \end{cases}$$

Let ξ_n, $n = 0, 1, 2, \ldots$, be functions from $(T \cup N)^* \times (T \cup N)^*$ into $\mathbb{R}^{\geq 0}$ defined inductively as follows

$$\xi_0(t \mid s) = \begin{cases} 1 & \text{if } s = t \in N, \\ 0 & \text{otherwise,} \end{cases}$$

and

$$\xi_{n+1}(t \mid s) = \begin{cases} \vee_{A \in N}[\xi_n(t \mid A)\xi(A \mid s)] & \text{if } s, t \in N, \\ 0 & \text{otherwise.} \end{cases}$$

Let ξ' be the function from $(T \cup N)^* \times (T \cup N)^*$ into $\mathbb{R}^{\geq 0}$ such that for all $s, t \in (T \cup N)^*$,

$$\xi'(\tau \mid \sigma) = \begin{cases} \vee\{\xi(t \mid s), \ \wedge \vee[\xi(t \mid A)\xi_n(A \mid s)]\} & \text{if } s \in N, \ t \in N, \\ 0 & \text{otherwise.} \end{cases}$$

Let $G = (T, N, P, A_0)$ be the CMG, where P is the collection of all fuzzy productions $s \to^p t$ in which $\xi'(t \mid s) = p > 0$. Clearly, G has the desired properties. Let $x \in T^*$. For all $w \in (P \times \mathbb{N})^+$ and $\varepsilon > 0$, there exists $w_0 \in (P_0 \times \mathbb{N})^+$ such that $f_w(x \mid A) > f_{w_0}(x \mid A) - \varepsilon$. Moreover, for every $w_0 \in (P_0 \times \mathbb{N})^+$, there exists $w \in (P \times \mathbb{N})^+$ such that $f_{w_0}(x \mid A) \leq f_w(x \mid A)$. Thus, $f = f_G$. The converse follows from the fact that

$$\begin{aligned} f_G(x) &= \wedge\{\vee\{\eta(A) f_w(x \mid A) \mid A \in N\} \ w \in (P \times \mathbb{N})^*, \\ &\quad len(w) \leq len(x)\}. \ \blacksquare \end{aligned}$$

In view of Theorem 5.4.10, the usual procedures [9] could be modified to obtain the following normal form theorems for CMG.

Theorem 5.4.11 ***(Chomsky Normal Form)*** *f is a finitary CFFL if and only if $f = f_G$ for some reduced CMG $G = (T, N, P, A_0)$, where p contains only fuzzy productions of the forms $A_0 \to^p \Lambda$, $A \to^p a$, and $A \to^p BC$,where $A, B, C \in N$, and $a \in T$.* $\blacksquare$

Theorem 5.4.12 ***(Greibach Normal Form)*** *f is finitary CFFL if and only if $f = f_G$ for some reduced CMG $G = (T, N, P, A_0)$, where P contains only fuzzy productions of the forms $A_0 \to^p \Lambda$ and $A \to^p az$, where $A \in N$, $a \in T$, $z \in N^*$, $len(z) \leq 2$.* $\blacksquare$

Remark 5.4.13 *Theorems* 5.4.11 *and* 5.4.12 *are, in general, not valid if f is not finitary, unless we extend the definition of fuzzy productions to include $A \to^p x$, where $p = \infty$.*

5.5 Context-Free Fuzzy Languages

In this section, we study the family of languages generated by CMGs with cut points. The results are from [23].

Let f be a fuzzy language over T and $r \in \mathbb{R}^{\geq 0}_{\infty}$. Let $L(f, r, >)$ denote the set

$$L(f, r, >) = \{x \in T^* \mid f(x) > r\}.$$

$L(f, r, \geq)$ and $L(f, r, =)$ are defined in a similar manner.

Let $G = (T, N, P, A_0)$ be a CMG. Then $x \in L(f_G, r, >)$ if and only if there exists $w \in (P \times \mathbb{N})^*$ such that $A_0 \rightrightarrows^p x \pmod{w}$ and $p > r$.

Definition 5.5.1 *A* ***phrase structure grammar*** *is a quadruple* $G = (T, N, P, A_0)$ *where* T *and* N *are the terminals and the nonterminals, respectively,* P *is a finite collection of productions, and* $A_0 \in N$ *is the start symbol. Derivations according to* G*, the language* $L(G)$ *generated by* G*, and the various types of grammars in the Chomsky hierarchy are defined in the usual manner [7].*

Theorem 5.5.2 L *is a context-free language (CFL) if and only if* $L = L(f, 0, >)$ *for some CFFL* f*.*

Proof. Suppose $f = f_G$, where $G = (T, N, P, A_0)$ is a CMG. Let $G' = (T, N, P', A_0)$ be a context-free grammar, where $P' = \{s \to t \mid (s \to^p t) \in P \text{ and } p > 0\}$. Then it follows that $L = L(G')$. The converse is straightforward. ■

Definition 5.5.3 *For each* $a \in T$*, let* T_a *be a finite nonempty set and* $\psi(a) \subseteq T_a^*$*. Let* $\psi(\Lambda) = \{\Lambda\}$ *and* $\psi(ax) = \psi(a)\psi(x)$ *for every* $a \in T$ *and* $x \in T^*$*. Then* ψ *is called a* ***substitution.*** *If* $L \subseteq T^*$*, let* $\psi(L) = \cup_{x \in L}\psi(x)$*.*

Definition 5.5.4 *An* a-***transducer*** *is a 6-tuple* $M = (U, S, V, H, q_0, F)$ *where* U*,* S*, and* V *are finite nonempty sets (of input, state, and output symbols, respectively),* H *is a finite subset* $S \times U^* \times V^* \times S$*,* $q_0 \in S$ *is the initial state, and* $F \subseteq S$ *is the set of accepting states.*

Let $M = (U, S, V, H, q_0, F)$ be an a-transducer. For each $x \in U^*$,

$$\begin{aligned} M(x) = \{ & y \in V^* \mid \exists x_1, x_2, \ldots, x_k \in U^*, y_1, y_2, \ldots, y_k \in V^*, \\ & \text{and } q_1, q_2, \ldots, q_k \in S \text{ such that } x = x_1 x_2 \cdots x_k, \\ & y = y_1 y_2 \cdots y_k, q_k \in F, \text{ and } (q_{i-1}, x_i, y_i, q_i) \in H \\ & \text{for all } i, 1 \leq i \leq k\}. \end{aligned}$$

For each $L \subseteq U^*$, let

$$M(L) = \cup_{x \in L} M(x).$$

It is well known that if L is a CFL and M is an a-transducer, then $M(L)$ is a CFL. It is also well known that if L is a CFL, and ψ is a substitution such that $\psi(a)$ is a CFL for all a, then $\psi(L)$ is also a CFL.

Theorem 5.5.5 L *is a* CFL *if and only if* $L = L(f_G, r, >)$ *for some strict* CMG G *and* $r \in \mathbb{R}^{\geq 0}$*.*

Proof. Let $L = L(f_G, r, >)$, where $G = (T, N, P, A_0)$ is a strict CMG and $r \in \mathbb{R}^{\geq 0}$. Without loss of generality, we may assume that $N = N_1 \cup N_2$, where

(i) $N_1 \cap N_2 = \emptyset$,
(ii) $A_0 \in N_1$,
(iii) if $(A \to^1 x) \in P$, then $A \in N_1$, and
(iv) if $(A \to^p x) \in P$ and $p < 1$, then $A \in N_2$.

Let P_1 be the set of all fuzzy productions in P of the form $A \to^1 x$, and let $P_2 = P \backslash P_1$. For every $\rho = (A \to^p s) \in P_2$, let $M_\rho = (U, S, V, H, S_0, \{q_1\})$ be an a-transducer, where $U = T \cup N_2$, $S = \{q_0, q_1\}$, $V = T \cup N$, and $H = \{(q, u, u, q) \mid q \in S, u \in T \cup N_2\} \cup \{(q_0, A, x, q_1)\}$. Note that if $L_0 \subseteq (T \cup N_2)^*$, then $M_\rho(L_0)$ is obtained from L by first omitting all words in L_0 not containing any occurrence of A, and then replacing exactly one occurrence of A by x in the remaining words of L_0. For each $A \in N_1$, let $L(A) = (f_{G_A}, 0, >)$, where

$$G_A = (T \cup N_2, N_1, P_1, A).$$

Clearly, $L(A)$ is a CFL for all $A \in N_1$. Let ψ be a substitution such that $\psi(A) = L(A)$ if $A \in N_1$, and $\psi(a) = a$ if $a \in T \cup N_2$. For each $L_1 L_2 \subseteq (T \cup N_2)^*$ and $k \in \mathbb{N}$, define $L_1 \to L_2$ if and only if there exists $\rho \in P_2$ such that $L_2 = (M_\rho(L_1))$; and $L_1 \to^k L_2$ if and only if there exists

$$L'_0, L'_1, ..., L'_k \subseteq (T \cup N_2)^*$$

such that $L_1 = L'_0$, $L_2 = L'_k$ and $L_{i-1} \to L_i$ for $i = 1, 2, \ldots, k$. Define $L_1 \rightrightarrows^k L_2$ if and only if there exists $L_3 \subseteq (T \cup N_2)^*$ such that $L_1 \rightrightarrows^k L_3$ and $L_2 = L_3 \cap T^*$. Clearly, if L_1 is a CFL and $L_1 \rightrightarrows^k L_2$ for some k, then L_2 is also a CFL. Since $r > 0$, there exists $n \in \mathbb{N}$ such that $x \in L$ implies $x \in L_2$ for some $L_2 \subseteq T^*$, where $\psi(A_0) \rightrightarrows^k L_2$ and $k \leq n$. Let $\Gamma = \{L_2 \subseteq T^* \mid \psi(A_0) \rightrightarrows^k L_2$ for some $k \leq n$ and $L_2 \cap L \neq \emptyset\}$. It is easy to verify that $L_2 \in \Gamma$ implies $L_2 \subseteq L$. Thus $L = \cup_{L_2 \in \Gamma} L_2$. Therefore, L is a CFL. The converse is straightforward. ∎

Theorem 5.5.6 *Let $G = (T, N, P, A_0)$ be a CMG, where there exists $c < 1$ such that $(A \to^p x) \in P$ implies $p \leq c$. Then $L(f_G, r, >)$ is finite for all $r \in \mathbb{R}^{\geq 0}$.*

Proof. Clearly, f_G is a finitary $CFFL$. Thus, by Theorem 5.4.12, $f_G = f_{G_1}$ for some CMG G_1 which is in Greibach normal form. Let n be the largest integer such that $c^n > r$. Then $x \in L(f_{G_1}, r, >)$ implies $len(x) \leq n$. Therefore, $L(f_G, r, >) = L(f_{G_1}, r, >)$ is finite. ∎

Note that Theorems 5.5.5 and 5.5.6 remain valid if $>$ is replaced by $\geq$.

Theorem 5.5.7 *Let f be a CFFL. Then $L(f, \infty, =)$ is a CFL.*

Proof. Let $f = f_G$, where $G = (T, N, P, A_0)$ is a CMG. By Remark 5.4.13, we may assume that G is in Greibach normal form as long as we allow fuzzy productions of the form $s \rightarrow^p t$, where $p = \infty$. For each $a \in T$, let $\bar{a}$ be a new symbol and let $\bar{T} = \{\bar{a} \in T\}$. For each $x \in (T \cup N)^*$, let $\bar{x} \in (\bar{T} \cup N)^*$, where $\bar{x}$ is obtained from x by replacing every $a \in T$ in x by $\bar{a}$. Let $G = (T \cup \bar{T}, N, P_0, A_0)$ be the context-free grammar where $P_0 = \{A \rightarrow x \mid (A \rightarrow^p x) \in P \text{ and } p < \infty\} \cup \{A \rightarrow \bar{x} \mid (A \rightarrow^p x) \in P$ and $p = \infty\}$. Let $M = (U, S, V, H, q_0, \{q_1\})$ be an a-transducer where $U = T \cup \bar{T}$, $S = \{q_0, q_1\}$, $V = T$, and

$$H = \{(q, u, u, q) \mid u \in T,\ q \in S\} \cup \{(q, \bar{a}, a, q_1) \mid a \in T,\ q \in S\}.$$

Note that for all $L_0 \subseteq (T \cup \bar{T})^*$, $M(L_0)$ is obtained from L_0 by first omitting all those words in L_0 which do not contain any $\bar{a} \in T$, and then replacing all $\bar{a} \in \bar{T}$ by the corresponding $a \in T$ in the remaining words. Since G is in Greibach normal form, $x \in L(f, \infty, =)$ if and only if x can be derived from A by using at least a fuzzy production of the form $(A \rightarrow^p x) \in P$ with $p = \infty$. Thus, $L = M(L(G_0))$. Hence, L is a CFL. ■

Let $r \in \mathbb{R}^{\geq 0}$. Let

$$\mathcal{L}_r = \{L(f, r, >) \mid f \text{ is a CFFL}\}$$

and

$$\mathcal{L} = \cup_{r \in \mathbb{R}^{\geq 0}} \mathcal{L}_r.$$

$\mathcal{C}$ is the family of all CFL and $\mathcal{R}$ is the family of all regular languages.

Theorem 5.5.8 $\mathcal{L}_0 = \mathcal{C}$.

Proof. The proof follows from Theorem 5.5.2. ■

Theorem 5.5.9 *Let $r \in \mathbb{R}^{\geq 0}$. Then $\mathcal{C} \subseteq \mathcal{L}_r$.*

Proof. Let $L \in \mathcal{C}$. Then $L = L(G)$ for some context-free grammar $G = (T, N, P, A_0)$. Let $p = 1 + r$ and $G' = (T, N, P', A_0)$ be a CMG, where $P' = \{A \rightarrow x \mid (A \rightarrow x) \in P\}$. It follows that $L = L(f_{G'}, r, >)$. Thus, $L \in \mathcal{L}_r$. Hence $\mathcal{C} \subseteq \mathcal{L}_r$. ■

Theorem 5.5.10 *Let $r \in \mathbb{R}$ and $r > 0$. Then $\mathcal{L} = \mathcal{C}_r$.*

Proof. Let $L \in \mathcal{L}$. Then $L = L(f_{G_1}, r, >)$ for some CMG $G_1 = (T, N, P, \eta_1)$ and $r_1 \geq 0$. By Theorems 5.5.7 and 5.5.8, it suffices to consider the case where $r_1 > 0$. Let

$$G = (T, N, P, \eta)$$

where $\eta(A) = (r/r_1)\ \eta_1(A)$ for all $A \in N$. Clearly, $L = L(f_G, r, >)$. Thus $L \in \mathcal{L}_r$. Hence $\mathcal{L} \subseteq \mathcal{L}_r$. Since $\mathcal{L}_r \subseteq \mathcal{L}$, $\mathcal{L} = \mathcal{L}_r$. ∎

Theorem 5.5.11 *$L \in \mathcal{L}$ if and only if $L = L(f, r, >)$ for some finitary CFFL f and $r \in \mathbb{R}^{\geq 0}$.*

Proof. Let $L \in \mathcal{L}$. By Theorem 5.5.10, $L = L(f_G, 1/2, >)$ for some CMG $G = (T, N_1, P_1, A_1)$. By Remark 5.4.13, we may assume that G is in Greibach normal form as long as we allow fuzzy productions of the form $A \rightarrow^p x$ where $p = \infty$. By Theorem 5.5.7,

$$L(f_{G_1}, \infty, =) = L(G_2)$$

for some context-free grammar $G_2 = (T, N_2, P_2, A_2)$. We assume that G_2 is also in Greibach normal form. We also assume that $N_1 \cap N_2 = \emptyset$ and let $A_0 \notin T \cup N_1 \cup N_2$. Let $G = (T, N, P, A_0)$ be a CMG, where $N = N_1 \cup N_2 \cup \{A_0\}$, and P consists of all fuzzy productions of the following forms:

(1) $A \rightarrow^p x$, where $(A \rightarrow^p x) \in P_1$ and $p < \infty$,
(2) $A_0 \rightarrow^p x$, where $(A \rightarrow^p x) \in P_1$ and $p < \infty$,
(3) $A \rightarrow^1 x$, where $(A \rightarrow x) \in P_2$, and
(4) $A_0 \rightarrow^1 x$, where $(A \rightarrow x) \in P_2$.

Clearly, G is in Greibach normal form. Thus by Theorems 5.4.12, f_G is finitary. Now it is easy to show that $L = L(f_G, 1/2, >)$. The converse is straightforward. ∎

Theorem 5.5.12 *$\mathcal{C}$ is a proper subfamily of $\mathcal{L}$.*

Proof. Let $L = \{a^n b^n c^m \mid n \geq m \geq 1\}$. It is well known that $L \notin \mathcal{C}$. Now let $G = (\{a, b, c\}, \{A_0, A_1, A_2\}, P, A_0)$ be a CMG where P consists of the fuzzy productions $A_0 \rightarrow^1 A_1 A_2$, $A_1 \rightarrow^2 aA_1b$, $A_1 \rightarrow^1 ab$, $A_2 \rightarrow^{0.5} cA_2$, and $A_2 \rightarrow^1 c$. It is easy to verify that $L = L(f_G, b, >)$. Thus $L \in \mathcal{L}$. ∎

We write

$$s \rightrightarrows^p t \text{ if } s \rightrightarrows^p t (\text{mod } w)$$

for some w.

Lemma 5.5.13 *For each L where $L = L(f_G, r, >)$ for some CMG G and $r \in \mathbb{R}^{\geq 0}$, there exists $n \in \mathbb{N}$ with the property that each word $x \in L$ and $len(x) > n$ is of the form $s_1 s_2 s_3 s_4 s_5$ where $s_2 \neq \Lambda$, and either $s_1 s_2^k s_3 s_4^k s_5 \in L$ for all $k \in \mathbb{N}$ or $s_1 s_3 s_5 \in L$.*

Proof. By Theorems 5.4.12, 5.5.10, and 5.5.11, $L = L(f_G, 1, >)$, where $G = (T, N, P, A_0)$ is a CMG in Greibach normal form. Let $|N| = m$. If $x = s_1s_2s_3s_4s_5 \in L$ and $len(x) > m$, then we can show that there exists an $A \in N$ such that $A_0 \Rightarrow^{p_1} s_1At_1 \Rightarrow^{p_2} s_1s_2At_2t_1 \Rightarrow^{p_3} s_1s_2s_3s_4s_5 = x$, where $s_i \in T^*$, $t_i \in N^*$, $s_2 \neq \Lambda$, $A \Rightarrow^{p_4} s_3$, $t_2 \Rightarrow^{p_5} s_4$, $t_1 \Rightarrow^{p_6} s_5$, $p_4p_5p_6 = p_3$, and $p_1p_2p_3 > 1$. It is easily shown that if $p_2p_5 < 1$, then $s_1s_3s_5 \in L$. ■

Theorem 5.5.14 $\{a^nb^nc^n \mid n \geq 1\} \notin \mathcal{L}$.

Proof. Let $L = \{a^nb^nc^n \mid n \geq 1\} \in \mathcal{L}$. By Theorems 5.4.12, 5.5.10, and 5.5.11, $L = L(f_G, 1, >)$, where $G = (T, N, P, A)$ is a CMG in Greibach normal form. Let $|N| = m$ and let $x = s_1s_2s_3s_4s_5s_6s_7 \in L$, where $len(x) > m^2$. Then there exists $A \in N$ such that

$$A_0 \longrightarrow s_1At_1 \longrightarrow s_1s_2At_2t_1 \longrightarrow s_1s_2s_3At_3t_2t_1 \longrightarrow s_1s_2s_3s_4s_5s_6s_7,$$

where $s_i \in T^*$, $t_i \in N^*$, $s_2, s_3 \neq e$, $A \Rightarrow^{p_5} s_3$, $t_3 \Rightarrow^{p_6} s_5$, $t_1 \Rightarrow^{p_7} s_7$, $p_5p_6p_7p_8 = p_4$ and $p_1p_2p_3p_4 > 1$. It follows from the proof of Lemma 5.5.13 that neither $p_2p_7 \geq 1$ nor $p_3p_6 \geq 1$. However, if $p_2p_7 < 1$ and $p_3p_6 < 1$, then by Lemma 5.5.13, $s_1s_2s_4s_6s_7$ and $s_1s_3s_4s_5s_7$ are in L. This is a contradiction. Hence, $L \notin \mathcal{L}$. ■

Since it is well known that $\{a^nb^nc^n \mid n \geq 1\}$ is a context-sensitive language, we have the following result.

Theorem 5.5.15 *There exists a context-sensitive language which is not in* $\mathcal{L}$. ■

It can be shown by Theorem 5.5.14 and the proof of Theorem 5.5.12 that $\mathcal{L}$ is not closed under intersection. However, it can be shown that $\mathcal{L}$ is closed under union, intersection with regular sets, concatenation with CFL, substitution by CFL, homomorphism, inverse homomorphism, reversal, *a*-transducer mapping and so on. This can be accomplished by modifying existing proofs of the closure properties of CFLs.

5.6 Deterministic Acceptors of Regular Fuzzy Languages

The results in this section are from [28]. It is well known that there is a one-to-one relationship between finite automata, regular (type-3) languages, and regular expressions. This relatonship demonstrates the use of regular expressions for describing deterministically finite fuzzy automata and regular fuzzy languages. It introduces a normal form for the production of a regular fuzzy grammar when the max-min rule is used.

Specific fuzzy system models based on fuzzy set theory include a description of decision making in a fuzzy environment [3], finite fuzzy automata as learning systems [29], and fuzzy grammars and languages [10, 11]. Fuzzy automata, grammars, and languages are leading to greater understanding of nondeterministic algorithms and of pattern recognition tasks employing syntactic pattern recognition techniques.

In this section, an algorithm is developed for constructing a deterministic finite automaton which classifies the strings of a language with a regular fuzzy grammar. The derivations of the grammar are governed by the max-min rule [10, 11]. An equivalent unambiguous regular fuzzy grammar with productions in a normal form is developed from an extension of this algorithm.

Definition 5.6.1 *A* ***regular fuzzy grammar*** (**RFG**) *is a four-tuple RFG* $= (V_N, V_T, S, P)$, *where* V_N *is a finite set of non-terminals,* V_T *is a finite set of terminals,* $S \in V_N$ *is the starting symbol,* P *is a finite set of productions,* $V_N \cap V_T = \emptyset$, *and the elements in* P *are of the form* $A \overset{\theta}{\rightarrow} aB$ *or* $A \overset{\theta}{\rightarrow} a, A, B \in V_N,\ a \in V_T,\ 0 < \theta \leq 1$.

Definition 5.6.2 *A* ***finite fuzzy automaton*** (**FFA**) *is a six-tuple* $FFA = (\Sigma, Z, Q, \delta, w, q_s)$, *where* Σ *is a finite input alphabet,* Z *is a finite output alphabet,* Q *is a finite set of states,* $\delta : \Sigma \times Q \times [0,1] \rightarrow Q$ *is the fuzzy state transition map,* $\omega : Q \rightarrow Z$ *is the output map, and* $q_s \in Q$ *is the starting state.*

A regular fuzzy grammar reduces to a conventional grammar when production weights are all equal to 1. Similarly, a finite fuzzy automaton reduces to a conventional finite-state Moore machine by restricting the transition weights to the value 1.

A conventional formal language is defined abstractly as a subset of V_T^*. A fuzzy language is a fuzzy subset of V_T^* [10, 11].

Definition 5.6.3 *A fuzzy subset of* V_T^* *is called a* ***fuzzy language (FL)*** *in the alphabet V.*

Definition 5.6.4 *Given a regular fuzzy grammar RFG, the* ***membership grade of a string*** x *of* V_T^* *in the regular fuzzy language L(RFG) is the maximum value of any derivation of* x, *where the value of a specific derivation of* x *is equal to the minimum weight of the productions used.*

From the max-min rule for a fuzzy language, each string of V_T^* will have its highest computable membership grade. It is known [10] that given a grammar RFG, a corresponding finite fuzzy automaton FFA can be constructed which "accepts" the language L(RFG). The following theorem describes the construction of a deterministic nonfuzzy, finite automation (DFA) which computes the membership function of L (RFG). Unless otherwise specified, we use the symbol $\sum$ to denote both the automaton input alphabet and the language terminals, i.e., $\sum = V_T$.

Theorem 5.6.5 *Let a regular fuzzy grammar RFG be given. Then there exits a deterministic Moore sequential machine DFA, with output Alphabet* $Z \subseteq \{\theta \mid \theta \text{ is a production weight}\} \cup \{0\}$, *which computes the membership function* $\mu: \sum^* \to [0,1]$ *of the language L(RFG).*

Proof. We give a five-step algorithm for constructing the DFA.

Step 1: Given the regular fuzzy grammar, obtain the corresponding FFA.

The FFA is obtained in the same way that a nonfuzzy finite automaton is obtained from a nonfuzzy regular grammar [4], with the exception that a production weight is assigned to the corresponding transition [10].

Step 2: Obtain the set M of possible nonzero membership grades of strings in the language L(RFG).

M is taken to be the finite set of distinct production weights or, equivalently, the weights of the transitions of the FFA. (The reasoning is as follows: (a) the max and min operations do not introduce a weight not already assigned to some production and (b) each production weight initially may be the membership grade of some string (or strings) in the language L(RFG).)

Step 3: For each $\theta \in M$, obtain the regular expression $F'(\theta)$ describing those $x \in \sum^*$ such that $\mu(x) \geq \theta$.

(It is known that given a regular fuzzy language and a threshold θ, $0 \leq \theta \leq 1$, the nonfuzzy "threshold language"

$$L(\theta) = \{x \mid x \in \textstyle\sum^*,\ \mu(x) \geq \theta\}$$

is regular.) The regular expression $F'(\theta)$ can be found in the following method. Examine the transition diagram of the finite fuzzy automaton FFA and retain without weight only those transitions whose weight is equal to or greater than θ. This yields a nondeterministic nonfuzzy machine which recognizes the language $L(\theta)$ and from which the regular expression $F'(\theta)$ can be obtained directly by standard techniques for nondeterministic transition graphs or by conversion to a deterministic finite automaton and solution of the descriptive equations [4].

Step 4: For each $\theta \in M$, obtain the regular expression $F(\theta)$ describing those strings x of $\sum^*$ such that $\mu(x) = \theta$. (It is known that if $F'(\theta_1)$ and $F'(\theta_2)$ are regular expressions, then so are the Boolean functions of $F'(\theta_1)$ and $F'(\theta_2)$. Specifically, if $F'(\theta_1)$ and $F'(\theta_2)$ define two threshold languages, then the new regular expression

$$F(\theta_2) = F'(\theta_2) \cap \overline{F'(\theta_1)}$$

defines the regular language consisting of those strings which are in $L(\theta_2)$ and not in $L(\theta_1)$.) Consider the finite set M of possible nonzero membership grades in L(RFG). Select two adjacent membership grades θ_1, θ_2 in M such that $\theta_1 > \theta_2$. Then the regular expression

$$F(\theta_2) = F'(\theta_2) \cap \overline{F'(\theta_1)}$$

defines the set of strings

$$\begin{aligned}\{x \mid x \in \sum^*, x \in L(\theta_2), x \in \overline{L(\theta_1)}\} &= \{x \mid x \in \sum^*,\ \mu(x) \geq \theta_2,\\ &\quad \mu(x) < \theta_1\}\\ &= \{x \mid x \in \sum^*,\ \mu(x) = \theta_2\}.\end{aligned}$$

$F(\theta_2)$ identifies one equivalence class of an equivalence relation E in $\sum^*$, where x is E-related to y if and only if $\mu(x) = \mu(y)$. This procedure, applied to all pairs of adjacent membership grades in M beginning with the lowest value, yields the disjoint regular expressions defining the deterministic finite automaton DFA.

Step 5: Use the regular expressions $F(\theta)$ to obtain the state transition diagram of the DFA, where $\theta \in M$. ∎

The procedure for obtaining a state transition diagram by taking derivatives of a regular expression is discussed in [4].

Another method of decomposing a fuzzy grammar into nonfuzzy grammars using the concept of level set can be found in [32].

Corollary 5.6.6 *Given a regular fuzzy grammar* RFG_1*, there is an equivalent unambiguous* RFG_2 *in which productions have the form* $A \xrightarrow{1} aB$ *or* $A \xrightarrow{\theta} a$*, where* $A, B \in V_N$*,* $a \in V_T$*, and* $0 < \theta \leq 1$*.*

Proof. Construct the DFA as described in proof of Theorem 5.6.5. Then there is in P, a production $A \xrightarrow{1} aB$ (or $A \rightarrow aB$, with weight 1 understood) for each transition $\delta(q_A, a) = q_B$. If $\omega(q_B) = \theta \neq 0$, then there is in P a production $A \xrightarrow{\theta} a$ for each transition $\delta(q_A, a) = q_B$. The starting symbol S of RFG$_2$ corresponds to the starting state q_s of the DFA. Suppose the terminal string $x = ab\ldots cd$ causes the DFA started in state q_s, to halt in a state with output θ. Then there is a derivation

$$S \rightarrow aA \rightarrow abB \rightarrow \cdots \rightarrow ab\cdots cD \xrightarrow{\theta} ab\cdots cd$$

in RFG$_2$. Conversely, such a derivation in RFG$_2$ yields a string which causes the DFA to terminate in a state with output θ. RFG$_2$ is unambiguous since it is obtained from a deterministic finite automaton. ∎

The following example illustrates Theorem 5.6.5 and Corollary 5.6.6.

Example 5.6.7 (28) *Consider the regular fuzzy grammar* $RFG_1 = (V_N, V_T, S, P)$*, where* $V_N = \{S, A, B\}$*,* $V_T = \{a, b\}$*, and the productions are as follows:*

$$\begin{array}{lll} S \xrightarrow{0.3} aS & S \xrightarrow{0.5} aA & S \xrightarrow{0.7} aB \\ S \xrightarrow{0.3} bS & S \xrightarrow{0.2} bA & A \xrightarrow{0.5} b \\ B \xrightarrow{0.4} b. & & \end{array}$$

Step 1: The corresponding fuzzy machine is shown in Figure 5.1.

Step 2: $M = \{0.7,\ 0.5,\ 0.4,\ 0.3,\ 0.2\}$.

Step 3:

$$\begin{array}{ll} \theta = 0.2, & F'(0.2) = (a+b)^*(ab+bb) \\ \theta = 0.3, & F'(0.3) = (a+b)^*ab \\ \theta = 0.4, & F'(0.4) = ab \\ \theta = 0.5, & F'(0.5) = ab \\ \theta = 0.7, & F'(0.7) = \emptyset. \end{array}$$

Step 4:

$$\begin{array}{lll} F(0.2) & = F'(0.2) \cap \overline{F'(0.3)} & = (a+b)^*bb \\ F(0.3) & = (a+b)(a+b)^*ab & \\ F(0.4) & = \emptyset & \\ F(0.5) & = ab & \\ F(0.7) & = \emptyset. & \end{array}$$

Step 5: The deterministic classifier of strings in the regular fuzzy language $L(RFG)$ *is shown in Figure 5.2. Using the method of Corollary 5.6.6, the productions of the equivalent grammar* RFG_2 *are as follows:*

$$\begin{array}{ll} S \rightarrow aA \mid bB & \\ A \rightarrow aC \mid dD & A \xrightarrow{0.5} b \\ B \rightarrow aC \mid bE & B \xrightarrow{0.2} b \end{array}$$

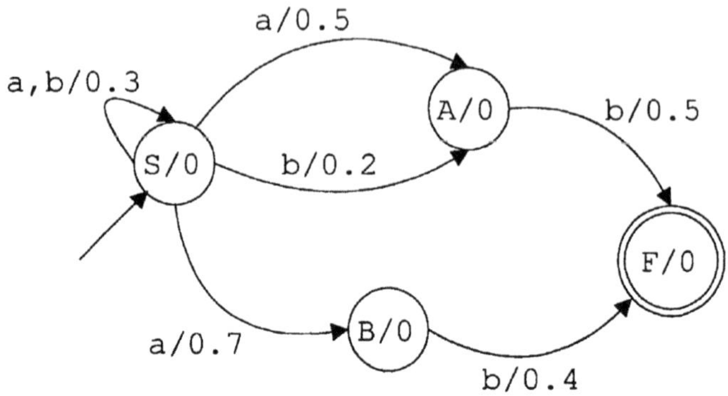

Figure 5.1[1]: FFA Obtained from RFG_1

$$\begin{array}{ll} C \to aC \mid bF & C \xrightarrow{0.3} b \\ D \to aC \mid bE & D \xrightarrow{0.2} b \\ E \to aC \mid bE & E \xrightarrow{0.2} b \\ F \to aC \mid bE & F \xrightarrow{0.2} b. \end{array}$$

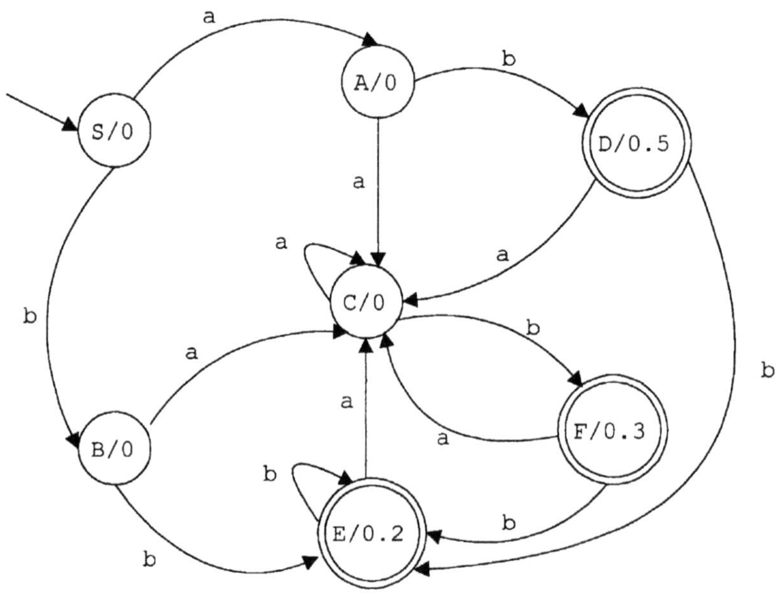

Figure 5.2[2]: DFA obtained from RFG_1

[1] Figure 5.1 is from [28], reprinted with permission by Copyright 1974 IEEE.
[2] Figure 5.2 is from [28], reprinted with permission by Copyright 1974 IEEE.

Example 5.6.8 (28) *Consider the string ab. Using RFG_1,*

$$S \xrightarrow{0.5} aA \xrightarrow{0.5} ab \text{ and } S \xrightarrow{0.7} ab \xrightarrow{0.4} ab,$$

we have that $\mu(ab) = \max\{\min\{0.5, 0.5\}, \min\{0.7, 0.4\}\} = 0.5$. *Using RFG_2, we have that* $S \rightarrow aA \xrightarrow{0.5} ab$.

5.7 Fuzzy Languages on a Free Monoid

In this section, we introduce the notions of fuzzy regular languages, adjunctive languages, and dense languages. We present their algebraic properties. The results are from [26].

The study of fuzzy grammars, the rules of fuzzy syntaxes, and the recognition ability of fuzzy automata extended the application area of fuzzy set theory and reduced the difference between formal languages and natural languages [12, 24, 11].

In the remainder of the chapter, we let $\sum$ denote an alphabet with $1 \leq |\sum| < \infty$ and $\sum^*$ be the free monoid generated from $\sum$ with the operation of concatenation. The empty string (or word) Λ is the identity of $\sum^*$. Recall that $\mathcal{FP}(X)$ denotes the set of all fuzzy subsets of X. We use the superscript T to denote the transpose of a matrix.

Definition 5.7.1 (12) *A **finite F-automaton** on an alphabet is a 5-tuple* $M = (X, Y, (T_u)_{u\in\Sigma}, \sigma_0, \sigma_1)$ *such that*

(i) $X = \{x_1, x_2, \dots, x_n\}$ *is the set of states,*

(ii) $Y = \{y_1, y_2, \dots, y_n\} \subseteq X$ *is the set of outputs,*

(iii) $(T_u)_{u\in\Sigma}$ *are F-transitive matrices, where* $\delta_{x_i x_j} : \sum \rightarrow [0,1]$ *and* $T_u = (\delta_{x_i x_j}(u))$, $i, j = 1, 2, \dots, n$,

(iv) $\sigma_0 = (i_1, i_2, \dots, i_n) \in \mathcal{FP}(X)$ *are F-initial states,*

(v) $\sigma_1 = (j_1, j_2, \dots, j_n)^T, \sigma_1^T \in \mathcal{FP}(X)$ *are F-terminal states.*

Let $M = (X, Y, (T_u)_{u\in\Sigma}, \sigma_0, \sigma_1)$ be a finite F-automaton. Define $\delta : X \times \sum \times X \rightarrow [0,1]$ by

$$\delta(x_i, u, x_j) = \delta_{x_i x_j}(u)$$

for $i, j = 1, 2, \dots, n$ and $\forall u \in \sum$. Then δ is an F-transitive function.

The extension on $\sum^*$ of δ is defined as follows:

$$\delta : X \times \sum^* \times X \rightarrow [0,1], \tag{5.10}$$

$$\delta(x, \Lambda, x') = \begin{cases} 1 \text{ if } x = x' \\ 0 \text{ if } x \neq x'. \end{cases}$$

For all $\theta = u_1 u_2 \cdots u_k \in \sum^*$, $\theta \neq \Lambda$, define

$$\delta(x,\theta,x') = \vee_{x_1,x_2,\ldots,x_{k-1}\in X}(\delta(x,u_1,x_1)\wedge\delta(x_1,u_2,x_2)\wedge\ldots\wedge\delta(x_{k-1},u_k,x'))$$

or

$$T_\theta = T_{u_1} \circ T_{u_2} \circ \cdots \circ T_{u_k},$$

where $\circ$ is the sup-min composition of fuzzy matrices.

Definition 5.7.2 (12) *Any member of $\mathcal{FP}(\sum^*)$ is called an **F-language** on the free monoid $\sum^*$. For all $u \in \sum$, define $\mathbf{u} \in \mathcal{FP}(\sum^*)$ by*

$$\mathbf{u}(\theta) = \begin{cases} 1 & \text{if } \theta = u, \\ 0 & \text{if } \theta \neq u, \end{cases}$$

*$\forall \theta \in \sum^*$. Then $\mathbf{u}$ is called the **basic F-language generated by** u. We set*

$$E = \{\mathbf{u} \mid u \in \textstyle\sum\} \cup \{\Lambda\}.$$

Definition 5.7.3 (12) *Let $U \subseteq \{\mu \in \mathcal{FP}(\sum^*) \mid \mu$ is an F-language$\}$ be such that*

(i) $\forall a \in [0,1]$, $\mu \in U \Rightarrow a \cdot \mu \in U$,

(ii) $\mu_1, \mu_2 \in U \Rightarrow \mu_1 \cup \mu_2 \in U$,

(iii) $\mu_1, \mu_2 \in U \Rightarrow \mu_1 \circ \mu_2 \in U$, where $(\mu_1 \circ \mu_2)(\theta) = \vee \{\mu_1(u) \wedge \mu_2(v) \mid uv = \theta\}$, $\forall \theta \in \sum^*$,

(iv) $\mu \in U \Rightarrow \widehat{\mu} \in U$, where $\widehat{\mu}$ is the Kleene's closure of μ, i.e.,

$$\widehat{\mu} = \mu_0 \cup \mu \cup \mu^2 \cup \cdots,$$

where $\mu_0(\theta) = 0$ $\forall \theta \in \sum^*$.

*Then U is called a **closed family of F-languages** on $\sum^*$.*

Let

$$\mathcal{F} = \{U \mid U \text{ is a closed family of } F\text{-languages on } \textstyle\sum^*\}.$$

Clearly, $\mathcal{FP}(\sum^*) \in \mathcal{F}$.

Definition 5.7.4 (12) *Let*

$$\mathbf{FR}(\textstyle\sum^*) = \displaystyle\bigcap_{E \subset M \subset \mathcal{F}} M.$$

*(i) $\mathbf{FR}(\sum^*)$ is called the **family of F-regular languages** on $\sum^*$.*

*(ii) If $\mu \in \mathbf{FR}(\sum^*)$, then μ is called an **F-regular language** on $\sum^*$.*

Clearly, $\mathbf{FR}(\sum^*)$ is a subalgebra of $\mathcal{FP}(\sum^*)$ generated by E with the four operations in Definition 5.7.3.

Definition 5.7.5 (12) *Let $M = (X, Y, T_u, \sigma_0, \sigma_1)$ be a finite F-automaton on $\sum$. Define $f_M : \sum^* \to [0,1]$ by $\forall u \in \sum^*$,*

$$f_M(u) = \sigma_0 \circ T_u \circ \sigma_1.$$

*Then f_M is called the **F-language determined** by the F-automaton M.*

From [12], $\mu \in \mathbf{FR}(\sum^*)$ if and only if there exists a finite F-automaton M such that $f_M = \mu$.

Definition 5.7.6 (12) *We say that P_L is a **main congruence** if $\forall x, y \in \sum^*$, $x \equiv y(P_L)$ if and only if $\forall u, v \in \sum^*$, $uxv \in L \Leftrightarrow uyv \in L$.*

Proposition 5.7.7 (27) *An ordinary language $L \subseteq \sum^*$ is regular if and only if the index of P_L is finite.* ∎

Proposition 5.7.8 (27) *An ordinary language $L \subseteq \sum^*$ is adjunctive if and only if $\forall x, y \in \sum^*$, $x \equiv y(P_L) \Rightarrow x = y$.* ∎

Definition 5.7.9 *Let $\mu \in \mathcal{FP}(\sum^*)$. We says that F_μ is an **F-main congruence** with respect to μ on $\sum^*$ if*

$$\forall x, y \in \textstyle\sum^*, x \equiv y(F_\mu) \Leftrightarrow \forall u, v \in \textstyle\sum^*,\ \mu(uxv) = \mu(uyv).$$

Proposition 5.7.10 *Let $x, y \in \sum^*$. Then $x \equiv y(F_\mu)$ if and only if $\forall a \in [0,1]$, $x \equiv y(P_{\mu_a})$, where $\mu_a = \{x \in \sum^* \mid \mu(x) \geq a\}$.*

Proof. Suppose that $x \equiv y(F_\mu)$. Then $\forall u, v \in \sum^*$, $\mu(uxv) = \mu(uyv)$. Thus $\forall a \in [0,1]$, $\mu(uxv) \geq a$ if and only if $\mu(uyv) \geq a$. Thus $\forall u, v \in \sum^*$, $\forall a \in [0,1]$, $uxv \in \mu_a$ if and only if $uyv \in \mu_a$. Hence $x \equiv y(P_{\mu_a})$.

Conversely, suppose that $\forall a \in [0,1]$, $x \equiv y(P_{\mu_a})$. Then $\forall u, v \in \sum^*$, $\mu(uxv) \geq a \Leftrightarrow \mu(uyv) \geq a$. Let $\mu(uxv) = c$ and $\mu(uyv) = b$. It follows easily that $c \geq b$ and $c \leq b$ and so $c = b$. This implies that $x \equiv y(F_\mu)$. ∎

Proposition 5.7.11 *Let $x, y \in \sum^*$. Then $x \equiv y(F_\mu)$ if and only if $\forall a \in [0,1]$, $x \equiv y(P_{\mu_{a+}})$, where $\mu_{a+} = \{x \in \sum^* \mid \mu(x) > a\}$.*

Proof. Suppose that $\forall a \in [0,1]$, $x \equiv y(P_{\mu_a})$. Then $\forall u, v \in \sum^*$, $\mu(uxv) > a \Leftrightarrow \mu(uyv) > a$. Let $\mu(uxv) = b$. Now for any $\varepsilon > 0$, we have $\mu(uxv) > b - \varepsilon$. This implies that $\mu(uyv) > b - \varepsilon$ for all $\varepsilon > 0$. Let $\varepsilon \to 0$. Then $\mu(uyv) \geq b = \mu(uxv)$. Similarly, $\mu(uxv) \geq \mu(uyv)$. Hence, $x \equiv y(F_\mu)$. The converse is straightforward. ∎

Corollary 5.7.12 $F_\mu = \bigcap_{a \in [0,1]} P_{\mu_a} = \bigcap_{a \in [0,1]} P_{\mu_{a+}}$. ∎

5.8 Algebraic Character and Properties of F-Regular Languages

Proposition 5.8.1 *Let $\mu \in \mathcal{FP}(\sum^*)$. Then $\mu \in \mathbf{FR}(\sum^*)$ if and only if the index of F_μ is finite.*

Proof. Suppose that $\mu \in \mathbf{FR}(\sum^*)$. Then there exists a finite F-automaton $M = (X, Y, T_u, \sigma_0, \sigma_1)$ such that $f_M = \mu$. Clearly, $f_M = \sigma_0 \circ T \circ \sigma_1$. Define a relation ρ on $\sum^*$ as follows: $\forall x, y \in \sum^*$, $x \equiv y(\rho) \Leftrightarrow T_x = T_y$. Clearly, ρ is an equivalence relation. Let $x, y \in \sum^*$. Now

$$\begin{aligned} x \equiv y(\rho) &\Rightarrow T_x = T_y \\ &\Rightarrow \forall u, v \in \textstyle\sum^*, T_u \circ T_x \circ T_v = T_u \circ T_y \circ T_v \\ &\Rightarrow \forall u, v \in \textstyle\sum^*, T_{uxv} = T_{uyv} \\ &\Rightarrow \forall u, v \in \textstyle\sum^*, \sigma_0 \circ T_{uxv} \circ \sigma_1 = \sigma_0 \circ T_{uyv} \circ \sigma_1 \\ &\Rightarrow \forall u, v \in \textstyle\sum^*, \mu(uxv) = \mu(uyv) \\ &\Rightarrow x \equiv y(F_\mu). \end{aligned}$$

Hence $\rho \subseteq F_\mu$. Since $\sum$ and X are finite, $\{T_w \mid w \in \sum^*\}$ is a finite set. Therefore, since the index of ρ is finite, the index of F_μ is also finite.

Conversely, suppose that the index of F_μ is finite. Let $[\Lambda], [u_1], \ldots, [u_s]$ be the distinct congruence classes of F_μ, where $u_i \in \sum^*$, $i = 1, 2, \ldots, s$. Let $M = (X, Y, T_u, \sigma_0, \sigma_1)$ be a finite F-automaton, where

$$X = Y = \{[\Lambda], [u_i], \ldots, [u_s]\},$$

the F-transitive matrix is defined as $\sigma_{[w]} \circ T_u = \sigma_{[wu]}$, and $\sigma_{[\Lambda]} = (1, 0, \ldots, 0)$, $\sigma_{[u_i]} = (0, \ldots, 0, 1, 0, \ldots, 0)$, $i = 1, 2, \ldots, s$. Set $\sigma_0 = \sigma_{[\Lambda]}$,

$$\sigma_1 = (\mu([\Lambda]), \mu([u_1]), \ldots, \mu([u_s]))^T.$$

Since $\forall u \in \sum^*$, $u \in [u]$, $\mu(u) = \mu([u])$. Also

$$f_M(u) = \sigma_0 \circ T_u \circ \sigma_1 = \sigma_{[\Lambda]} \circ T_u \circ \sigma_1 = \sigma_{[u]} \circ \sigma_1 = \mu([u]).$$

Thus, $\forall u \in \sum^*$, $\mu(u) = f_M(u)$. Hence $\mu = f_M$ and so $\mu \in \mathbf{FR}(\sum^*)$. ∎

Proposition 5.8.2 *Let $\mu \in \mathcal{FP}(\sum^*)$. Then $\mu \in \mathbf{FR}(\sum^*)$ if and only if $\forall a \in [0, 1]$, μ_a is regular and $|\operatorname{Im}(\mu)| < \infty$.*

Proof. Suppose that $\mu \in \mathbf{FR}(\sum^*)$. Then the index of F_μ is finite. Let $a \in [0, 1]$. By Corollary 5.7.12, $P_{\mu_a} \supseteq F_\mu$, $\forall a \in [0, 1]$. Hence the index of P_{μ_a} is finite and so μ_a is regular. We may assume that the congruence classes of F_μ are $[u_1], [u_2], \ldots, [u_s]$ since the index of F_μ is finite. Clearly, $\forall i$, $i = 1, 2, \ldots, s$, $\forall u, v \in \sum^*$, if $u, v \in [u_i]$, then $\mu(u) = \mu(v)$. Therefore, $\operatorname{Im}(\mu) = \{\mu(u_1), \mu(u_2), \ldots, \mu(u_s)\}$ and so $|\operatorname{Im}(\mu)| < \infty$.

Conversely, suppose that $\forall a \in [0, 1]$, μ_a is regular and $|\operatorname{Im}(\mu)| < \infty$. Then the characteristic function $\mathcal{K}_{\mu_a}$ of μ_a is an F-regular language. By the resolution theorem of F-sets (see [12]), $\mu = \bigcup_{a\in[0,1]} a\ \mathcal{K}_{\mu_a}$. Let $\mathrm{Im}(\mu) = \{a_1, a_2, \ldots, a_k\}$. Then $\mu = a_1\mathcal{K}_{\mu_{a_1}} \cup a_2\mathcal{K}_{\mu_{a_2}} \cup \cdots \cup a_k\mathcal{K}_{\mu_{a_k}}$. Hence by Definition 5.7.4, $\mu \in \mathbf{FR}(\sum^*)$. ■

Proposition 5.8.3 $(\mathbf{FR}(\sum^*), \cap, \cup, c)$ *is a de Morgan algebra, where* c *denotes complement of fuzzy subsets.*

Proof. Since $\mathbf{FR}(\sum^*) \subset \mathcal{FP}(\sum^*)$ and $(\mathcal{FP}(\sum^*), \cup, \cap, c)$ is a de Morgan algebra, it suffices to show that $(\mathbf{FR}(\sum^*), \cup, \cap, c)$ is a subalgebra of $(\mathcal{FP}(\sum^*), \cup, \cap, c)$. By Definition 5.7.4, it follows that $\forall \mu, \nu \in \mathbf{FR}(\sum^*)$, $\mu \cup \nu \in \mathbf{FR}(\sum^*)$. Also, since $\mu,\nu \in \mathbf{FR}(\sum^*)$, $\forall a \in [0,1]$, μ_a, ν_a are regular, and $|\mathrm{Im}(\mu)|$ and $|\mathrm{Im}(\nu)|$ are finite. Thus $\forall a \in [0,1]$, $(\mu\cap\nu)_a = \mu_a \cap \nu_a$ is regular and $|\mathrm{Im}(\mu\cap\nu)| \leq |\mathrm{Im}(\mu)| + |\mathrm{Im}(\nu)|$ is finite. Hence by Proposition 5.8.3, $\mu \cap \nu \in \mathbf{FR}(\sum^*)$. Let $x, y \in \sum^*$. Then

$$\begin{aligned} x \equiv y(F_\mu) \quad &\Leftrightarrow \quad \forall u, v \in \textstyle\sum^*, \mu(uxv) = \mu(uyv) \\ &\Leftrightarrow \quad \forall u, v \in \textstyle\sum^*, 1 - \mu(uxv) = 1 - \mu(uyv) \\ &\Leftrightarrow \quad \forall u, v \in \textstyle\sum^*, \overline{\mu}(uxv) = \overline{\mu}(uyv) \\ &\Leftrightarrow \quad x \equiv y(F_{\overline{\mu}}). \end{aligned}$$

Thus since the index of F_μ is finite, the index of $F_{\overline{\mu}}$ is finite. Hence, $\overline{\mu} \in \mathbf{FR}(\sum^*)$. ■

Proposition 5.8.4 *Let* $\mu,\nu \in \mathbf{FR}(\sum^*)$. *Then*
(i) $\mu \circ \nu \in \mathbf{FR}(\sum^*)$,
(ii) $\nu^{-1}\mu$ *(or* $\mu\nu^{-1}$*)* $\in \mathbf{FR}(\sum^*)$,
where $\nu^{-1}\mu$ *(or* $\mu\nu^{-1}$*) is called the* ***F-left (-right, respectively) quotient*** *of* μ *with respect to* ν *and is defined as follows:*

$$(\nu^{-1}\mu)(u) = \vee\{\mu(xu) \wedge \nu(x) \mid x \in \Sigma^*), \quad \forall u \in \textstyle\sum^*,$$

$$(\mu\nu^{-1})(u) = \vee\{\mu(ux) \wedge \nu(x) \mid x \in \Sigma^* \}, \quad \forall u \in \textstyle\sum^*.$$

Proof. (i) Let $a \in [0,1]$. Since $(\mu\circ\nu)_a = \mu_a\nu_a$ and μ and ν are F-regular, μ_a and ν_a are regular. Also since $(\mu\circ\nu)_a = \mu_a\nu_a$ is an adjoin of two regular languages, $(\mu\circ\nu)_a$ is regular. Now the degrees of membership of elements in $\mu\circ\nu$ are obtained from the degrees of membership of elements in μ and ν with operations max and min. Thus $\mathrm{Im}(\mu \circ \nu) \subseteq \mathrm{Im}(\mu) \cup \mathrm{Im}(\nu)$ and so $\mathrm{Im}(\mu \circ \nu)$ is finite. Hence $\mu \circ \nu \in \mathbf{FR}(\sum^*)$.

(ii) For all $s, t \in \sum^*$, $s \equiv t(F_\mu)$

$$\begin{aligned} &\Rightarrow \quad \forall u, v \in \textstyle\sum^*, \mu(usv) = \mu(utv) \\ &\Rightarrow \quad \forall u, v, x \in \textstyle\sum^*, \mu(xusv) = \mu(xutv) \\ &\Rightarrow \quad \forall u, v, x \in \textstyle\sum^*, \ \mu(xusv) \wedge \nu(x) = \ \mu(xutv) \wedge \nu(x) \\ &\Rightarrow \quad \forall u, v \in \textstyle\sum^*, \ \vee_{x\in\Sigma^*}(\mu(xusv) \wedge \nu(x)) = \ \vee_{x\in\Sigma^*} (\mu(xutv) \wedge \nu(x)) \\ &\Rightarrow \quad \forall u, v \in \Sigma^*, \ (\nu^{-1}\mu)(usv) = (\nu^{-1}\mu)(utv) \\ &\Rightarrow \quad s \equiv t(F_{\nu^{-1}\mu}). \end{aligned}$$

Thus, $F_\mu \subseteq F_{\nu^{-1}\mu}$. Since $\mu \in \mathbf{FR}(\sum^*)$, the index of F_μ is finite and so the index of $F_{\nu^{-1}\mu}$ is also finite Consequently, $\nu^{-1}\mu \in \mathbf{FR}(\sum^*)$. ■

Remark 5.8.5 *ν in Proposition 5.8.4(ii) may be any language.*

Proposition 5.8.6 *Let f be a homomorphism from $\sum^*$ onto $\sum_1^*$, where $\sum_1$ is a nonempty set. Then*
(i) $\mu \in \mathbf{FR}(\sum^*) \Rightarrow f(\mu) \in \mathbf{FR}(\sum_1^*)$;
(ii) $\nu \in \mathbf{FR}(\sum_1^*) \Rightarrow f^{-1}(\nu) \in \mathbf{FR}(\sum^*)$.

Proof. (i) Let $\mu \in \mathbf{FR}(\sum^*)$. Let $a \in [0,1]$ and $v_1 \in \sum_1^*$. Now $v_1 \in (f(\mu))_a$ if and only if $(f(\mu))(v_1) \geq a$ if and only if $\vee\{\mu(v) \mid f(v) = v_1\} \geq a$ if and only if $\exists v', f(v') = v_1$ and $\mu(v') \geq a$ (since $|\operatorname{Im}(\mu)| < \infty$) if and only if $f(v') = v_1,\ v' \in \mu_a$ if and only if $v_1 \in f(\mu_a)$. Hence $(f(\mu))_a = f(\mu_a)$. Since μ_a is regular, $f(\mu_a)$ is regular and so $(f(\mu))_a$ is regular. Since $\operatorname{Im}(\mu)$ is finite, it follows that $\operatorname{Im}(f(\mu)) \subseteq \operatorname{Im}(\mu)$ and so $\operatorname{Im}(f(\mu))$ is finite. Hence $f(\mu) \in \mathbf{FR}(\sum_1^*)$.

(ii) Let $\nu \in \mathbf{FR}(\sum_1^*)$ and $x, y \in \sum^*$. Then

$$\begin{aligned} x \equiv y(F_{f^{-1}(\nu)}) &\Leftrightarrow (f^{-1}(\nu))(uxv) = (f^{-1}(\nu))(uyv) \quad \forall u, v \in \textstyle\sum^* \\ &\Leftrightarrow \nu(f(uxv)) = \nu(f(uyv)) \quad \forall u, v \in \textstyle\sum^* \\ &\Leftrightarrow \nu(f(u)f(x)f(v)) = \nu(f(u)f(y)f(v)) \quad \forall u, v \in \textstyle\sum^* \\ &\Leftrightarrow \nu(u_1 f(x) v_1) = \nu(u_1 f(y) v_1) \quad \forall u_1, v_1 \in \textstyle\sum_1^* \\ &\Leftrightarrow f(x) \equiv f(y)(F_\nu). \end{aligned}$$

This implies that the index of $F_{f^{-1}(\nu)}$ in $\sum^*$ equals the index of F_ν in $\sum_1^*$. Since the index of F_ν is finite, the index of $F_{f^{-1}(\nu)}$ is finite and so $f^{-1}(\nu) \in \mathbf{FR}(\sum^*)$. ■

Definition 5.8.7 *Let $\mu \in \mathcal{FP}(\sum^*)$ and let $h(\mu) = \vee\{\mu(u) | u \in \Sigma^*\}$.*
*(i) Then $h(\mu)$ is called the **height** of μ.*
*(ii) If there exists $u_0 \in \sum^*$ such that $h(\mu) = \mu(u_0)$, then u_0 is called a **saddle point** of μ.*

Proposition 5.8.8 *Let $\mu \in \mathbf{FR}(\sum^*)$. Then there exists $n \in \mathbb{N}$ such that μ has a saddle point with length not greater than $n-1$.*

Proof. Since $\mu \in \mathbf{FR}(\sum^*)$, there exists a finite automaton $M = (X, Y, T_u, \sigma_0, \sigma_1)$ such that $\mu = f_M = \sigma_0 \circ T \circ \sigma_1$, where $|X| = n$, $\sigma_0 = (a_1, a_2, \ldots, a_n)$, and $\sigma_1 = (b_1, b_2, \ldots, b_n)^T$. Clearly,

$$h(\mu) = \vee_{u \in \Sigma^*} \mu(u) = \vee \{\vee_{u \in \cup_{k=0}^{n-1} \Sigma^k} \mu(u), \vee_{u \in \Sigma^* \setminus \cup_{k=0}^{n-1} \Sigma^k} \mu(u)\},$$

where $\sum^0 = \{\Lambda\}$. We now show that

$$\vee_{u \in \cup_{k=0}^{n-1} \Sigma^k} \mu(u) \geq \vee_{u \in \Sigma^* \setminus \cup_{k=0}^{n-1} \Sigma^k} \mu(u).$$

Let $v \in \sum^* \setminus \cup_{k=0}^{n-1} \sum^k$ be such that $len(v) \geq n$. Let $v = \theta_1\theta_2 \cdots \theta_m$, $m \geq n$, $\theta_i \in \sum$, $i = 1, \ldots, m$. Then

$$\begin{aligned}\mu(v) &= \sigma_0 \circ T_v \circ \sigma_1 \\ &= \sigma_0 \circ T_{\theta_1} \circ T_{\theta_2} \circ \cdots \circ T_{\theta_m} \circ \sigma_1 \\ &= \vee_{1 \leq i_1, \ldots, i_m \leq n}(a_i \wedge r_{ii_1}^{(1)} \wedge r_{i_1 i_2}^{(2)} \wedge \ldots \wedge r_{i_{m-1} i_m}^{(m)} \wedge b_{i_m}),\end{aligned}$$

where $T_{\theta_k} = (r_{ij}^{(k)})$, $k = 1, 2, \ldots, m$. Consider the term

$$d = a_i \wedge r_{ii_1}^{(1)} \wedge r_{i_1 i_2}^{(2)} \wedge \ldots \wedge r_{i_{m-1} i_m}^{(m)} \wedge b_{i_m}.$$

The number of different elements in the indices of d is not greater than n. Since $m \geq n$, there exists p, q with $p < q$ and $i_p = i_q$. Now

$$d \leq d' = a_i \wedge r_{ii_1}^{(1)} \wedge \ldots \wedge r_{i_{p-1} i_p}^{(p)} \wedge r_{i_q i_{q+1}}^{(q+1)} \wedge \ldots \wedge r_{i_{m-1} i_m}^{(m)} \wedge b_{i_m}.$$

If the number $m+p-q+1$ of indices $i, i_1, \ldots, i_p, i_{q+1}, \ldots, i_m$ is still greater than n, then we repeat the above process and eliminate a part of r until the number of indices is not greater than n. Thus we may assume that $m + p - q + 1 \leq n$. Now

$$\begin{aligned}d &\leq \vee_{1 \leq i, i_1, \ldots, i_p, i_{q+1}, \ldots, i_m \leq n}(a_i \wedge r_{ii_1}^{(1)} \wedge \ldots \wedge r_{i_{p-1} i_p}^{(p)} \wedge r_{i_q i_{q+1}}^{(q+1)} \wedge \ldots \\ &\quad \wedge r_{i_{m-1}}^{(m)} {}_{i_m} \wedge b_{i_m}) \\ &= \sigma_0 \circ T_{\theta_1} \circ \cdots \circ T_{\theta_p} \circ T_{\theta_{q+1}} \circ \cdots \circ T_{\theta_m} \circ \sigma_1.\end{aligned}$$

Let $v' = \theta_1\theta_2 \cdots \theta_p\theta_{q+1} \cdots \theta_m$. Clearly, $len(v') = m + p - q \leq n - 1$. Furthermore, $d \leq \sigma_0 \circ T_{v'} \circ \sigma_1 = \mu(v') \leq \vee_{u \in \cup_{k=0}^{n-1} \Sigma^k} \mu(u)$. Since d was an arbitrary term in $\mu(v)$, it follows that $\mu(v) \leq \vee_{u \in \cup_{k=0}^{n-1} \Sigma^k} \mu(u)$. Thus

$$\vee_{u \in \Sigma^* \setminus \cup_{k=0}^{n-1} \Sigma^k} \mu(u) \leq \vee_{u \in \cup_{k=0}^{n-1} \Sigma^k} \mu(u).$$

Hence, $h(\mu) = \vee_{u \in \Sigma^*} \mu(u) = \vee_{u \in \cup_{k=0}^{n-1}} \mu(u)$. Since $\cup_{k=0}^{n-1} \sum^k$ is a finite set, there exists $u_0 \in \sum^*$ such that $len(u_0) \leq n - 1$ and $h(\mu) = \mu(u_0)$. ∎

Proposition 5.8.9 *(Action lemma of the F-pump) If $\mu \in \mathbf{FR}(\sum^*)$, then there exists a number n such that for any $u \in \sum^*$, $u = xyw$ with $len(xy) \leq n$, $len(y) \geq 1$, and $\forall j \geq 0$, $\mu(xy^j w) \geq \mu(u)$ provided $len(u) \geq n$.*

Proof. As the proof of Proposition 5.8.8, let $\mu = f_M$ and $n = |X|$. Let $u \in \sum^*$ and let $u = \theta_1\theta_2 \cdots \theta_m$, $m \geq n$. Now

$$\mu(u) = \vee_{1 \leq i, i_1, \ldots, i_m \leq n}(a_i \wedge r_{ii_1}^{(1)} \wedge \ldots \wedge r_{i_{m-1} i_m}^{(m)} \wedge b_{i_m}).$$

Since m and n are finite, there exists a term d in the above equality such that

$$\mu(u) = d = a_i \wedge r^{(1)}_{ii_1} \wedge \ldots \wedge r^{(m)}_{i_{m-1}i_m} \wedge b_{i_m}.$$

Since the number of indices $i, i_1, \ldots, i_m$ is greater than n, there exist, as in the proof of Proposition 5.8.8, p, q with $p < q$, $i_p = i_q$, and

$$d' = a_i \wedge r^{(1)}_{ii_1} \wedge \ldots \wedge r^{(p)}_{i_{p-1}i_p} \wedge r^{(q+1)}_{i_q i_{q+1}} \wedge \ldots \wedge r^{(m)}_{i_{m-1}} i_m \wedge b_{i_m} \geq \mu(u),$$

where $m + p - q < n$. Furthermore,

$$\begin{aligned}&\vee_{i \leq i, i_1, \ldots, i_p, i_{qm}, \ldots, i_m \leq n} a_i \wedge r^{(1)}_{ii_1} \wedge \ldots \wedge r^{(p)}_{i_{p-1}i_p} \wedge r^{(q+1)}_{i_q i_{q+1}} \wedge \ldots \\ &\wedge r^{(m)}_{i_{m-1}i_m} \wedge b_{i_m} \geq \mu(u).\end{aligned}$$

That is, $\mu(\theta_1 \cdots \theta_p \theta_q \cdots \theta_m) \geq \mu(u)$. Let $x = \theta_1 \cdots \theta_p, = y = \theta_{p+1} \cdots \theta_q$, and $w = \theta_{q+1} \cdots \theta_m$. Clearly, $len(xy) \leq n$, $len(y) \geq 1$, and $\mu(xy^0w) \geq \mu(u)$.

Also,

$$\begin{aligned} d_{(i)} &= a_i \wedge r^{(1)}_{ii_1} \wedge \ldots \wedge r^{(p)}_{i_{p-1}i_p} \wedge \\ &\quad \underbrace{r^{(p+1)}_{i_p i_{p+1}} \wedge \ldots \wedge r^{(q)}_{i_{q-1}i_q} \wedge \ldots \wedge r^{(p+1)}_{i_p i\, p+1} \wedge \ldots \wedge r^{(q)}_{i_{q-1}i_q}}_{\left(r^{(p+1)}_{i_p i_{p+1}}, \ldots, r^{(q)}_{i_{q-1}, i_q}\right) j\text{-times}} \\ &\quad \wedge r^{(q+1)}_{i_q i_{q+1}} \wedge \ldots \wedge r^{(m)}_{i_{m-1}i_m} \wedge b_{i_m} \\ &= d \\ &= \mu(u), \end{aligned}$$

where (i) stands for the set of all indices in the above formula. Thus $\mu(\theta_1 \cdots \theta_p (\theta_{p+1} \cdots \theta_q)^j \theta_{q+1} \cdots \theta_m) = \vee\{ d_{(i)} \mid 1 \leq (i) \leq n\} \geq \mu(u)$, i.e., $\mu(xy^jw) \geq \mu(u)$, $\forall j \geq 1$. Consequently, $\forall j \geq 0$, $\mu(xy^jw) \geq \mu(u)$. ∎

Corollary 5.8.10 *Let* $\mu \in \mathbf{FR}(\sum^*)$, $\forall a \in [0,1]$. *Then there exists a constant* n *such that for any* $u \in \mu_a$, *we have* $u = xyw$ *with* $len(xy) \leq n$, $len(y) \geq 1$, *and* $\forall j \geq 0$, $xy^jw \in \mu_a$ *provided* $len(u) \geq n$. ∎

If A is ordinary regular language, then its characteristic function $\mathcal{K}_A \in \mathbf{FR}(\sum^*)$ and Corollary 5.8.10 becomes the ordinary action lemma of the pump (see [11]). Hence, Proposition 5.8.9 is a generalization of the ordinary action lemma of the pump and for this reason is called the action lemma of the F-pump.

F-Adjunctive Languages

Definition 5.8.11 *Let* $\lambda \in \mathcal{FP}(\sum^*)$. *If* $\forall x, y \in \sum^*$, $x \equiv y(F_\lambda) \Rightarrow x = y$, *then* λ *is called an* ***F-adjunctive language***.

Proposition 5.8.12 *The following statements are equivalent:*

(i) $\lambda \in \mathcal{FP}(\sum^*)$ is an F-adjunctive language.
(ii) F_λ is the identity relation.
(iii) $\forall x \in \sum^*$, $[x]_{F_\lambda} = \{x\}$.
(iv) $\forall x, y \in \sum^*$, if $x \neq y$, then there exist $u, v \in \sum^*$ such that $\lambda(uxv) \neq \lambda(uyv)$. ■

Proposition 5.8.13 *Let $\lambda \in \mathcal{FP}(\sum^*)$. Then λ is an F-adjunctive language if there exist $a \in [0,1]$ such that λ_a is an ordinary adjunctive language.* ■

Proposition 5.8.14 *Let $\lambda \in \mathcal{FP}(\sum^*)$. Then λ is an F-adjunctive language $\Leftrightarrow \forall x, y \in \sum^*$, $\{x \equiv y(\bigcap_{a \in (0,1)} P_{\lambda_a}) \Rightarrow x = y\} \Leftrightarrow \forall x, y \in \sum^*$, $\{x \equiv y(\bigcap_{a \in (0,1)} P_{\lambda_a}) \Rightarrow x = y\}$.* ■

Lemma 5.8.15 *Let f be a homomorphism from $\sum^*$ onto $\sum^*$. Then*
(i) $\forall \mu \in \mathcal{FP}(\sum^*)$, $(f(\mu))_a = f(\mu)_a$, $\forall a \in [0,1]$;
(ii) $\forall \nu \in \mathcal{FP}(\sum^*)$, $(f^{-1}(\nu))_a = f^{-1}(\nu)_a$, $\forall a \in [0,1]$. ■

Proposition 5.8.16 *Let f be a homomorphism from $\sum^*$ onto $\sum_1^*$. Let $\lambda \in \mathcal{FP}(\sum^*)$ be an F-adjunctive language. Then $f(\lambda) \in \mathcal{FP}(\sum_1^*)$ is F-adjunctive.*

Proof. Since λ is F-adjunctive, $\forall x, y \in \sum^*, x \equiv y(\bigcap_a P_{\lambda_a}) \Rightarrow x = y$. That is, $\forall a$, $\forall u, v \in \sum^*$, $uxv \in \lambda_a \Leftrightarrow uyv \in \lambda_a$ implies that $x = y$. Let $f(x), f(y) \in f(\sum^*) = \sum_1^*$ and $f(x) \equiv f(y)(\bigcap_a P_{(f(\lambda)_a)})$. Then $\forall a \in [0,1]$, $\forall u, v \in \sum^*$, $f(u)f(x)f(v) \in (f(\lambda))_a \Leftrightarrow f(u)f(y)f(v) \in (f(\lambda))_a$. This implies that $f(uxv) \in (f(\lambda))_a \Leftrightarrow f(uyv) \in (f(\lambda))_a$, $\forall a$, $\forall u, v \in \sum^*$, which in turn imples that $f(uxv) \in f(\lambda_a) \Leftrightarrow f(uyv) \in f(\lambda_a)$, $\forall a$, $\forall u, v \in \sum^*$. This is equivalent to $uxv \in \lambda_a \Leftrightarrow uyv \in \lambda_a$, $\forall a, \forall u, v \in \sum^*$. Thus $x \equiv y(P_{\lambda_a}), \forall a$ and so $x \equiv y\,(\cap_a P_{\lambda_a}\,)$. Hence $x = y$ and so $f(x) = f(y)$. Consequently, $f(\lambda)$ is F-adjunctive. ■

Proposition 5.8.17 *Let $f : \sum^* \to \sum_1^*$ be an isomorphism. Suppose that $\nu \in \mathcal{FP}(\sum_1^*)$ is F-adjunctive language. Then $f^{-1}(\nu) \in \mathcal{FP}(\sum^*)$ is F-adjunctive.*

Proof. Let $x, y \in \sum^*$ and $x \equiv y(\bigcap_a P_{(f^{-1}(\nu))_a})$. Then $uxv \in (f^{-1}(\nu))_a \Leftrightarrow uyv \in (f^{-1}(\nu))_a$ $\forall a, \forall u, v \in \sum^*$, i.e., $uxv \in f^{-1}(\nu_a) \Leftrightarrow uyv \in f^{-1}(\nu_a)$ $\forall a, \forall u, v \in \sum^*$. This implies that $f(uxv) \in \nu_a \Leftrightarrow f(uyv) \in \nu_a$ $\forall a, \forall u, v \in \sum^*$. Thus $f(x) \equiv f(y)(P_{\nu_a})$ $\forall a$ and so $f(x) \equiv f(y)\,(\cap_a P_{\nu_a})$. This implies that $f(x) = f(y)$. Since f is one-one, $x = y$. Hence $f^{-1}(\nu)$ is F-adjunctive. ■

Proposition 5.8.18 *Let $\lambda \in \mathcal{FP}(\sum^*)$. If there exists $\nu \in \mathcal{FP}(\sum^*)$ such that $\nu^{-1}\lambda$ is adjunctive, then λ is F-adjunctive.*

Proof. Since $\nu^{-1}\lambda$ is adjunctive, $\forall s,t \in \sum^*$, $s \equiv t(F_{\nu^{-1}\lambda}) \Rightarrow s = t$. Suppose that $s \equiv t(F_\lambda)$. Then $\forall u,v \in \sum^*$, $\lambda(usv) = \lambda(utv)$. Thus $\lambda(xusv) = \lambda(xutv)$ $\forall x,u,v \in \sum^*$. This implies that $\lambda(xusv) \wedge \nu(x) = \lambda(xutv) \wedge \nu(x)$ $\forall x,u,v \in \sum^*$ and so

$$\vee\{\lambda(xusv) \wedge \nu(x) \mid x \in \Sigma^*\} = \vee\{\lambda(xuyv) \wedge \nu(x) \mid x \in \Sigma^*\}$$

$\forall u,v \in \sum^*$. Thus $(\nu^{-1}\lambda)(usv) = (\nu^{-1}\lambda)(utv)$, $\forall u,v \in \sum^*$ and so $s \equiv t(F_{\nu^{-1}\lambda})$. Hence $s = t$. Consequently, λ is F-adjunctive. ■

Proposition 5.8.19 *Let $\lambda \in \mathcal{FP}(z^*)$. Then λ is F-adjunctive if and only if λ is not F-regular.*

Proof. If λ is F-regular, then the index of F_λ is finite. This contradicts the fact that λ is F-adjunctive and z^* is divided into infinite classes that contain only one element by F_λ.

Conversely, if λ is not F-adjunctive, then there exist i,j with $i \neq j$ and $z^i \equiv z^j(F_\lambda)$. Now z^* is divided into at most j classes $[\Lambda], [z], [z^2], \ldots, [z^{j-1}]$. Hence the index of F_λ is finite, i.e., λ is F-regular. ■

Proposition 5.8.20 *Let $\lambda \in F(z^*)$. If there exists $a \in [0,1]$ such that*
(i) $\forall m \geq 1$, $\exists n \in \mathbb{N}$, $\lambda(z^{m+n}) \geq a$,
(ii) $\forall m \geq 1$, $\exists l \in \mathbb{N}$, $\lambda(z^{m+l}) < a$,
(iii) $\forall m \geq 1$, $\forall t \in \mathbb{N}, \exists s \geq t$, $(\lambda(z^{s+i}) < a$ or $\lambda(z^{s+i}) \geq a$, $\forall i = 1,2,\ldots,m)$,
then λ is an F-adjunctive language.

Proof. It suffices to show that $\forall r \geq 0$, $k \geq 1$, $z^r \not\equiv z^{r+k}(F_\lambda)$. In fact, if $k+1 = m \geq 1$, then we set $s \geq 1$ when $r = 0$ and set $s \geq r$ when $r \geq 1$. From (iii), it follows that $\forall i = 1,2,\ldots,m$, $\lambda(z^{s+i}) < a$ or $\lambda(z^{s+1}) \geq a$.

Case 1: $\lambda(z^{s+i}) < a$, $i = 1,2,\ldots,k+1$. From (i), we have that it is possible to make m_i the smallest integer with $\lambda(z^{s+k+1+m_1}) \geq a$. Thus

$$\lambda(z^{s-r+1}z^r z^m) < a, \qquad \lambda(z^{s-r+1}z^{r+k}z^{m_1}) \geq a,$$

i.e., there exist $z^{s-r+1}, z^{m_1} \in z^*$ such that

$$\lambda(z^{s-r+1}z^r z^{m_1}) \neq \lambda(z^{s-r+1}z^{r+k}z^{m_1}).$$

Hence $z^r \not\equiv z^{r+k}(F_\lambda)$.

Case 2: $\lambda(z^{s+i}) \geq a$, $i = 1,2,\ldots,k+1$. From (ii), we have that it is possible to make m_2 the smallest integer with $\lambda(z^{s+k+1+m_2}) < a$. Similarly, it is easy to show that $z^r \not\equiv z^{r+k}(F_\lambda)$. ■

Definition 5.8.21 *Let $\lambda \in \mathcal{FP}(\sum^*)$ and $a \in [0,1]$. Then λ is called a a-**discrete language** if $\forall x,y \in \sum^*$, $x \neq y$, $\lambda(x) \geq a$ and $\lambda(y) \geq a$ implies that $len(x) \neq len(y)$.*

Lemma 5.8.22 *Let* $|\sum| \geq 2$ *and* $\lambda \in \mathcal{FP}(\sum^*)$. *Then* λ *is an* F*-adjunctive language if and only if* $\forall u, v \in \sum^*$, $len(u) = len(v)$ *and* $u \equiv v(F_\lambda)$ *implies that* $u = v$. ■

Proposition 5.8.23 *Let* $\lambda \in \mathcal{FP}(\sum^*)$. *Suppose that* λ *is a-discrete and* $\forall w \in \sum^*$, *there exist* $u, v \in \sum^*$ *such that* $\lambda(uwv) \geq a$. *Then* λ *is an* F*-adjunctive language. Conversely if* λ *is an* F*-adjunctive language, then* $\forall w \in \sum^*$ *there exist* $u, v \in \sum^*$ *such that* $\lambda(uwv) > 0$.

Proof. If $u \equiv v(F_\lambda)$ and $len(u) = len(v)$, then by the hypothesis there exist $x, y \in \sum^*$ such that $\lambda(xuy) > a$. Since $u \equiv v(F_\lambda)$, $\lambda(xuy) = \lambda(xvy)$ and so $\lambda(xvy) > a$. Now, since $len(xuy) = len(xvy)$ and λ is a-discrete, it follows that $xuy = xvy$ and so $u = v$. Hence λ is F-adjunctive.

Conversely, if there exists $w \in \sum^*$ such that $\forall u, v \in \sum^*$, $\lambda(uwv) = 0$, then $\forall u, v \in \sum^*$, $\lambda(uwv) = \lambda(uw^2v) = 0$, i.e., $w \equiv w^2(F_\lambda)$, which contradicts the fact that λ is an F-adjunctive language. ■

Proposition 5.8.24 *If* $\lambda \in \mathcal{FP}(\sum^*)$ *is an* F*-adjunctive language, then* $\forall w \in \sum^*$, $|\lambda^w| = \infty$, *where* $\lambda^w = \{x \mid x \in \sum^* w \sum^*,\ \lambda(x) > 0\}$.

Proof. Since λ is F-adjunctive, $\forall w \in \sum^*$, $|\lambda^w| \neq 0$. Suppose that $|\lambda^w| < \infty$. Let $u \in \lambda^w$ be such that $|u| = \vee\{|v| \mid v \in \lambda^w\}$. Clearly, $|\lambda^{wu}| = 0$. This contradicts the fact that λ is an F-adjunctive language. ■

Proposition 5.8.25 *Let* $|\sum| \geq 2$. *Then* λ *is an* 0*-discrete* F*-adjunctive language if and only if Supp*(λ) *is an ordinary discrete adjunctive language.*

Proof. If Supp(λ) is an ordinary discrete adjunctive language, then λ is clearly a 0-discrete F-adjunctive language.

Conversely, suppose that λ is a 0-discrete F-adjunctive language and Supp(λ) is not adjunctive. By Lemma 5.8.22, there exist $x, y \in \sum^*$ such that $x \neq y$, $len(x) = len(y)$, and $x \equiv y(P_{\text{Supp}(\lambda)})$. This implies that $\forall u, v \in \sum^*$, $uxv \in \text{Supp}(\lambda) \Leftrightarrow uyv \in \text{Supp}(\lambda)$. Consequently, $\forall u, v \in \sum^*$, $\lambda(uxv) > 0 \Leftrightarrow \lambda(uyv) > 0$. Since λ is an F-adjunctive language, by Proposition 5.8.23, it follows that $\forall x \in \sum^*$ there exist $u_0, v_0 \in \sum^*$ with $\lambda(u_0xv_0) > 0$. This implies that $\lambda(u_0yv_0) > 0$. Since $len(u_0xv_0) = len(u_0yv_0)$ and λ is 0-discrete, $u_0xv_0 = u_0yv_0$ and so $x = y$, a contradiction. Hence Supp(λ) must be an adjunctive language. If $\forall x, y \in \sum^*$ with $x \neq y$, $x, y \in \text{Supp}(\lambda)$, then it is clear that $\lambda(x) > 0$ and $\lambda(y) > 0$. Since λ is 0-discrete, $len(x) \neq len(y)$. Consequently, Supp(λ) is discrete. ■

F-Dense Languages

Definition 5.8.26 *Let* $\mu \in \mathcal{FP}(\sum^*)$ *and* $w \in \sum^*$. *Let*

$$\mu_a^w = \{x \mid x \in \sum^* w \sum^*, \mu(x) > a\}, \quad a \in [0, 1].$$

If $\forall w \in \sum^*$, $|\mu_a^w| \neq 0$, *then* μ *is called an* a-**dense language**. *In particular, a* 0*-dense language is called an* F**-dense language** *and we write* $\mu_0^w = \mu^w$.

Proposition 5.8.27 *Let* $\omega \in \mathcal{FP}(\sum^*)$. *The following statements are equivalent.*

(i) ω *is an* F*-dense language.*
(ii) $\forall w \in \sum^*$, $|\omega^w| = \infty$.
(iii) There exists an F*-adjunctive language* λ *with* $\omega \supseteq \lambda$.

Proof. (i) $\Rightarrow$ (ii) Immediate from Proposition 5.8.24.

(ii) $\Rightarrow$ (iii) Define an ordering relation on $\sum^*$ as follows: if $len(x) < len(y)$, then $x < y$; if $len(x) = len(y)$, then $x < y$ means that x, y is in the dictionary ordering of elements of $\sum$. Now

$$\sum^* = \{\Lambda < w_1 < w_2 < \cdots < w_n < \cdots\}.$$

Since $|\omega^w| = \infty$, $\forall w \in \sum^*$, $|\mathrm{Supp}(\omega)| = \infty$ and the number of elements in $\{[x]_\omega\}$ is greater than 1.

Let $x_1, x_2, \ldots, x_n, \ldots$ be the representative elements from these classes. Define $\lambda \in \mathcal{FP}(\sum^*)$ as follows:

$$\lambda(x) = \begin{cases} \omega(x) & \text{if } x = x_i \text{ for some } i \\ 0 & \text{if } x \neq x_i \text{ for all } i. \end{cases}$$

Clearly, $\lambda \subseteq \omega$. Now $\forall x, y \in \sum^*$, $x \neq y$ if $\lambda(x) > 0$, $\lambda(y) > 0$, then $len(x) \neq len(y)$ and so λ is 0-discrete. Since $|\omega^{w_1}| = \infty$, there exists an element $u_1 w_1 v_1$ with the shortest length in ω^{w_1}. We select $u_1 w_1 v_1$ as the representative element. Hence $\lambda(u_1 w_1 v_1) = \omega(u_1 w_1 v_1) > 0$. Similarly, since $|\omega^{w_2}| = \infty$, we can choose $u_2 w_2 v_2 \in \omega^{w_2}$ such that $\lambda(u_2 w_2 v_2) = \omega(u_2 w_2 v_2) > 0$ and $len(u_1 w_1 v_1) < len(u_2 w_2 v_2)$. Hence since $\forall i$ $|\omega^{w_i}| = \infty$, we can choose $u_i w_i v_i \in \omega^{w_i}$ such that $\lambda(u_i w_i v_i) = \omega(u_i w_i v_i) > 0$ and $len(u_{i-1} w_{i-1} v_{i-1}) < len(u_i w_i v_i)$. Hence $|\lambda^{w_i}| \neq 0$, $\forall i$. By Proposition 5.8.23, λ is an F-adjunctive language and $\omega \supseteq \lambda$.

(iii) $\Rightarrow$ (i) Now $\omega \supseteq \lambda$ implies that $|\omega^w| > |\lambda^w|$, $\forall w \in \sum^*$. Since λ is F-adjunctive, $|\lambda^w| \neq 0$ $\forall w \in \sum^*$ by Proposition 5.8.23. Thus $|\omega^w| \neq 0$ $\forall w \in \sum^*$. Hence ω is an F-dense language. ■

Proposition 5.8.28 *Let* $\mu \in \mathcal{FP}(\sum^*)$. *Then either* μ *or* $\overline{\mu}$ *must be an* F*-dense language.*

Proof. If $\forall w \in \sum^*$, $|\mu^w| = \infty$, then μ is F-dense. Suppose that there exists $w \in \sum^*$ such that $|\mu^w| < \infty$. Then $|\{x \mid x \in \sum^* w \sum^*, \mu(x) > 0\}| < \infty$. Thus $|\{x \mid x \in \sum^* wu \sum^*, \mu(x) > 0\}| < \infty$ $\forall u \in \sum^*$ and so $|\{x \mid x \in \sum^* wu \sum^*, \mu(x) = 0\}| = \infty$ $\forall u \in \sum^*$. This implies that $|\{x \mid x \in \sum^* wu \sum^*, \overline{\mu}(x) = 1 > 0\}| = \infty$. Hence $|(\overline{\mu})^{wu}| = \infty$ $\forall u \in \sum^*$.

Since $\sum^* wu \sum^* \subseteq \sum^* u \sum^*$, $\forall u \in \sum^*$, $|(\overline{\mu})^{wu}| \leq |(\overline{\mu})^u| = \infty$. Thus $\overline{\mu}$ is F-dense. ■

Proposition 5.8.29 *Let $\omega = \mu \cup \nu$. Then ω is a-dense if and only if μ or ν is a-dense.*

Proof. Suppose that ω is a-dense and ν is not a-dense. Then there exists $w_0 \in \sum^*$ such that

$$|\{x \mid x \in \textstyle\sum^* w_0 \sum^*, \nu(x) > a\}| < \infty.$$

This implies that $\forall u \in \sum^*$, $|\{x \mid x \in \sum^* w_0 u \sum^*,\ \nu(x) > a\}| < \infty$. Since ω is a-dense, $\forall w \in \sum^*$, $|\{x \mid x \in \sum^* w \sum^*,\ \omega(x) > a\}| = \infty$, i.e., $|\{x \mid x \in \sum^* w \sum^*,\ (\mu \cup \nu)(x) > a\}| = \infty\ \forall w \in \sum^*$. This is equivalent to $|\{x \mid x \in \sum^* w \sum^*,\ \mu(x) \vee \nu(x) > a\}| = \infty\ \forall w \in \sum^*$. This implies that $|\{x \mid x \in \sum^* w_0 u \sum^*,\ \mu(x) \vee \nu(x) > a\}| = \infty\ \forall u \in \sum^*$. From this, it follows that $|\{x \mid x \in \sum^* w_0 u \sum^*,\ \mu(x) > a\}| = \infty\ \forall u \in \sum^*$. Hence $\forall u \in \sum^*$, $|\mu_a^{w_0 u}| = \infty$ and so $|\mu_a^u| = \infty\ \forall u \in \sum^*$. Thus μ is a-dense.

Conversely, suppose that μ is a-dense. Now $\forall w \in \sum^*$, $|\{x \mid x \in \sum^* w \sum^*,\ \mu(x) > 0\}| = \infty$ and so $|\{x \mid x \in \sum^* w \sum^*,\ \mu(x) \vee \nu(x) > a\}| = \infty\ \forall w \in \sum^*$. This implies that $|\{x \mid x \in \sum^* w \sum^*,\ \omega(x) > a\}| = \infty$ $\forall w \in \sum^*$. Hence ω is a-dense. ■

Definition 5.8.30 *Let $\mu \in \mathcal{FP}(\sum^*)$ and $\Omega \subseteq \sum^*$. Then $\mu|_\Omega \in \mathcal{FP}(\sum^*)$ is called the **restriction** of μ on Ω if $\forall x \in \sum^*$,*

$$\mu|_\Omega(x) = \begin{cases} \mu(x), & x \in \Omega \\ 0, & x \notin \Omega. \end{cases}$$

Proposition 5.8.31 *Let $\omega \in \mathcal{FP}(\sum^*)$. Then ω is a-dense if and only if $\omega|_{\Sigma^* w \Sigma^*}$ is a-dense, $w \in \sum^*$.*

Proof. Suppose that ω is a-dense. Then $\forall w, u \in \sum^*$, $|\{x \mid x \in \sum^* wu \sum^*,$ $\omega(x) > a\}| \neq 0$. Thus there exists $x \in \sum^* wu \sum^* \subseteq \sum^* w \sum^* \cap \sum^* u \sum^*$ such that $\omega(x) > a$. Hence $\forall u \in \sum^*$, there exists $x \in \sum^* u \sum^*$ such that $\omega|_{\Sigma^* w \Sigma^*}(x) > a$. This imples that $\forall u \in \sum^*$,

$$|\{x \mid x \in \textstyle\sum^* u \sum^*, \omega|_{\Sigma^* w \Sigma^*}(x) > a\}| \neq 0.$$

Hence $\omega|_{\Sigma^* w \Sigma^*}$ is a-dense, $w \in \sum^*$.

Conversely, suppose that $\forall w \in \sum^*$, $\omega|_{\Sigma^* w \Sigma^*}$ is a-dense. Then $\forall w \in \sum^*$, $|\{x \mid x \in \sum^* w \sum^*,\ \omega|_{\Sigma^* w \Sigma^*}(x) > a\}| \neq 0$. Thus $\forall w \in \sum^*, |\{x \mid x \in \sum^* w \sum^*,\ \omega(x) > a\}| \neq 0$. Hence ω is a-dense. ■

Proposition 5.8.32 *Let* $|\sum| \geq 2$, $\omega = \mu \cup \nu$, *and* ω *be an* F*-adjunctive language. Then one of the following statements hold.*
(i) μ *or* ν *is an* F*-adjunctive language.*
(ii) μ *and* ν *are* F*-dense languages.*

Proof. Suppose that μ,ν are not F-adjunctive and ν is not F-dense. Then there exists $w \in \sum^*$ such that

$$|\{x \mid x \in \sum^* w \sum^*, \nu(x) > 0\}| < \infty.$$

Since μ is not F-adjunctive, there exist $u, v \in \sum^*$ such that $u \neq v$, $len(u) = len(v)$, and $u \equiv v(F_\mu)$. Now $|\nu^w| < \infty$ and so we may choose u such that $len(u) > \vee\{len(z) \mid z \in \nu^w\}$. Since $uw \equiv vw(F_\mu)$, $\forall x, y \in \sum^*$, $\mu(xuwy) = \mu(xvwy)$, and $len(xuwy) = len(xvwy) > len(u)$. Thus $xuwy$, $xuvy \notin \nu^w$ and so $\nu(xuwy) = \nu(xvwy) = 0$. Now $\forall x, y \in \sum^*$,

$$\begin{aligned} \omega(xuwy) &= \mu(xuwy) \vee \nu(xuwy) \\ &= \mu(xuwy) \\ &= \mu(xvwy) \\ &= \mu(xvwy) \vee \nu(xvwy) \\ & \quad \omega(xvwy). \end{aligned}$$

Hence $uw \equiv vw(F_\omega)$. Since ω is an F-adjunctive language, $uw = vw$, which contradicts the fact that $u \neq v$. Thus ν is F-dense. Similarly, it can be shown that μ is F-dense. ■

Proposition 5.8.33 *Let* λ *be an* F*-adjunctive language and* $w \in \sum^*$. *Then* $\lambda|_{\Sigma^* w \Sigma^*}$ *is an* F*-adjunctive language.*

Proof. Clearly, $\lambda = \lambda|_{\Sigma^* w \Sigma^*} \cup \lambda|_{\overline{\Sigma^* w \Sigma^*}}$. Let $x \in \sum^*$. Then

$$\begin{aligned} \lambda|_{\overline{\Sigma^* w \Sigma^*}}(x) &= \begin{cases} \lambda(x) & x \in \overline{\Sigma^* w \Sigma^*} \\ 0 & x \notin \overline{\Sigma^* w \Sigma^*}, \end{cases} \\ &= \begin{cases} \lambda(x) & x \notin \sum^* w \sum^* \\ 0 & x \in \sum^* w \sum^* \end{cases} . \end{aligned}$$

Clearly, $(\lambda|_{\overline{\Sigma^* w \Sigma^*}})^w$ is the empty set and so $\lambda|_{\overline{\Sigma^* w \Sigma^*}}$ is not F-dense. By Proposition 5.8.32, $\lambda|_{\Sigma^* w \Sigma^*}$ is an F-adjunctive language. ■

Proposition 5.8.34 *Let* $|\sum| \geq 2$, λ *be a* 0*-discrete* F*-adjunction language, and* $\lambda = \mu \cup \nu$. *Then* μ *or* ν *is an* F*-adjunctive language.*

Proof. If μ and ν are not F-adjunctive, then there exist $x_1, x_2, y_1, y_2 \in \sum^*$ such that $x_1 \neq x_2$, $y_1 \neq y_2$, $len(x_1) = len(x_2)$, $len(y_1) = len(y_2)$, and $x_1 \equiv x_2(F_\mu)$, $y_1 \equiv y_2(F_\nu)$. Clearly, we have $x_1 y_1 \equiv x_2 y_1 (F_\mu)$ and $x_1 y_1 \equiv x_1 y_2 (F_\nu)$. Since λ is F-adjunctive, there exist $u, v \in \sum^*$ such that $\lambda(ux_1y_1v) = \mu(ux_1y_1v) \vee \nu(ux_1y_1v) > 0$ by Proposition 5.8.23. If $\mu(ux_1y_1v) \geq \nu(ux_1y_1v)$, then

$$\lambda(ux_1y_1v) = \mu(ux_1y_1v) = \mu(ux_2y_1v) > 0$$

and furthermore $\lambda(ux_2y_1v) = \mu(ux_2y_1v) \vee \nu(ux_2y_1v) > 0$. Since λ is 0-discrete, $len(ux_1y_1v) = len(ux_2y_1v)$ implies that $ux_1y_1v = ux_2y_1v$, which contradicts that $x_1 \neq x_2$. Similarly, if $\mu(ux_1y_1v) \leq \nu(ux_1y_1v)$, then we can show that $y_1 = y_2$, a contradiction. Hence μ or ν is an F-adjunctive language. ■

Proposition 5.8.35 *Every 0-discrete F-adjunctive language may be resolved into the adjoint union of two 0-discrete F-adjunctive languages.*

Proof. Let ω be an 0-discrete F-adjunctive language. Let $L \subseteq \sum^*$ be an adjunctive language and let

$$B_1 = \text{Supp}(\omega) \cap \textstyle\sum^* L \sum^*,$$

$$B_2 = \text{Supp}(\omega) \cap \textstyle\sum^* \overline{L} \sum^*.$$

Then $B_1 \cup B_2 = \text{Supp}(\omega)$, $B_1 \cap B_2 = \phi$, and $\omega = \omega|_{B_1} \cup \omega|_{B_2}$, $\omega|_{B_1} \cap \omega|_{B_2} = \phi$. Now since ω is 0-discrete, $\omega|_{B_1}$ and $\omega|_{B_2}$ are also 0-discrete. Since L is an adjunctive language, $\forall w \in \sum^*$ there exist $u, v \in \sum^*$ such that $uwv \in L$. Also since ω is F-adjunctive and $uwv \in \sum^*$, there exist $s, t \in \sum^*$ with $\omega(suwvt) > 0$. This implies that $x = suwvt \in \text{Supp}(\omega) \cap \sum^* L \sum^* = B_1$. Thus $\forall w \in \sum^*$, $|(\omega|_{B_1})^w| \neq 0$. Hence $\omega|_{B_1}$ is F-dense. By Proposition 5.8.23, $\omega|_{B_1}$ is F-dense. Similarly, $\omega|_{B_2}$ is also F-dense since the adjunctivity of L implies that $\overline{L}$ is adjunctive. ■

Proposition 5.8.36 *Every F-dense language may be resolved into the adjoint union of an F-adjunctive language and an 0-discrete F-adjunctive language.*

Proof. Let ω be an F-dense. First suppose that ω is not an F-adjunctive.

Case 1: $|\sum| > 2$. Let $\sum^* = \{u_1, u_2, \ldots, u_n, \ldots\}$. Since ω is F-dense, $\forall u_i$, $|\{x \mid x \in \sum^* u_i \sum^*, \omega(x) > 0\}| = \infty$. Choose $x_i = s_i u_i t_i \in \omega^{u_i}$, $i = 1, 2, \ldots$ such that

$$len(x_1) < len(x_2) < \cdots < len(x_n) < \cdots.$$

Let $B = \{x_1, x_2, \ldots, x_n, \ldots\}$. By Definition 5.8.21 and Proposition 5.8.23, $\omega|_B$ is an 0-discrete F-adjunctive language. Next, we show that $\omega|_{\overline{B}}$ is F-adjunctive. Supppose that $\omega|_{\overline{B}}$ is not F-adjunctive.

Since ω is not F-adjunctive, there exist v_1, v_2 such that $v_1 \neq v_2$, $len(v_1) = len(v_2)$, and $v_1 \equiv v_2(F_\omega)$. Also, since $\omega|_{\overline{B}}$ is not F-adjunctive, there exist

w_1, w_2 such that $w_1 \neq w_2$, $len(w_1) = len(w_2)$, and $w_1 \equiv w_2(F_{\omega|_{\overline{B}}})$. Clearly, $v_1w_1 \equiv v_1w_2(F_{\omega|_{\overline{B}}})$. Since $\omega|_B$ is F-adjunctive, $v_1w_1 \not\equiv v_1w_2(F_{\omega|_B})$. Thus there exist $s, t \in \sum^*$ such that $\omega|_B(sv_1w_1t) \neq \omega|_B(sv_1w_2t)$. Now either $\omega|_B(sv_1w_1t) > \omega|_B(sv_1w_2t)$ or $\omega|_B(sv_1w_1t) < \omega|_B(sv_1w_2t)$. To be specific, suppose that $\omega|_B(sv_1w_1t) > \omega|_B(sv_1w_2t)$. Clearly, $sv_1w_1t \in B$. If $\omega|_B(sv_1w_2t) > 0$, then $\omega(sv_1w_1t) > \omega(sv_1w_2t) > 0$, i.e., $v_1w_1 \equiv v_1w_2(F_\omega)$. If $\omega|_B(sv_1w_2t) = 0$ and $sv_1w_2t \in B$, then $\omega(sv_1w_2t) = 0$ and $\omega(sv_1w_1t) > \omega(sv_1w_2t)$, i.e., $v_1w_1 \not\equiv v_1w_2(F_\omega)$. If $\omega|_B(sv_1w_2t) = 0$ and $sv_1w_2t \notin B$, then $sv_1w_2t \in \overline{B}$. From $v_1w_1 \equiv v_1w_2(F_{\omega|_{\overline{B}}})$ and $sv_1w_1t \notin \overline{B}$ we obtain that $\omega(sv_1w_2t) = \omega|_{\overline{B}}(sv_1w_2t) = \omega|_{\overline{B}}(sv_1w_1t) = 0$. Hence $v_1w_1 \not\equiv v_1w_2(F_\omega)$. Therefore, it always holds that $v_1w_1 \not\equiv v_1w_2(F_\omega)$, i.e.,

$$\omega(sv_1w_1t) > \omega(sv_1w_2t).$$

However, $v_1 \equiv v_2(F_\omega)$, $v_1w_1 \equiv v_2w_1(F_\omega)$, $v_1w_2(F_\omega)$, and furthermore, $\omega(sv_2w_1t) > \omega(sv_2w_2t)$. Since $sv_1w_1t \neq sv_2w_1t$, $len(sv_1w_1t) = len(sv_2w_1t)$, $sv_1w_1t \in B$, and by the definition of B, $sv_2w_1t \notin B$, i.e., $sv_2w_1t \in \overline{B}$. Consequently,

$$\omega(sv_2w_1t) = \omega|_{\overline{B}}(sv_2w_1t) > \omega(sv_2w_2t) \geq \omega|_{\overline{B}}(sv_2w_2t),$$

i.e., $w_1 \not\equiv w_2(F_{\omega|_{\overline{B}}})$, a contradiction. Hence, $\omega|_{\overline{B}}$ is an F-adjunctive language, $\omega = \omega|_B \cup \omega|_{\overline{B}}$, and $\omega|_B \cap \omega|_{\overline{B}} = \phi$.

Case 2: $|\sum| = 1$. Since ω is not F-adjunctive, by Definition 5.7.6, ω is an F-regular language. If $\omega|_{\overline{B}}$ is not F-adjunctive, then $\omega|_{\overline{B}}$ is F-regular. Since $\omega = \omega|_{\overline{B}} \cup \omega|_B, \omega_a = (\omega|_B)_a \cup (\omega|_{\overline{B}})_a, \forall a \in [0,1]$. Clearly, ω_a and $(\omega|_{\overline{B}})_a$ are regular languages, $\forall a \in [0,1]$. Since $(\omega|_B)_a \cap (\omega|_{\overline{B}})_a = \phi, \forall a \in [0,1]$, $(\omega|_B)_a$ is a regular language. Also since ω is F-regular, $|\{\omega(x) | \ x \in \sum^*\}| < \infty$ and $|\{(\omega|_B)(x) | \ x \in \sum^*\}| < \infty$. By Proposition 5.8.2, $\omega|_B$ is an F-regular language. This contradicts that $\omega|_B$ is an F-adjunctive language. Hence $\omega|_{\overline{B}}$ is also an F-adjunctive language.

Suppose that ω is an F-adjunctive language. Now $\omega = \omega|_B \cup \omega|_{\overline{B}}$, where $\omega|_B$ is an 0-discrete F-adjunctive language. By Proposition 5.8.14, $\omega|_B = \omega|_{B_1} \cup \omega|_{B_2}$ and $\omega|_{B_1} \cap \omega|_{B_2} = \phi$, where $\omega|_{B_1}$ and $\omega|_{B_2}$ are two 0-discrete F-adjunctive languages. Thus

$$\omega = \omega|_{B_1} \cup \omega|_{B_2} \cup \omega|_{\overline{B}}.$$

If $\omega|_{B_2} \cup \omega|_{\overline{B}}$ is an F-adjunctive language, then the result is true. Suppose that $\omega|_{B_2} \cup \omega|_{B^c}$ is not F-adjunctive. Then $\omega|_{B_2}$ is F-adjunctive. By Proposition 5.8.27, $\omega|_{B_2} \cup \omega|_{\overline{B}}$ is an F-dense language. From the preceding process we know that $\omega|_{\overline{B}}$ must be an F-adjunctive language. Since $\omega = \omega|_B \cup \omega|_{\overline{B}}$, the desired result follows. ■

5.9 References

1. Bacon, G. C., Minimal-state stochastic finite systems, *IEEE Trans. Computers,* 11 (1964), 107-108.

2. Bellman, R.E., Kalaba, R., and Zadeh, L.A., Abstraction and pattern classification, *J. Math Appl.,* 13 (1966) 1-7.

3. Bellman, R.E. and Zadeh, L.A., Decision-making in a fuzzy environment, *Management Sci.,* 17 (1970) 141-164.

4. Booth, T.L., *Sequential Machines and Automata Theory,* Wiley, New York, 1968.

5. Carlyle, J.W., Reduced forms for stochastic sequential machines, *J. Math. Anal. Appl.,* 7 (1963), 167-175.

6. Fu, K. S. and Huang, T., Stochastic grammars and languages, *J. Comput. Inform. Sci.,* 1 (1972) 135-170.

7. Ginsburg, S., *The Mathematical Theory of Context-free Languages,* McGraw-Hill, New York, 1966.

8. Ginsburg, S. and Greibach, S., Abstract family of languages, In: Studies in Abstract Families of Languages, *Amer. Math. Soc. Mem.,* 87 (1969) 1-32.

9. Hopcroft, J. E. and Ullman, J. E., *Formal Languages and Their Relations to Automata,* Addison-Wesley, Reading, MA, 1969.

10. Lee, E.T. and Zadeh, L.A., Fuzzy languages and their acceptance by automata, In: 4th Princeton Conf. Information Science and Systems, 1970, p.399.

11. Lee, E.T. and Zadeh, L.A., Note on fuzzy languages, *Inform. Sci.,* 1 (1969) 421-434.

12. Negoita, C.V. and Ralescu, D.A., *Application of Fuzzy Sets to Systems Analysis.* Halsted Press, New York, 1975.

13. Paz, A., Homomorphisms between stochastic sequential machines and related problems, *Math Systems Theory,* 2 (1968) 223-245.

14. Salomaa, A., Probabilistic and weighted grammars, *Inform. Contr.,* 15 (1969) 529-544.

15. Santos, E.S., Maximin automata, *Information Control,* 13 (1968) 363-377.

16. Santos, E.S., Maximin, minimax, and composite sequential machines, *J. Math. Anal. Appl.*, 24 (1968), 246-259.

17. Santos, E.S., Maximin sequential-like machines and chains, *Math. System Theory*, 3 (1969) 300-309.

18. Santos, E.S., Max-product machines, *J. Math. Anal. Appl.* 37 (1972), 677-686.

19. Santos, E.S., On reductions of maxmin machines, *J. Math. Anal. Appl.*, 40 (1972) 60-78.

20. Santos, E. S., Probabilistic grammars and automata, *Inform. Contr.*, 21 (1972) 27-47.

21. Santos, E. S., Realizations of fuzzy languages by probabilistic, max-product and maxmin automata, *Inform. Sci.*, 8 (1975).

22. Santos, E. S., Max-product grammars and languages, *Inform. Sci.*, 9 (1975) 1-23.

23. Santos, E.S., Context-free fuzzy languages, *Inform. Cont.*, 26 (1974) 1-11.

24. Santos, E.S., Fuzzy automata and languages. *Inform. Sci.* 10 (1976) 193-197.

25. E. S. Santos and Wee, W. G., General formulation of sequential machines, *Information Control*, 12 (1968) 5-10.

26. Shen, J. Fuzzy language on free monoid, *Inf. Sci.*, 88 (1996) 149-168.

27. Shyr, H.J., *Free Monoids and Languages.* Taipei, Taiwan, 1979.

28. Thomason, M.G. and Marinos, P.N., Deterministic acceptors of regular fuzzy languages, *IEEE Trans. Sys., Man, and Cyber.*, (1974) 228-230.

29. Wee,W.G., and Fu, K.S., A formulation of fuzzy automata and its applications as a model of learning systems, *IEEE Trans. Syst. Sci. Cybern.*, 5 (1969) 215-223.

30. Zadeh, L.A., Fuzzy sets and systems, In: Symposium of System Theory, Polytechnic Institute of Brooklyn, New York, 1965, 29-37.

31. Zadeh, L.A., Fuzzy sets, *Information Control*, 8 (1965) 338-353.

32. Zadeh, L.A., Fuzzy languages and their relation to human and machine intelligence, Electron. Res. Lab., Univ. California, Berkeley, Rep. ERL-M302, Aug. 1971.

6
Algebraic Fuzzy Automata

A **fuzzy finite state machine (ffsm)** is a triple $M = (Q, X, \mu)$, where Q and X are finite nonempty sets and μ is a fuzzy subset of $Q \times X \times Q$, i.e., $\mu : Q \times X \times Q \to [0, 1]$. Let X^* denote the set of all words of elements of X of finite length .

Q is called the set of **states** and X is called the set of **input symbols**. Let Λ denote the empty word in X^* and $len(x)$ denote the length of x $\forall x \in X^*$. As previously noted, X^* is a semigroup with identity Λ with respect to the binary operation concatenation of two words.

6.1 Semigroups of Fuzzy Finite State Machines

Definition 6.1.1 *Let $M = (Q, X, \mu)$ be a ffsm. Define $\mu^* : Q \times X^* \times Q \to [0, 1]$ by*

$$\mu^*(q, \Lambda, p) = \begin{cases} 1 \text{ if } q = p \\ 0 \text{ if } q \neq p \end{cases}$$

and

$$\mu^*(q, xa, p) = \vee\{\mu^*(q, x, r) \wedge \mu(r, a, p) | r \in Q\}$$

$\forall\ x \in X^*, a \in X.$

Lemma 6.1.2 *Let $M = (Q, X, \mu)$ be a ffsm. Then*

$$\mu^*(q, xy, p) = \vee\{\mu^*(q, x, r) \wedge \mu^*(r, y, p) | r \in Q\}$$

$\forall q, p \in Q$ *and* $\forall x, y \in X^*$.

Proof. Let $q, p \in Q$ and $x, y \in X^*$. We prove the result by induction on $len(y) = n$. If $n = 0$, then $y = \Lambda$ and hence $xy = x\Lambda = x$. Thus $\vee\{\mu^*(q,x,r) \wedge \mu^*(r,y,p) | r \in Q\} = \vee\{\mu^*(q,x,r) \wedge \mu^*(r,\Lambda,p) | r \in Q\} = \mu^*(q,x,p) = \mu^*(q,xy,p)$ by the definition of μ^*. Thus the result is true for $n = 0$. Suppose the result is true for all $u \in X^*$ such that $len(u) = n-1, n > 0$. Let $y = ua$ where $u \in X^*$, $a \in X$ and $len(u) = n-1, n > 0$. Now $\mu^*(q,xy,p) = \mu^*(q,xua,p) = \vee\{\mu^*(q,xu,r) \wedge \mu(r,a,p) | r \in Q\} = \vee\{(\vee\{\mu^*(q,x,s) \wedge \mu^*(s,u,r) | s \in Q\}) \wedge \mu(r,a,p) | r \in Q\} = \vee\{\vee\{\mu^*(q,x,s) \wedge \mu^*(s,u,r) \wedge \mu(r,a,p)\} | \, r, s \in Q\} = \vee\{\mu^*(q,x,s) \wedge (\vee\{\mu^*(s,u,r) \wedge \mu(r,a,p) | r \in Q\}) | s \in Q\} = \vee\{\mu^*(q,x,s) \wedge \mu^*(s,ua,p) | s \in Q\} = \vee\{\mu^*(q,x,s) \wedge \mu^*(s,y,r) | \, s \in Q\}$. Thus the result is true for $len(y) = n$. ■

Define a relation $\equiv$ on X^* by $\forall x, y \in X^*$, $x \equiv y$ if and only if $\mu^*(q,x,p) = \mu^*(q,y,p) \; \forall \; q, p \in Q$.

Clearly $\equiv$ is an equivalence relation on X^*. Let $z \in X^*$ and let $x \equiv y$. Then $\forall p, q \in Q$, $\mu^*(q,xz,p) = \vee\{\mu^*(q,x,r) \wedge \mu^*(r,z,p) | r \in Q\} = \vee\{\mu^*(q,y,r) \wedge \mu^*(r,z,p) | r \in Q\} = \mu^*(q,yz,p)$. Thus $xz \equiv yz$. Similarly $zx \equiv zy$. Thus $\equiv$ is a congruence relation on the semigroup X^*. We have thus proved the following result.

Theorem 6.1.3 *Let* $M = (Q, X, \mu)$ *be a ffsm. Define a relation* $\equiv$ *on* X^* *by* $\forall x, y \in X^*$, $x \equiv y$ *if and only if* $\mu^*(q,x,p) = \mu^*(q,y,p) \; \forall \; q, p \in Q$. *Then* $\equiv$ *is a congruence relation on* X^*. ■

Let $x \in X^*$, $[x] = \{y \in X^* | x \equiv y\}$ and $E(M) = \{[x] | x \in X^*\}$.

Theorem 6.1.4 *Let* $M = (Q, X, \mu)$ *be a ffsm. Define a binary operation* $*$ *on* $E(M)$ *by* $\forall \; [x], [y] \in E(M), [x] * [y] = [xy]$. *Then* $(E(M), *)$ *is a finite semigroup with identity.*

Proof. Clearly $*$ is well defined and associative. Now $[x] * [\Lambda] = [x\Lambda] = [x] = [\Lambda x] = [\Lambda] * [x] \; \forall \; [x] \in E(M)$. Thus $[\Lambda]$ is the identity of $(E(M), *)$. Hence $(E(M), *)$ is a semigroup with identity. Let $x \in X^*$ and let $x = x_1 x_2 \ldots x_n$, where $x_1, x_2, \ldots, x_n \in X$. Then $\forall \; q, p \in Q$,

$$\mu^*(q,x,p) = \vee\{\mu(q,x_1,q_1) \wedge \mu(q_1,x_2,q_2) \wedge \ldots \wedge \mu(q_{n-1},x_n,p) \mid q_1, q_2, \ldots, q_{n-1} \in Q\}.$$

Hence since $\text{Im}(\mu)$ is finite, $\text{Im}(\mu^*)$ is finite. Thus $(E(M), *)$ is a finite semigroup with identity. ■

We now define another type of congruence relation on X^*. Let $x, y \in X^*$. Define $x \approx y$ if and only if $(\forall s, t \in Q, \mu^*(s,x,t) > 0 \Leftrightarrow \mu^*(s,y,t) > 0)$. Clearly $\approx$ is an equivalence relation on X^*. Let $z \in X^*$ and $x \approx y$. Then $\forall s, t \in Q$, $\mu^*(s,zx,t) = \vee\{\mu^*(s,z,r) \wedge \mu^*(r,x,t) | r \in Q\} > 0$ if and only if

$\exists\, r \in Q$ such that $\mu^*(s,z,r) \wedge \mu^*(r,x,t) > 0$ if and only if $\exists\, r \in Q$ such that $\mu^*(s,z,r) \wedge \mu^*(r,y,t) > 0$ if and only if $\mu^*(s,zy,t) = \vee\{\mu^*(s,z,r) \wedge \mu^*(r,y,t) | r \in Q\} > 0$. Hence $zx \approx zy$. Similarly $xz \approx yz$. Hence $\approx$ is a congruence relation on X^*. So, we have the following theorem.

Theorem 6.1.5 *Let $M = (Q, X, \mu)$ be a ffsm. Let $x, y \in X^*$. Define a relation $\approx$ on X^* by $x \approx y$ if and only if $\forall s,t \in Q, \mu^*(s,x,t) > 0 \Leftrightarrow \mu^*(s,y,t) > 0$. Then $\approx$ is a congruence relation on X^*.* ■

Let $x \in X^*$ and let $[\![x]\!] = \{y \in X^* | x \approx y\}$. Let $\widetilde{E(M)} = \{[\![x]\!] | x \in X^*\}$.

Theorem 6.1.6 *Let $M = (Q, X, \mu)$ be a ffsm. Define a binary operation $\tilde{*}$ on $\widetilde{E(M)}$ by $\forall\, [\![x]\!], [\![y]\!] \in \widetilde{E(M)}$, $[\![x]\!] \tilde{*} [\![y]\!] = [\![xy]\!]$. Then $(\widetilde{E(M)}, \tilde{*})$ is a finite semigroup with identity and $[x] \to [\![x]\!]$ is a homomorphism of $E(M)$ onto $\widetilde{E(M)}$.*

Proof. Clearly $(\widetilde{E(M)}, \tilde{*})$ is a semigroup with identity. Define $f : E(M) \to \widetilde{E(M)}$ by $f([x]) = [\![x]\!]\ \forall\ [x] \in E(M)$. Let $x, y \in X^*$ and $[x] = [y]$. Then $\forall s,t \in Q, \mu^*(s,x,t) = \mu^*(s,y,t)$. Thus $\forall s,t \in Q, \mu^*(s,x,t) > 0 \Leftrightarrow \mu^*(s,y,t) > 0$. Hence $x \approx y$ or $[\![x]\!] = [\![y]\!]$. Thus f is well defined. Clearly f is an onto homomorphism. Now since $E(M)$ is finite $\widetilde{E(M)}$ is finite. ■

Definition 6.1.7 *Let $M = (Q, X, \mu)$ be a ffsm. For all $x \in X^*$ define the fuzzy subset x^M of $Q \times Q$ by $x^M(s,t) = \mu^*(s,x,t)\ \forall s,t \in Q$.*

Definition 6.1.8 *Let P, R and T be nonempty sets. Let f be a fuzzy subset of $P \times R$ and g be a fuzzy subset of $R \times T$. Define the fuzzy subset $f \circ g$ of $R \times T$ by $(f \circ g)(p,t) = \vee\{f(p,r) \wedge g(r,t) | r \in R\}\ \forall p \in P$ and $\forall t \in T$.*

Theorem 6.1.9 *Let $M = (Q, X, \mu)$ be a ffsm. Let $S_M = \{x^M | x \in X^*\}$. Then*

(i) $x^M \circ y^M = (xy)^M\ \forall x, y \in X^*$,

(ii) $(S_M, \circ)$ is a finite semigroup with identity, where $\circ$ is defined as in Definition 6.1.8.

Proof. (i) Let $s,t \in Q$. Then $(xy)^M(s,t) = \mu^*(s,xy,t) = \vee\{\mu^*(s,x,q) \wedge \mu^*(q,y,t) | q \in Q\} = \vee\{x^M(s,q) \wedge y^M(q,t) | q \in Q\} = (x^M \circ y^M)(s,t)$. Thus $(xy)^M = x^M \circ y^M$.

(ii) Clearly $(S_M, \circ)$ is a finite semigroup with identity, where Λ^M is the identity element. S_M is finite since Q and $\mathrm{Im}(\mu)$ are finite. ■

Theorem 6.1.10 *Let $M = (Q, X, \mu)$ be a ffsm. Then $S_M \simeq E(M)$.*

Proof. Define $f : S_M \to E(M)$ by $f(x^M) = [x]\ \forall x^M \in S_M$. Let $x^M, y^M \in S_M$. Then $x^M = y^M$ if and only if $x^M(s,t) = y^M(s,t)\ \forall s,t \in Q$

if and only if $\mu^*(s, x, t) = \mu^*(s, y, t)\ \forall s, t \in Q$ if and only if $[x] = [y]$. Thus f is single valued and one-one. Now

$$f(x^M \circ y^M) = f((xy)^M) = [xy] = [x] * [y] = f(x^M) * f(y^M).$$

Thus f is a homomorphism. Clearly f is onto. Hence $S_M \simeq E(M)$. ∎

Let $M = (Q, X, \mu)$ be a ffsm. The index of an equivalence relation is the number of distinct equivalence classes. Let $\sim$ be a congruence relation of finite index on X^*. Let $x \in X^*$ and $\prec x \succ = \{y \in X^* | x \sim y\}$. Let $\widetilde{Q} = \{\prec x \succ\ | x \in X^*\}$. Define $\sigma : \widetilde{Q} \times X \times \widetilde{Q} \to [0, 1]$ by $\forall \prec x \succ \in \widetilde{Q}$ and $\forall\ a \in X$, $\sigma(\prec x \succ, a, \prec xa \succ)$ an arbitrary fixed element in $(0, 1]$ and $\forall \prec x \succ, \prec w \succ \in \widetilde{Q}$,

$$\sigma(\prec x \succ, a, \prec w \succ) = \begin{cases} \sigma(\prec x \succ, a, \prec xa \succ) \text{ if } w \sim xa \\ 0 \text{ otherwise} \end{cases}$$

Let $\prec x \succ, \prec y \succ, \prec u \succ, \prec v \succ \in \widetilde{Q}$ and $a, b \in X$. Suppose that $(\prec x \succ, a, \prec u \succ) = (\prec y \succ, b, \prec v \succ)$. Then $\prec x \succ = \prec y \succ$, $a = b$, $\prec u \succ = \prec v \succ$. Now $u \sim xa$ if and only if $v \sim ya$. Thus $\sigma(\prec x \succ, a, \prec u \succ) = \sigma(\prec y \succ, b, \prec v \succ)$. Hence σ is single valued. Thus $\widetilde{M} = (\widetilde{Q}, X, \sigma)$ is ffsm. Now extend σ to σ^* as μ was extended to μ^* in Definition 6.1.1.

Lemma 6.1.11 *Let $\widetilde{M}$ be as above and let $\prec z \succ, \prec w \succ \in \widetilde{Q}$. Then the following assertions hold.*

(i) $\forall x \in X^*$, if $\sigma^*(\prec z \succ, x, \prec w \succ) > 0$, then $\prec zx \succ = \prec w \succ$.

(ii) $\sigma^*(\prec z \succ, x, \prec zx \succ) > 0\ \forall\ z, x \in X^*$.

Proof. (i) Let $x \in X^*$ and $len(x) = n$. If $n = 0$, then $x = \Lambda$. Hence if $\sigma^*(\prec z \succ, x, \prec w \succ) > 0$, then $\prec z \succ = \prec w \succ$ or $\prec zx \succ = \prec w \succ$. Suppose the result is true $\forall\ y \in X^*$ such that $len(y) = n - 1$, $n > 0$. Let $x = ya$ where $y \in X^*$, $a \in X$ and $len(y) = n - 1$. Let

$$\sigma^*(\prec z \succ, ya, \prec w \succ) = \sigma^*(\prec z \succ, x, \prec w \succ) > 0.$$

Now $\sigma^*(\prec z \succ, ya, \prec w \succ) = \vee\{\sigma^*(\prec z \succ, y, \prec q \succ) \wedge \sigma(\prec q \succ, a, \prec w \succ) \mid \prec q \succ \in \widetilde{Q}\} > 0$. Hence $\sigma^*(\prec z \succ, y, \prec q \succ) > 0$ and $\sigma^*(\prec q \succ, a, \prec w \succ) > 0$ for some $\prec q \succ \in \widetilde{Q}$. Thus by the induction hypothesis, $\prec zy \succ = \prec q \succ$ and $\prec qa \succ = \prec w \succ$. Hence $\prec zx \succ = \prec zya \succ = \prec qa \succ = \prec w \succ$. The result now follows by induction.

(ii) Let $z, x \in X^*$ and $len(x) = n$. If $n = 0$, then $x = \Lambda$. Hence $\sigma^*(\prec z \succ, x, \prec zx \succ) > 0$. Suppose the result is true $\forall\ y \in X^*$ such that $len(y) = n - 1, n > 0$. Let $x = ya$ where $y \in X^*$, $a \in X$ and $len(y) = n - 1$. Now $\sigma^*(\prec z \succ, x, \prec zx \succ) = \sigma^*(\prec z \succ, ya, \prec zya \succ) = \vee\{\sigma^*(\prec z \succ, y, \prec q \succ) \wedge \sigma(\prec q \succ, a, \prec zya \succ) | \prec q \succ \in \widetilde{Q}\} \geq \sigma^*(\prec z \succ, y, \prec zy \succ) \wedge \sigma(\prec zy \succ, a, \prec zya \succ) > 0$. The result now follows by induction. ∎

Let $x, y \in X^*$. Suppose $x \sim y$. Let $\prec z \succ, \prec w \succ \in \widetilde{Q}$. Suppose $\sigma^*(\prec z \succ, x, \prec w \succ) > 0$. Then $\prec zx \succ = \prec w \succ$. Since $x \sim y$ and $\sim$ is a congruence relation, $\prec zx \succ = \prec zy \succ$. Thus $\prec zy \succ = \prec w \succ$. Hence $\sigma^*(\prec z \succ, y, \prec w \succ) > 0$. Similarly if $\sigma^*(\prec z \succ, y, \prec w \succ) > 0$, then $\sigma^*(\prec z \succ, x, \prec w \succ) > 0$. Hence $x \approx y$. Conversely, suppose $x \approx y$. Let $\prec z \succ \in \widetilde{Q}$. Now $\sigma^*(\prec z \succ, x, \prec zx \succ) > 0$. Hence $\sigma^*(\prec z \succ, y, \prec zx \succ) > 0$. Thus $\prec zy \succ = \prec zx \succ$. Choose $z = \Lambda$. Then $\prec x \succ = \prec y \succ$ or $x \sim y$. Thus $x \sim y$ if and only if $x \approx y$. We summarize the above discussion in the following theorem.

Theorem 6.1.12 *A fuzzy finite state machine M can be constructed from a given congruence relation $\sim$ on X^* of finite index in such a way that $\sim$ is the same congruence relation $\approx$ as on M.* ■

6.2 Homomorphisms

Definition 6.2.1 *Let $M_1 = (Q_1, X_1, \mu_1)$ and $M_2 = (Q_2, X_2, \mu_2)$ be two ffsms. A pair (α, β) of mappings, $\alpha : Q_1 \to Q_2$ and $\beta : X_1 \to X_2$, is called a* ***homomorphism****, written $(\alpha, \beta) : M_1 \to M_2$, if $\mu_1(q, x, p) \leq \mu_2(\alpha(q), \beta(x), \alpha(p))\ \forall\, q, p \in Q_1$ and $\forall\, x \in X_1$.*

The pair (α, β) is called a ***strong homomorphism*** *if*

$$\mu_2(\alpha(q), \beta(x), \alpha(p)) = \vee\{\mu_1(q, x, t) | t \in Q_1,\ \alpha(t) = \alpha(p)\}$$

$\forall\, q, p \in Q_1$ *and* $\forall\, x \in X_1$.

A homomorphism (strong homomorphism) $(\alpha, \beta) : M_1 \to M_2$ is called an **isomorphism (strong isomorphism)** if α and β are both one-one and onto.

Remark 6.2.2 *(i) In Definition 6.2.1, if $X_1 = X_2$ and β is the identity map, then we simply write $\alpha : M_1 \to M_2$ and say that α is a homomorphism or strong homomorphism accordingly.*

(ii) If (α, β) is a strong homomorphism with α one-one, then

$$\mu_2(\alpha(q), \beta(x), \alpha(p)) = \mu_1(q, x, p)$$

$\forall\, q, p \in Q_1$ *and* $\forall\, x \in X_1$.

Lemma 6.2.3 *Let $M_1 = (Q_1, X_1, \mu_1)$ and $M_2 = (Q_2, X_2, \mu_2)$ be two ffsms. Let $(\alpha, \beta) : M_1 \to M_2$ be a strong homomorphism. Then $\forall q, r \in Q_1$, $\forall x \in X_1$, if $\mu_2(\alpha(q), \beta(x), \alpha(r)) > 0$, then $\exists\, t \in Q_1$ such that $\mu_1(q, x, t) > 0$ and $\alpha(t) = \alpha(r)$. Furthermore, $\forall p \in Q$ if $\alpha(p) = \alpha(q)$, then $\mu_1(q, x, t) \geq \mu_1(p, x, r)$.*

Proof. Let $p, q, r \in Q_1$, $x \in X_1$ and $\mu_2(\alpha(q), \beta(x), \alpha(r)) > 0$. Then

$$\vee\{\mu_1(q, x, s) | s \in Q_1,\ \alpha(s) = \alpha(r)\} > 0.$$

Since Q_1 is finite, $\exists\ t \in Q_1$ such that $\alpha(t) = \alpha(r)$ and $\mu_1(q, x, t) = \vee\{\mu_1(q, x, s) | s \in Q_1,\ \alpha(s) = \alpha(r)\} > 0$. Suppose $\alpha(p) = \alpha(q)$. Then

$$\mu_1(q, x, t) = \mu_2(\alpha(q), \beta(x), \alpha(r)) = \mu_2(\alpha(p), \beta(x), \alpha(r)) \geq \mu_1(p, x, r). \blacksquare$$

Definition 6.2.4 *Let $M_1 = (Q_1, X_1, \mu_1)$ and $M_2 = (Q_2, X_2, \mu_2)$ be two ffsms. Let $(\alpha, \beta) : M_1 \to M_2$ be a homomorphism. Define $\beta^* : X_1^* \to X_2^*$ by $\beta^*(\Lambda) = \Lambda$ and $\beta^*(ua) = \beta^*(u)\beta(a)\ \forall\ u \in X_1^*,\ a \in X_1$.*

Lemma 6.2.5 *Let M_1, $M_2, (\alpha, \beta)$ and β^* be as above. Then $\beta^*(uv) = \beta^*(u)\beta^*(v)\ \forall\ u, v \in X_1^*$.*

Proof. Let $u, v \in X_1^*$ and $len(v) = n$. If $n = 0$, then $v = \Lambda$ and hence $\beta^*(uv) = \beta^*(u) = \beta^*(u)\beta^*(v)$. Suppose now the result is true $\forall\ y \in X_1^*$ such that $len(y) = n - 1, n > 0$. Let $v = ya$ where $y \in X_1^*, a \in X_1$ and $len(y) = n - 1$. Then $\beta^*(uv) = \beta^*(uya) = \beta^*(uy)\beta(a) = \beta^*(u)\beta^*(y)\beta(a)$ $= \beta^*(u)\beta^*(ya) = \beta^*(u)\beta^*(v)$. The result now follows by induction. $\blacksquare$

Theorem 6.2.6 *Let M_1, M_2 be as above. Let $(\alpha, \beta) : M_1 \to M_2$ be a homomorphism. Then $\mu_1^*(q, x, p) \leq \mu_2^*(\alpha(q), \beta^*(x), \alpha(p))\ \forall\ q, p \in Q_1$ and $x \in X_1^*$.*

Proof. Let $q, p \in Q_1$ and $x \in X_1^*$. We prove the result by induction on $len(x) = n$. If $n = 0$, then $x = \Lambda$ and $\beta^*(x) = \beta^*(\Lambda) = \Lambda$. Now if $q = p$, then $\mu_1^*(q, \Lambda, p) = 1 = \mu_2^*(\alpha(q), \Lambda, \alpha(p))$. If $q \neq p$, then $\mu_1^*(q, \Lambda, p) = 0 \leq \mu_2^*(\alpha(q), \Lambda, \alpha(p))$. Suppose now the result is true $\forall\ y \in X^*$ such that $len(y) = n-1, n > 0$. Let $x = ya$ where $y \in X_1^*$, $a \in X_1$ and $len(y) = n-1$. Now

$$\begin{aligned}
\mu_1^*(q, x, p) &= \mu_1^*(q, ya, p) \\
&= \vee\{\mu_1^*(q, y, r) \wedge \mu_1^*(r, a, p) | r \in Q_1\} \\
&\leq \vee\{\mu_2^*(\alpha(q), \beta^*(y), \alpha(r)) \wedge \mu_2^*(\alpha(r), \beta(a), \alpha(p)) | r \in Q_1\} \\
&\leq \vee\{\mu_2{}^*(\alpha(q),\ \beta^*(y),\ r') \wedge \mu_2{}^*(r',\ \beta^*(a),\ \alpha(p) \mid r' \in Q_2\} \\
&= \mu_2^*(\alpha(q), \beta^*(y)\beta(a), \alpha(p)) \\
&= \mu_2^*(\alpha(q), \beta^*(ya), \alpha(p)) \\
&= \mu_2^*(\alpha(q), \beta^*(x),\ \alpha(p)). \blacksquare
\end{aligned}$$

Theorem 6.2.7 *Let M_1, M_2 be as above. Let $(\alpha, \beta) : M_1 \to M_2$ be a strong homomorphism. Then α is one-one if and only if $\mu_1^*(q, x, p) = \mu_2^*(\alpha(q), \beta^*(x), \alpha(p))\ \forall\ q, p \in Q_1$ and $x \in X_1^*$.*

Proof. Suppose α is one-one. Let $p, q \in Q_1$ and $x \in X_1^*$. Let $len(x) = n$. We prove the result by induction on n. Let $n = 0$. Then $x = \Lambda$ and

$\beta^*(\Lambda) = \Lambda$. Now $\alpha(q) = \alpha(p)$ if and only if $q = p$. Hence $\mu_1^*(q, \Lambda, p) = 1$ if and only if $\mu_2^*(\alpha(q), \beta^*(\Lambda), \alpha(p)) = 1$. Suppose the result is true $\forall\ y \in X_1^*$, $len(y) = n - 1$, $n > 0$. Let $x = ya$, $len(y) = n - 1$, $y \in X_1^*$, $a \in X_1$. Then

$$\begin{aligned}
\mu_2^*(\alpha(q), \beta^*(x), \alpha(p)) &= \mu_2^*(\alpha(q), \beta^*(ya), \alpha(p)) \\
&= \mu_2^*(\alpha(q), \beta^*(y)\beta(a), \alpha(p)) \\
&= \vee\{\mu_2^*(\alpha(q), \beta^*(y), \alpha(r)) \wedge \mu_2(\alpha(r), \beta(a), \alpha(p)) \\
&\quad |r \in Q_1\} \\
&= \vee\{\mu_1^*(q, y, r) \wedge \mu_1(r, a, p)|r \in Q_1\}, \\
&\qquad \text{by Remark 6.2.2(2)} \\
&= \mu_1^*(q, ya, p) \\
&= \mu_1^*(q, x, p).
\end{aligned}$$

Conversely, let $q, p \in Q_1$ and let $\alpha(q) = \alpha(p)$. Then $1 = \mu_2^*(\alpha(q), \Lambda, \alpha(p)) = \mu_1^*(q, \Lambda, p)$. Hence $q = p$, i.e., α is one-one. ■

6.3 Admissible Relations

Definition 6.3.1 *Let $M = (Q, X, \mu)$ be a ffsm and let $\sim$ be an equivalence relation on Q. Then $\sim$ is called an* **admissible relation** *if and only if $\forall$ $p, q, r \in Q, \forall a \in X$, if $p \sim q$ and $\mu(p, a, r) > 0$, then $\exists\ t \in Q$ such that $\mu(q, a, t) \geq \mu(p, a, r)$ and $t \sim r$.*

Theorem 6.3.2 *Let $M = (Q, X, \mu)$ be a ffsm and let $\sim$ be an equivalence relation on Q. Then $\sim$ is an admissible relation if and only if $\forall\ p, q, r \in Q, \forall x \in X^*$, if $p \sim q$ and $\mu^*(p, x, r) > 0$, then $\exists\ t \in Q$ such that $\mu^*(q, x, t) \geq \mu^*(p, x, r)$ and $t \sim r$.*

Proof. Suppose $\sim$ is admissible. Let $p, q \in Q$ be such that $p \sim q$. Let $x \in X^*$, $r \in Q$ be such that $\mu^*(p, x, r) > 0$. Suppose $len(x) = n$. If $n = 0$, then $x = \Lambda$. Thus $\mu^*(p, x, r) > 0 \Rightarrow p = r$ and $\mu^*(p, x, p) = 1$. Now $\mu^*(q, x, q) = 1 = \mu^*(p, x, p)$ and $q \sim p$. Thus the result is true for $n = 0$. Suppose now the result is true $\forall\ y \in X^*$ such that $len(y) = n - 1$, $n > 0$. Let $x = ya$ where $y \in X_1^*$, $a \in X_1$ and $len(y) = n - 1$. Now $\mu^*(p, x, r) = \mu^*(p, ya, r) = \vee\{\mu^*(p, y, q_1) \wedge \mu^*(q_1, a, r)|q_1 \in Q\} > 0$. Let $s \in Q$ be such that $\mu^*(p, y, s) \wedge \mu^*(s, a, r) = \vee\{\mu^*(p, y, q_1) \wedge \mu^*(q_1, a, r) \mid q_1 \in Q\}$. Then $\mu^*(p, y, s) > 0$ and $\mu^*(s, a, r) > 0$. By the induction hypothesis, $\exists\ t_s \in Q$ such that $\mu^*(q, y, t_s) \geq \mu^*(p, y, s)$ and $t_s \sim s$. Now $\mu(s, a, r) > 0$ and $t_s \sim s$. Since $\sim$ is admissible, $\exists\ t \in Q$ such that $\mu(t_s, a, t) \geq \mu(s, a, r)$ and $t \sim r$. Thus $\exists\ t \in Q$ such that $t \sim r$ and $\mu^*(q, y, t_s) \wedge \mu(t_s, a, t) \geq$

$\mu^*(p, y, s) \wedge \mu^*(s, a, r)$. Thus

$$\begin{aligned}\mu^*(p,x,r) &= \mu^*(p,y,s) \wedge \mu^*(s,a,r) \\ &\leq \mu^*(q,y,t_s) \wedge \mu(t_s,a,t) \\ &\leq \vee\{\mu^*(q,y,r_1) \wedge \mu(r_1,a,t) | r_1 \in Q\} \\ &= \mu^*(q,ya,t) \\ &= \mu^*(q,x,t)\end{aligned}$$

and $t \sim r$. The result now follows by induction. The converse is trivial. ∎

Let $M = (Q, X, \mu)$ be a ffsm and let $\sim$ be an admissible relation on Q. For $q \in Q$, let $[q]$ denote the equivalence class of q. Let $\widetilde{Q} = Q/\sim$ $= \{[q] | q \in Q\}$. Define the fuzzy subset $\widetilde{\mu}$ of $\widetilde{Q} \times X \times \widetilde{Q}$ by

$$\widetilde{\mu}([q], x, [p]) = \vee\{\mu(q, x, t) | t \in [p]\}$$

$\forall q, p \in Q, x \in X$. Suppose that $[q] = [q']$, $x = y$, and $[p] = [p']$, $q, q', p, p' \in Q$ and $x, y \in X$. Then $q \sim q'$. Now

$$\widetilde{\mu}([q], x, [p]) = \vee\{\mu(q, x, r) | r \in [p]\}$$

and

$$\widetilde{\mu}([q'], y, [p']) = \widetilde{\mu}([q'], x, [p']) = \vee\{\mu(q', x, t) | t \in [p']\}.$$

Let $r \in [p]$ be such that $\mu(q, x, r) > 0$. Then since $\sim$ is admissible, $\exists$ $t \in Q$ such that $\mu(q', x, t) \geq \mu(q, x, r) > 0$ and $t \sim r$. Now since $t \sim r$, $t \in [p] = [p']$. Thus $\exists$ $t \in [p']$ such that $\mu(q', x, t) \geq \mu(q, x, r) > 0$. Similarly if $\mu(q', x, t) > 0$ for some $t \in [p']$, then $\exists$ $r \in [p]$ such that $\mu(q, x, r) \geq \mu(q', x, t) > 0$. Hence

$$\widetilde{\mu}([q], x, [p]) = \widetilde{\mu}([q'], x, [p']).$$

Thus $\widetilde{\mu}$ is single-valued. Hence $(\widetilde{Q}, X, \widetilde{\mu})$ is a ffsm. Define $\underline{\alpha} : Q \to \widetilde{Q}$ by $\underline{\alpha}(q) = [q]$ $\forall$ $q \in Q$. Clearly, $\underline{\alpha}$ maps Q onto $\widetilde{Q}$. Let $\beta : X \to X$ be the identity map. Let $q, t \in Q$ and $x \in X$. Then $\widetilde{\mu}(\underline{\alpha}(q), x, \underline{\alpha}(t)) = \widetilde{\mu}([q], x, [t])$ $= \vee\{\mu(q, x, r) | r \in [t]\} \geq \mu(q, x, t)$. Hence $(\underline{\alpha}, \beta)$ is a homomorphism.

Definition 6.3.3 *Let $M_1 = (Q_1, X, \mu_1)$ and $M_2 = (Q_2, X, \mu_2)$ be ffsms. Let $\alpha : (Q_1, X, \mu_1) \to (Q_2, X, \mu_2)$ be a strong homomorphism. The kernel of α, denoted Ker α, is defined to be the set*

$$Ker\ \alpha = \{(p, q) | \alpha(p) = \alpha(q)\}.$$

Lemma 6.3.4 *Let α be as defined in Definition 6.3.3. Then Ker α is an admissible relation.*

Proof. Now clearly Ker α is an equivalence relation. Let $p, q \in Q_1$ and $(p, q) \in$ Ker α. Then $\alpha(p) = \alpha(q)$. Let $a \in X$, $r \in Q_1$ and $\mu_1(p, a, r) > 0$. Then $\mu_2(\alpha(q), a, \alpha(r)) = \mu_2(\alpha(p), a, \alpha(r)) \geq \mu_1(p, a, r) > 0$. By Lemma 6.2.3, $\exists\, t \in Q_1$ such that $\mu_1(q, a, t) \geq \mu_1(p, a, r) > 0$ and $\alpha(t) = \alpha(r)$. Since $\alpha(t) = \alpha(r)$, $(t, r) \in$ Ker α.Thus Ker α is admissible. ∎

Theorem 6.3.5 *Let $M_1 = (Q_1, X, \mu_1)$ and $M_2 = (Q_2, X, \mu_2)$ be two ffsms and let $\alpha : (Q_1, X, \mu_1) \to (Q_2, X, \mu_2)$ be an onto strong homomorphism. Then $\exists$ an isomorphism*

$$\gamma : (Q_1/(Ker\ \alpha),\ X,\ \widetilde{\mu_1}) \ \to (Q_2, X, \mu_2)$$

such that $\alpha = \gamma \circ \underline{\alpha}$.

Proof. Define $\gamma : Q_1/(Ker\ \alpha) \to Q_2$ by $\gamma([q]) = \alpha(q)$. Let $p, q \in Q_1$ be such that $[p] = [q]$. Then $(p, q) \in$ Ker α and hence $\alpha(p) = \alpha(q)$ or $\gamma([q]) = \gamma([p])$. Now, let $q, p \in Q_1$ and $x \in X$. Then $\widetilde{\mu_1}([q], x, [p]) = \vee\{\mu_1(q, x, r) | r \in [p]\} = \vee\{\mu_1(q, x, r) | \alpha(r) = \alpha(p), r \in Q_1\} = \mu_2(\alpha(q), x, \alpha(p)) = \mu_2(\gamma([q]), x, \gamma([p]))$. Thus γ is a homomorphism. Clearly γ maps $(Q_1/(Ker\ \alpha), X, \widetilde{\mu_1})$ one-to-one onto (Q_2, X, μ_2). ∎

6.4 Fuzzy Transformation Semigroups

Definition 6.4.1 *A **fuzzy transformation semigroup (fts)** is a triple (Q, S, ρ), where Q is a finite nonempty set, S is a finite semigroup and ρ is a fuzzy subset of $Q \times S \times Q$ such that*

(i) $\rho(q, uv, p) = \vee\{\rho(q, u, r) \wedge \rho(r, v, p) | r \in Q\}$ $\forall u, v \in S$ and $\forall q, p \in Q$;

(ii) If S contains the identity e, then $\rho(q, e, p) = 1$ if $q = p$ and $\rho(q, e, p) = 0$ if $q \neq p$, $\forall\ q, p \in Q$.

*If, in addition, the following property holds, then (Q, S, ρ) is called **faithful**.*

(iii) Let $u, v \in S$. If $\rho(q, u, p) = \rho(q, v, p)$ $\forall q, p \in Q$, then $u = v$.

Let $M = (Q, S, \rho)$ be a fts. This fts may not be faithful. Define a relation R on S by $\forall\ u, v \in S$, uRv if and only if $\forall\ q, p \in Q$, $\rho(q, u, p) = \rho(q, v, p)$. Clearly R is an equivalence relation on S. Suppose that $u, v, x \in S$ and uRv. Then

$$\begin{aligned} \rho(q, ux, p) &= \vee\{\rho(q, u, r) \wedge \rho(r, x, p) \mid r \in Q\} \\ &= \vee\{\rho(q, v, r) \wedge \rho(r, x, p) \mid r \in Q\} \\ &= \rho(q, vx, p) \end{aligned}$$

$\forall\ q, p \in Q$. Similarly, $\rho(q, xu, p) = \rho(q, xv, p)$ $\forall\ q, p \in Q$. Hence R is a congruence relation on S. Let $[u]$ denote the equivalence class of R induced by u. Let $S/R = \{[u] |\ u \in S\}$. Define

$$\overline{\rho} : Q \times S/R \times Q \to [0, 1]$$

by

$$\overline{\rho}(q,[x],p) = \rho(q,x,p)$$

$\forall\, q,p \in Q$ and $\forall\, [x] \in S/R$. Clearly, $\overline{\rho}$ is single-valued. Now

$$\begin{aligned}\overline{\rho}(q,[x][y],p) &= \overline{\rho}(q,[xy],p)\\ &= \rho(q,xy,p)\\ &= \vee\{\rho(q,x,r)\wedge\rho(r,y,p)\mid r\in Q\}\\ &= \vee\{\overline{\rho}(q,[x],r)\wedge\overline{\rho}(r,[y],p)\mid r\in Q\}.\end{aligned}$$

Also

$$\overline{\rho}(q,[e],p) = \begin{cases} 1 \text{ if } p=q\\ 0 \text{ otherwise.}\end{cases}$$

Suppose that $\overline{\rho}(q,[x],p) = \overline{\rho}(q,[y],p)\ \forall q,p \in Q$. Then $\rho(q,x,p) = \rho(q,y,p)\ \forall q,p \in Q$. Hence xRy and so $[x] = [y]$. Thus $(Q, S/R, \overline{\rho})$ is a faithful fts. We call $(Q, S/R, \overline{\rho})$ the **faithful fuzzy transformation semigroup represented** by the triple (Q, S, ρ).

Theorem 6.4.2 *Let $M = (Q, X, \mu)$ be a ffsm. Let $E(M)$ be defined as before. Then $(Q, E(M), \rho)$ is a faithful fts where $\rho(q,[x],p) = \mu^*(q,x,p)\ \forall\ q,p \in Q,\ x \in X^*$.*

Proof. By Theorem 6.1.4, $E(M)$ is a finite semigroup with identity $[\Lambda]$. Clearly ρ is single-valued. Let $q,p \in Q$ and $[x],[y] \in E(M)$. Then

$$\begin{aligned}\rho(q,[x]*[y],p) &= \rho(q,[xy],p)\\ &= \mu^*(q,xy,p)\\ &= \vee\{\mu^*(q,x,r)\wedge\mu^*(r,y,p)|r\in Q\}\\ &= \vee\{\rho(q,[x],r)\wedge\rho(r,[y],p)|r\in Q\}.\end{aligned}$$

Now

$$\rho(q,[\Lambda],p) = \begin{cases} \mu^*(q,\Lambda,p) = 1 & \text{if } p=q\\ \mu^*(q,\Lambda,p) = 0 & \text{if } q\neq p,\end{cases}$$

by the definition of μ^*. Suppose $\rho(q,[x],p) = \rho(q,[y],p)\ \forall\ q,p \in Q$. Then $\mu^*(q,x,p) = \mu^*(q,y,p)\ \forall q,p \in Q$. Thus $x \equiv y$ or $[x] = [y]$. Hence $(Q, E(M), \rho)$ is a faithful fuzzy transformation semigroup. ∎

Let $M = (Q, X, \mu)$ be a ffsm. Then by Theorem 6.4.2 $(Q, E(M), \rho)$ is a fuzzy transformation semigroup which we denote by $FTS(M)$. We call $FTS(M)$ the fuzzy transformation semigroup associated with M.

Definition 6.4.3 *Let (Q, S, ρ) be a fts. Let $\sim$ be an equivalence relation on Q. Then $\sim$ is called an* ***admissible relation*** *if and only if $\forall p,q,r \in Q$, $\forall u \in S$, if $p \sim q$ and $\rho(p,u,r) > 0$, then $\exists\ t \in Q$ such that $\rho(q,u,t) \geq \rho(p,u,r)$ and $t \sim r$.*

Theorem 6.4.4 *Let $M = (Q, X, \mu)$ be a ffsm and let $\sim$ be an equivalence relation on Q. Then $\sim$ is an admissible relation for M if and only if $\sim$ is an admissible relation for the fuzzy transformation semigroup, $FTS(M) = (Q, E(M), \rho)$.*

Proof. Suppose $\sim$ is admissible for M. Let $p, q \in Q$ be such that $p \sim q$ and $[u] \in E(M)$. Let $\rho(p, [u], r) > 0$ for some $r \in Q$. Then $\mu^*(p, u, r) > 0$. Hence by Theorem 6.3.2, $\exists\ t \in Q$ such that $\mu^*(q, u, t) \geq \mu^*(p, u, r)$ and $t \sim r$. Thus $\rho(q, [u], t) = \mu^*(q, u, t) \geq \mu^*(p, u, r) = \rho(p, [u], r)$. Hence $\sim$ is admissible for $FTS(M)$. Conversely, suppose that $\sim$ is admissible for $FTS(M)$. Let $p, q \in Q$ be such that $p \sim q$ and $u \in X$. Let $\mu^*(p, u, r) > 0$ for some $r \in Q$. Then $\rho(p, [u], r) > 0$. Then $\exists\ t \in Q$ such that $\rho(q, [u], t) \geq \rho(p, [u], r)$ and $t \sim r$. Now $\mu^*(q, u, t) = \rho(q, [u], t) \geq \rho(p, [u], r) = \mu^*(p, u, r)$ and $t \sim r$. Hence $\sim$ is admissible for M. ■

Definition 6.4.5 *Let (Q_1, S_1, ρ_1) and (Q_2, S_2, ρ_2) be two ftss. A pair (f, g) of mappings, where $f : Q_1 \to Q_2$ and $g : S_1 \to S_2$, is said to be a* ***homomorphism*** *from (Q_1, S_1, ρ_1) to (Q_2, S_2, ρ_2) if*
(i) $g(xy) = g(x)g(y)\ \forall\ x, y \in S_1$,
(ii) If e_1 is the identity of S_1 and e_2 is the identity of S_2, then $g(e_1) = e_2$,
(iii) $\rho_1(q, x, p) \leq \rho_2(f(q), g(x), f(p))\ \forall q, p \in Q_1, x \in S_1$.
(f, g) is called a ***strong homomorphism*** *if it satisfies (i), (ii) and*

$$\rho_2(f(q), g(x), f(p)) = \vee\{\rho_1(q, x, t) | t \in Q_1,\ f(t) = f(p)\}$$

$\forall q, p \in Q_1, x \in S_1$.

A homomorphism (strong homomorphism) $(f, g) : (Q_1, S_1, \rho_1) \to (Q_2, S_2, \rho_2)$ is called an ***isomorphism (strong isomorphism)*** *if f and g are both one-one and onto.*

Let S be a semigroup with identity. Let $\mathcal{A} = (Q, S, \delta)$ be a faithful fts. Define the ffsm $M = (Q, S, \mu)$ by taking $\mu = \delta$. Consider FTS$(M) = (Q, E(M), \rho)$, where $E(M) = S^*/\sim$ and $\rho(q, [u], p) = \mu^*(q, u, p)$. Let e be the identity element of S and Λ the empty word in S^*. Now $\rho(q, [e], p) = \mu^*(q, e, p) = \delta(q, e, p) = 1$ if $p = q$ and 0 if $p \neq q$. Hence $\rho(q, [e], p) = \rho(q, [\Lambda], p)\ \forall p, q \in Q$. Thus $[e] = [\Lambda]$.

Theorem 6.4.6 *Let S be a semigroup with identity. Then $FTS(M)$ is isomorphic to $\mathcal{A} = (Q, S, \delta)$.*

Proof. Define $f : Q \to Q$ by $f(q) = q\ \forall q \in Q$ and $g : S \to E(M)$ by $g(x) = [x]\ \forall\ x \in S$. Let $x, y \in S$ be such that $g(x) = g(y)$. Then $[x] = [y]$. Thus $\mu^*(q, x, p) = \mu^*(q, y, p)\ \forall q, p \in Q$. Hence $\mu(q, x, p) = \mu(q, y, p)\ \forall\ q, p \in Q$. This implies that $\delta(q, x, p) = \delta(q, y, p)\ \forall q, p \in Q$. Since $\mathcal{A}$ is faithful, we find that $x = y$. Hence g is injective. Let $\cdot$ denote the binary operation of the semigroup S. Let $a, b \in S$. Then $a \cdot b \in S$ and $ab \in S^*$. Let

$q, p \in Q$. Now

$$\begin{aligned}\mu^*(q, a \cdot b, p) &= \mu(q, a \cdot b, p) \\ &= \delta(q, a \cdot b, p) \\ &= \vee\{\delta(q, a, r) \wedge \delta(r, b, p) | r \in Q\} \\ &= \vee\{\mu(q, a, r) \wedge \mu(r, b, p) | r \in Q\} \\ &= \mu^*(q, ab, p).\end{aligned}$$

Hence $[a \cdot b] = [ab]$. Thus $g(ab) = [a \cdot b] = [ab] = [a][b] = g(a)g(b)$. By induction it can be shown that if $c_i \in S$, $1 \leq i \leq n$, then $[c_1 \cdot c_2 \cdot \ldots \cdot c_n] = [c_1 c_2 \ldots c_n]$. Let $[u] \in E(M)$. If $u = \Lambda$, then $[\Lambda] = [e]$ and $g(e) = [\Lambda]$. Suppose $u = a_1 a_2 \ldots a_n$, $a_i \in S$, $1 \leq i \leq n$. Then $g(a_1 \cdot a_2 \cdot \ldots \cdot a_n) = [a_1 \cdot a_2 \cdot \ldots \cdot a_n] = [a_1 a_2 \ldots a_n] = [u]$. Thus g is surjective. Finally, $\rho(f(q), g(x), f(p)) = \rho(q, [x], p) = \mu^*(q, x, p) = \mu(q, x, p) = \delta(q, x, p)$. ■

Theorem 6.4.7 *Let $M_1 = (Q_1, X_1, \mu_1)$ and $M_2 = (Q_2, X_2, \mu_2)$ be two ffsms and let $(\alpha, \beta) : M_1 \to M_2$ be a strong homomorphism with α one-one and onto. Then $\exists$ a strong homomorphism (f_α, g_β) from $FTS(M_1)$ to $FTS(M_2)$.*

Proof. Define $f_\alpha : Q_1 \to Q_2$ by $f_\alpha(q) = \alpha(q)$ $\forall$ $q \in Q_1$ and $g_\beta : E(M_1) \to E(M_2)$ by $g_\beta([x]) = [\beta^*(x)]$ $\forall$ $[x] \in E(M_1)$. Let $[x], [y] \in E(M_1)$ and $[x] = [y]$. Then $\mu^*(q, x, p) = \mu^*(q, y, p)$ $\forall$ $q, p \in Q$. Now

$$\begin{aligned}\mu_2^*(\alpha(q), \beta^*(x), \alpha(p)) &= \mu_1^*(q, x, p) \\ &= \mu_1^*(q, y, p) \\ &= \mu_2^*(\alpha(q), \beta^*(y), \alpha(p))\end{aligned}$$

$\forall$ $q, p \in Q_1$. Thus since α is onto, $[\beta^*(x)] = [\beta^*(y)]$. Hence g_β is well defined. Now $g_\beta([x] * [y]) = g_\beta([xy]) = [\beta^*(xy)] = [\beta^*(x)\beta^*(y)] = [\beta^*(x)] * [\beta^*(y)] = g_\beta([x]) * g_\beta([y])$ and $g_\beta([\Lambda]) = [\beta^*(\Lambda)] = [\Lambda]$. Also

$$\begin{aligned}\rho_1(q, [x], p) &= \mu_1^*(q, x, p) \\ &= \mu_2^*(\alpha(q), \beta^*(x), \alpha(p)) \\ &= \rho_2(f_\alpha(q),\ g_\beta([x]), f_\alpha(p)).\end{aligned}$$

Hence by definition (f_α, g_β) is a strong homomorphism. ■

Definition 6.4.8 *A* ***polytransformation semigroup*** *is a triple (Q, S, ν), where Q is a finite nonempty set, S is a finite semigroup, and $\nu : Q \times S \to \mathcal{P}(Q) \backslash \{\phi\}$ such that*

(i) $\nu(\nu(q, u), v) = \nu(q, uv)$ $\forall$ $q \in Q$, $u, v \in S$. It is being understood for any subset P of Q, $\nu(P, u) = \cup_{u \in P} \nu(p, u)$.

(ii) If S contains identity e, then $\nu(q, e) = \{q\}$ $\forall$ $q \in Q$.

If, in addition, the following holds, then (Q, S, ν) is called ***faithful.***

(iii) Let $u, v \in S$. If $\nu(q, u) = \nu(q, v)$ $\forall$ $q \in Q$, then $u = v$.

Let $M = (Q, X, \mu)$ be a ffsm. For $q \in Q$ and $x \in X^*$, we define

$$S_x(q) = \{p \in Q | \mu^*(q, x, p) > 0\}.$$

Theorem 6.4.9 *Let $M = (Q, X, \mu)$ be a ffsm. Let $\widetilde{E(M)}$ be defined as before. If $\forall q \in Q$, $S_x(q) \neq \phi$, then $(Q, \widetilde{E(M)}, \nu)$ is a faithful polytransformation semigroup with identity.*

Proof. Define $\nu : Q \times \widetilde{E(M)} \to \mathcal{P}(Q) \backslash \{\phi\}$ by $\nu(q, \prec x \succ) = \{p \in Q \mid \mu^*(q, x, p) > 0\}$. Suppose that $\prec x \succ = \prec y \succ$. Then $x \approx y$. Thus

$$\begin{aligned} \nu(q, \prec x \succ) &= \{p \in Q | \mu^*(q, x, p) > 0\} \\ &= \{p \in Q | \mu^*(q, y, p) > 0\} \\ &= \nu(q, \prec y \succ). \end{aligned}$$

Thus ν is single-valued. We write $q \prec x \succ$ for $\nu(q, \prec x \succ)$. Then, by definition, $(q \prec x \succ) \prec y \succ = \cup_{p \in q \prec x \succ} p \prec y \succ$. Let $p \in q \prec xy \succ = \{p \in Q | \mu^*(q, xy, p) > 0\}$. Now $\mu^*(q, xy, p) > 0 \Rightarrow \mu^*(q, x, r) \wedge \mu^*(r, y, p) > 0$ for some $r \in Q$. Thus $\mu^*(q, x, r) > 0$ and $\mu^*(r, y, p) > 0$. Thus $r \in q \prec x \succ$ and $p \in r \prec y \succ$. Hence $p \in \cup_{t \in q \prec x \succ} t \prec y \succ$. Thus $q \prec xy \succ \subseteq (q \prec x \succ)(\prec y \succ)$. Let $p \in (q \prec x \succ)(\prec y \succ) = \cup_{t \in q \prec x \succ} t \prec y \succ$. Then $p \in t \prec y \succ$ for some $t \in q \prec x \succ$. Thus $\mu^*(t, y, p) > 0$ and $\mu^*(q, x, t) > 0$. Hence $\mu^*(q, xy, p) > 0$. Thus $p \in q \prec xy \succ$. Hence $(q \prec x \succ)(\prec y \succ) \subseteq q \prec xy \succ$. Therefore, $q \prec xy \succ = (q \prec x \succ)(\prec y \succ)$. Hence (i) of Definition 6.4.8 holds. It follows easily that $q \prec \Lambda \succ = \{q\}$ and so (ii) of Definition 6.4.8 holds. Suppose that $q \prec x \succ = q \prec y \succ$ $\forall q \in Q$. Then $\mu^*(q, x, p) > 0$ if and only if $\mu^*(q, y, p) > 0$ $\forall\, p \in Q$. Thus $\prec x \succ = \prec y \succ$. Hence (iii) of Definition 6.4.8 holds. ∎

6.5 Submachines

The study of submachines of fuzzy finite state machines is particularly well-suited to the use of algebraic techinques. In the next three sections, we examine connectedness and retrievability properties of finite state machines. We also show that a fuzzy finite state machine can be decomposed in terms of primary submachines.

Definition 6.5.1 *Let $M = (Q, X, \mu)$ be a ffsm. Let $p, q \in Q$. p is called an* ***immediate successor*** *of q if $\exists a \in X$ such that $\mu(q, a, p) > 0$. p is called a* ***successor*** *of q if $\exists x \in X^*$ such that $\mu^*(q, x, p) > 0$.*

Proposition 6.5.2 *Let $M = (Q, X, \mu)$ be a ffsm. Let $q, p, r \in Q$. Then the following assertions hold.*

(i) q is a successor of q.

(ii) if p is a successor of q and r is a successor of p, then r is a successor of q.

Proof. (i) Since $\mu^*(q, \Lambda, q) = 1 > 0$, q is a successor of q.

(ii) Now $\exists x, y \in X$ such that $\mu^*(q, x, p) > 0$ and $\mu^*(p, y, r) > 0$. Thus

$$\mu^*(q, xy, r) \geq \mu^*(q, x, p) \wedge \mu^*(p, y, r) > 0,$$

by Lemma . Hence r is a successor of q. ■

Definition 6.5.3 *Let $M = (Q, X, \mu)$ be a ffsm and let $q \in Q$. We denote by $S(q)$ the set of all successors of q.*

Definition 6.5.4 *Let $M = (Q, X, \mu)$ be a ffsm and let $T \subseteq Q$. The set of all successors of T, denoted by $S_Q(T)$ in Q, is defined to be the set*

$$S_Q(T) = \cup\{S(q) | q \in T\}.$$

If no confusion arises, then we write $S(T)$ for $S_Q(T)$.

Theorem 6.5.5 *Let $M = (Q, X, \mu)$ be a ffsm. Let $A, B \subseteq Q$. Then the following assertions hold.*

(i) If $A \subseteq B$, then $S(A) \subseteq S(B)$.
(ii) $A \subseteq S(A)$.
(iii) $S(S(A)) = S(A)$.
(iv) $S(A \cup B) = S(A) \cup S(B)$.
(v) $S(A \cap B) \subseteq S(A) \cap S(B)$.

Proof. The proofs of (i), (ii), (iv) and (v) are straightforward.

(iii) Clearly $S(A) \subseteq S(S(A))$. Let $q \in S(S(A))$. Then $q \in S(p)$ for some $p \in S(A)$. Thus $p \in S(r)$ for some $r \in A$. Now q is a successor of p and p is a successor of r. Hence by Proposition 6.5.2, q is a successor of r. Thus $q \in S(r) \subseteq S(A)$. Hence $S(S(A)) = S(A)$. ■

Exchange Property: Let $M = (Q, X, \mu)$ be a ffsm. Let $q, p \in Q$ and let $T \subseteq Q$. Suppose that if $p \in S(T \cup \{q\})$, $p \notin S(T)$, then $q \in S(T \cup \{p\})$. Then we say that M satisfies the **exchange property**.

Proposition 6.5.6 *Let $M = (Q, X, \mu)$ be a ffsm. Then the following assertions are equivalent.*

(i) M satisfies the exchange property.
(ii) $\forall p, q \in Q$, $q \in S(p)$ if and only if $p \in S(q)$.

Proof. (i)⇒(ii): Let $p, q \in Q$ and $p \in S(q)$. Now $p \notin S(\phi)$. Hence $q \in S(p)$. Similarly, if $q \in S(p)$ then $p \in S(q)$.

(ii)⇒(i): Let $T \subseteq Q$, $p, q \in Q$. Suppose $p \in S(T \cup \{q\})$, $p \notin S(T)$. Then $p \in S(q)$. Hence $q \in S(p) \subseteq S(T \cup \{p\})$. ■

Definition 6.5.7 *Let $M = (Q, X, \mu)$ be a ffsm. Let $T \subseteq Q$. Let ν be a fuzzy subset of $T \times X \times T$ and let $N = (T, X, \nu)$. The fuzzy finite state machine N is called a **submachine** of M if*

(i) $\mu|_{T\times X\times T}=\nu$ *and*
(ii) $S_Q(T)\subseteq T$.

We assume that $\phi=(\phi,X,\nu)$ is a submachine of M. Clearly, if K is a submachine of N and N is a submachine of M, then K is a submachine of M.

Theorem 6.5.8 *Let* $M=(Q,X,\mu)$ *be a ffsm. Let* $M_i=(Q_i,X,\mu_i)$, $i\in I$, *be a family of submachines of* M, *where* $Q_i\subseteq Q$. *Then the following assertions hold.*
(i) $\cap_{i\in I}M_i=(\cap_{i\in I}Q_i,X,\cap_{i\in I}\mu_i)$ *is a submachine of* M.
(ii) $\cup_{i\in I}M_i=(\cup_{i\in I}Q_i,X,\nu)$ *is a submachine of* M, *where*

$$\nu=\mu|_{\cup_{i\in I}Q_i\times X\times\cup_{i\in I}Q_i}.$$

Proof. (i) Let $(q,x,p)\in\cap_{i\in I}Q_i\times X\times\cap_{i\in I}Q_i$. Then

$$(\cap_{i\in I}\mu_i)(q,x,p)=\wedge_{i\in I}\mu_i(q,x,p)=\wedge_{i\in I}\mu(q,x,p)=\mu(q,x,p).$$

Also

$$S(\cap_{i\in I}Q_i)\subseteq\cap_{i\in I}S(Q_i)\subseteq\cap_{i\in I}Q_i.$$

Thus $\cap_{i\in I}M_i$ is a submachine of M.
(ii) Since

$$S(\cup_{i\in I}Q_i)=\cup_{i\in I}S(Q_i)=\cup_{i\in I}Q_i,$$

$\cup_{i\in I}M_i$ is a submachine of M. ■

Definition 6.5.9 *Let* $M=(Q,X,\mu)$ *be a ffsm. Then* M *is called* ***strongly connected*** *if* $\forall p,q\in Q$, $p\in S(q)$.

Definition 6.5.10 *Let* $M=(Q,X,\mu)$ *be a ffsm and let* $N=(T,X,\nu)$ *be a submachine of* M. N *is called* ***proper*** *if* $T\neq Q$ *and* $T\neq\phi$.

Theorem 6.5.11 *Let* $M=(Q,X,\mu)$ *be a ffsm. Then* M *is strongly connected if and only if* M *has no proper submachines.*

Proof. Suppose M is strongly connected. Let $N=(T,X,\nu)$ be a submachine of such that $T\neq\phi$. Then $\exists q\in T$. Let $p\in Q$. Since M is strongly connected $p\in S(q)$. Hence $p\in S(q)\subseteq S(T)\subseteq T$. Thus $T=Q$ and so $M=N$. Conversely, suppose M has no proper submachine. Let $p,q\in Q$ and let $N=(S(q),X,\nu)$ where $\nu=\mu|_{S(q)\times X\times S(q)}$. Then N is a submachine of M and $S(q)\neq\phi$. Hence $S(q)=Q$. Thus $p\in S(q)$. Hence M is strongly connected. ■

Proposition 6.5.12 *Let* $M=(Q,X,\mu)$ *be a ffsm. Let* $R\subseteq Q$. *Then* $N=(S(R),X,\mu_R)$ *is a submachine of* M *where* $\mu_R=\mu|_{S(R)\times X\times S(R)}$. ■

Definition 6.5.13 *Let $M = (Q, X, \mu)$ be a ffsm. Let $R \subseteq Q$ and $\{N_i \mid i \in I\}$ be the collection of all submachines of M whose state set contains R. Define $< R > = \cap_{i \in I}\{N_i | i \in I\}$. Then $< R >$ is called the submachine* ***generated*** *by R.*

In Definition 6.5.13, it is clear that $< R >$ is the smallest submachine of M whose state set contains R.

Proposition 6.5.14 *Let $M = (Q, X, \mu)$ be a ffsm. Let $R \subseteq Q$. Then*

$$< R > \ = (S(R), X, \mu_R).$$

Proof. Now $< R > = (\cap_{i \in I} Q_i, X, \cap_{i \in I} \mu_i)$, where $\{N_i | i \in I\}$ is the collection of all submachines of M whose state set contains R and $N_i = (Q_i, X, \mu_i)$, $i \in I$. It suffices to show that $S(R) = \cap_{i \in I} Q_i$. Since $(S(R), X, \mu_R)$ is a submachine of M such that $R \subseteq S(R)$, we have that $S(R) \supseteq \cap_{i \in I} Q_i$. Let $p \in S(R)$. Then $\exists\ r \in R$ and $x \in X^*$ such that $\mu^*(r, x, p) > 0$. Now $r \in \cap_{i \in I} Q_i$, and since $< R >$ is a submachine of M, $p \in \cap_{i \in I} Q_i$. Thus $S(R) \subseteq \cap_{i \in I} Q_i$. Hence $S(R) = \cap_{i \in I} Q_i$. ■

Definition 6.5.15 *Let $M = (Q, X, \mu)$ be a ffsm. M is called* ***singly generated*** *if $\exists q \in Q$ such that $M = < \{q\} >$. In this case q is called a* ***generator*** *of M and we say that M is generated by q.*

Theorem 6.5.16 *Let $M = (Q, X, \mu)$ be a ffsm. Let $R, T \subseteq Q$. Then the following assertions hold.*

(i) $< R \cup T > = < R > \cup < T >$.

(ii) $< R \cap T > \subseteq < R > \cap < T >$.

Proof. (i) By Theorem 6.5.5, $S(R \cup T) = S(R) \cup S(T)$. Now

$$\begin{aligned} \mu_{S(R) \cup S(T)} &= \mu|_{(S(R) \cup S(T)) \times X \times (S(R) \cup S(T))} \\ &= \mu|_{S(R \cup T) \times X \times S(R \cup T)} \\ &= \mu_{S(R \cup T)}. \end{aligned}$$

Hence $< R \cup T > = < R > \cup < T >$.

(ii) By Theorem 6.5.5, $S(R \cap T) \subseteq S(R) \cap S(T)$. Now

$$\mu_{S(R \cap T)} = \mu|_{S(R \cap T) \times X \times S(R \cap T)}$$

and

$$\mu_{S(R) \cap S(T)} = \mu|_{(S(R) \cap S(T)) \times X \times (S(R) \cap S(T))}.$$

Hence

$$\mu_{S(R \cap T)} = \mu_{S(R) \cap S(T)}|_{S(R \cap T) \times X \times S(R \cap T)}.$$

Hence $< R \cap T > \subseteq < R > \cap < T >$. ■

Definition 6.5.17 *Let $M = (Q, X, \mu)$ be a ffsm. Let $M_i = (Q_i, X, \mu_i)$, $i = 1, 2$, be submachines of M. If $M = < Q_1 \cup Q_2 >$, then we say that M is the* **union** *of M_1 and M_2 and we write $M = M_1 \cup M_2$. If $M = M_1 \cup M_2$ and $Q_1 \cap Q_2 = \phi$, then we say that M is the* **(internal) direct union** *of M_1 and M_2 and we write $M = M_1 \dot{\cup} M_2$.*

Suppose $M = M_1 \cup M_2$. Then $S(Q_i) = Q_i$ in M, $i = 1, 2$, since M_i is a submachine of M, $i = 1, 2$. Now $S(Q_1 \cup Q_2) = S(Q_1) \cup S(Q_2) = Q_1 \cup Q_2$.

Definition 6.5.18 *Let $M = (Q, X, \mu)$ be a ffsm. Let $T \subseteq Q$. T is called* **free** *if $\forall t \in T$, $t \notin S(T \backslash \{t\})$.*

Definition 6.5.19 *Let $M = (Q, X, \mu)$ be a ffsm. Let $T \subseteq Q$. If T is free and $M = < T >$, then T is called a* **basis** *of M.*

Theorem 6.5.20 *Let $M = (Q, X, \mu)$ be a ffsm. Let $T \subseteq Q$. Then the following assertions are equivalent.*
(i) T is a minimal system of generators of M.
(ii) T is a maximally free subset of Q.
(ii) T is a basis of M. ■

Theorem 6.5.21 *Let $M = (Q, X, \mu)$ be a ffsm. Suppose that M satisfies the exchange property. Then M has a basis and the cardinality of a basis is unique.* ■

Theorem 6.5.22 *Let $M = (Q, X, \mu)$ be a ffsm. Suppose that M satisfies the exchange property. Let $\{q_1, q_2, \ldots, q_n\}$ be a basis of M. Then*

$$M = \ < q_1 > \dot{\cup} < q_2 > \dot{\cup} \cdots \dot{\cup} < q_n > .$$

Proof. We have $< q_i > = (S(q_i), X, \mu_i)$, where $\mu_i = \mu|_{S(q_i) \times X \times S(q_i)}$. Now if $i \neq j$, then $S(q_i) \cap S(q_j) = \phi$ since the exchange property is equivalent to the statement that $\forall p, q \in Q$, $p \in S(q)$ if and only if $q \in S(p)$. Since $M = < q_1, q_2, \ldots, q_n >$, it follows that $M = < q_1 > \dot{\cup} < q_2 > \dot{\cup} \ldots \dot{\cup} < q_n >$. ■

6.6 Retrievability, Separability and Connectivity

Definition 6.6.1 *Let $M = (Q, X, \mu)$ be a ffsm. M is said to be* **retrievable** *if $\forall q \in Q$, $\forall y \in X^*$ if $\exists t \in Q$ such that $\mu^*(q, y, t) > 0$, then $\exists x \in X^*$ such that $\mu^*(t, x, q) > 0$.*

Definition 6.6.2 *Let $M = (Q, X, \mu)$ be a ffsm. M is said to be* **quasi-retrievable** *if $\forall q \in Q$, $\forall y \in X^*$ if $\exists t \in Q$ such that $\mu^*(q, y, t) > 0$, then $\exists x \in X^*$ such that $\mu^*(q, yx, q) > 0$.*

Definition 6.6.3 *Let $M = (Q, X, \mu)$ be a ffsm. Let $q, r, s \in Q$. Then r and s are said to be **q-related** if $\exists y \in X^*$ such that $\mu^*(q, y, r) > 0$ and $\mu^*(q, y, s) > 0$. If r and s are q-related, then r and s are said to be **q-twins** if $S(s) = S(r)$.*

Remark 6.6.4 *Let $M = (Q, X, \mu)$ be a ffsm. The following assertions are equivalent.*

(i) $\forall q, r, p \in Q$, $\forall x, y \in X^*$ if $\mu^*(q, y, r) > 0$ and $\mu^*(q, yx, p) > 0$, then $p \in S(r)$.

(ii) $\forall q, r, s \in Q$, if r and s are q-related, then r and s are q-twins.

Proof. (i)⇒(ii): Let $q, r, s \in Q$, be such that r and s are q-related. Then $\exists y \in X^*$ such that $\mu^*(q, y, r) > 0$ and $\mu^*(q, y, s) > 0$. Let $p \in S(s)$. Then $\exists x \in X^*$ such that $\mu^*(s, y, p) > 0$. Then $\mu^*(q, yx, p) > 0$. Thus by the hypothesis $p \in S(r)$. Similarly if $p \in S(r)$ then $p \in S(s)$.

(ii)⇒(i): Let $q, r, p \in Q$, $x, y \in X^*$ be such that $\mu^*(q, y, r) > 0$ and $\mu^*(q, yx, p) > 0$. Now $\mu^*(q, yx, p) = \vee\{\mu^*(q, y, s) \wedge \mu^*(s, x, p) | s \in Q\} > 0$. Hence $\exists s \in Q$ such that $\mu^*(q, y, s) > 0$ and $\mu^*(s, x, p) > 0$. Then r and s are q-twins and $p \in S(s)$. Thus by the hypothesis $p \in S(r)$. ■

Proposition 6.6.5 *Let $M = (Q, X, \mu)$ be a ffsm. Then the following assertions are equivalent.*

(i) M is retrievable

(ii) M is quasi-retrievable and $\forall\, q, r, s \in Q$, if r and s are q-related, then r and s are q-twins.

Proof. (i)⇒(ii): It is immediate that retrievability implies quasi-retrievability. Let $q, r, p \in Q$ and $x, y \in X^*$. Suppose that $\mu^*(q, y, r) > 0$ and $\mu^*(q, yx, p) > 0$. Since M is retrievable, $\exists z \in X^*$ such that $\mu^*(p, z, q) > 0$. Thus $p \in S(r)$. Hence (ii) holds by Remark 6.6.4.

(ii)⇒(i): Let $q \in Q$ and $y \in X^*$. Suppose $\exists t \in Q$ such that $\mu^*(q, y, t) > 0$. Then $\exists x \in X^*$ such that $\mu^*(q, yx, q) > 0$ since M is quasi-retrievable. By Remark 6.6.4, $q \in S(t)$. ■

Theorem 6.6.6 *Let $M = (Q, X, \mu)$ be a ffsm. The following assertions are equivalent.*

(i) M satisfies the exchange property,

(ii) M is the union of strongly connected submachines,

(iii) M is retrievable.

Proof. (i)⇒(ii): By (i), $M = \cup_{i=1}^{n} < q_i >$, where $\{q_1, q_2, \ldots, q_n\}$ is a basis of M. Also $S(q_i) \cap S(q_j) = \phi$ if $i \neq j$. Let $p, q \in S(q_i)$. Then $q_i \in S(p)$ and so $q \in S(p)$. Thus $< q_i >$ is strongly connected.

(ii)⇒(i): Now $M = \cup_{i=1}^{n} M_i$, where each $M_i = (Q_i, X, \mu_i)$ is strongly connected. Let $p, q \in Q$. Suppose $p \in S(q)$. Now $\exists i$ such that $q \in Q_i$. Then $p \in S(q) \subseteq S(Q_i) = Q_i$. Thus $p, q \in Q_i$. Since M_i is strongly connected, $q \in S(p)$. Hence M satisfies the exchange property by Proposition 6.5.6.

(ii)⇒(iii): Now $M = \cup_{i=1}^{n} M_i$, where each $M_i = (Q_i, X, \mu_i)$ is strongly connected. Let $q \in Q$, $y \in X^*$ be such that $\mu^*(q, y, t) > 0$ for some $t \in Q$. Now $q \in Q_i$ for some i. Thus $t \in S(q) \subseteq S(Q_i)$. Since M_i is strongly connected, $q \in S(t)$. Hence $\exists x \in X^*$ such that $\mu^*(t, x, q) > 0$. Thus M is retrievable. .

(iii)⇒(ii): Let $q \in Q$ and let $r, t \in S(q)$. Then $\exists y, z \in X^*$ such that $\mu^*(q, y, r) > 0$ and $\mu^*(q, z, t) > 0$. Since M is retrievable $\exists x \in X^*$ such that $\mu^*(r, x, q) > 0$. Hence $q \in S(r)$. Thus $t \in S(r)$. Hence $< q >$ is strongly connected. Now $M = \cup_{q \in Q} < q >$. ■

Definition 6.6.7 *Let $M = (Q, X, \mu)$ be a ffsm. Let $N = (T, X, \nu) \neq \phi$ be a submachine of M. Then N is said to be* ***separated*** *if $S(Q \backslash T) \cap T = \phi$.*

Theorem 6.6.8 *Let $M = (Q, X, \mu)$ be a ffsm. Let $N = (T, X, \nu) \neq \phi$ be a submachine of M. Then N is separated if and only if $S(Q \backslash T) = Q \backslash T$.*

Proof. Suppose N is separated. Let $q \in S(Q \backslash T)$. Now $S(Q \backslash T) \cap T = \phi$. Hence $q \notin T$. Thus $q \in Q \backslash T$. Hence $S(Q \backslash T) \subseteq Q \backslash T$. Thus $S(Q \backslash T) = Q \backslash T$. Conversely, suppose that $S(Q \backslash T) = Q \backslash T$. Clearly then $S(Q \backslash T) \cap T = \phi$. Thus N is separated. ■

Theorem 6.6.9 *Let $M = (Q, X, \mu)$ be a ffsm. Let $N = (T, X, \nu) \neq \phi$ be a submachine of M. If N is separated, then so is $C = (Q \backslash T, X, \sigma)$ where $\sigma = \mu|_{(Q \backslash T) \times X \times (Q \backslash T)}$.*

Proof. Now $\phi \neq T \neq Q$ and so $Q \backslash T \neq \phi$. By Theorem 6.6.8, $S(Q \backslash T) = Q \backslash T$. Hence C is a submachine of M. Now $S(Q \backslash (Q \backslash T)) = S(T) = T$. Hence $S(Q \backslash (Q \backslash T)) \cap (Q \backslash T) = T \cap (Q \backslash T) = \phi$. Hence C is separated. ■

Definition 6.6.10 *Let $M = (Q, X, \mu)$ be a ffsm. Then M is said to be* ***connected*** *if M has no separated proper submachine.*

Theorem 6.6.11 *Let $M = (Q, X, \mu)$ be a ffsm. Then the following assertions are equivalent.*

(i) M is strongly connected,

(ii) M is connected and retrievable.

(iii) Every submachine of M is strongly connected.

Proof. (i)⇒(ii): By Theorem 6.5.11, M does not have any proper submachines and so M has no proper separated submachines. Thus M is connected. We now show that M is retrievable. Let $q, t \in Q$ and $y \in X^*$ be such that $\mu^*(q, y, t) > 0$. Since M is strongly connected, $q \in S(t)$. Then $\exists x \in X^*$ such that $\mu^*(t, x, q) > 0$. Hence M is retrievable.

(ii)⇒(iii): Let $N = (T, X, \nu)$ be a submachine of M. Suppose $p, q \in T$ are such that $p \notin S(q)$. Then $S(q) \neq Q$ and so $K = (S(q), X, \mu|_{S(q) \times X \times S(q)})$ is a proper submachine of M. Since M is connected, $S(Q \backslash S(q)) \cap S(q) \neq \phi$. Let $r \in S(Q \backslash S(q)) \cap S(q)$. Then $r \in S(t)$ for some $t \in Q \backslash S(q)$ and $r \in S(q)$. Now $\exists\ y \in X^*$ such that $\mu^*(t, y, r) > 0$. Since M is retrievable,

$\exists z \in X^*$ such that $\mu^*(r, z, t) > 0$. Thus $t \in S(r)$. Hence $t \in S(r) \subseteq S(q)$, a contradiction. Thus $p \in S(q)$ $\forall p, q \in T$. Hence N is strongly connected.

(iii)⇒(i): Obvious. ■

6.7 Decomposition of Fuzzy Finite State Machines

Definition 6.7.1 *Let $M = (Q, X, \mu)$ be a ffsm. Let P be a submachine of M. Then P is called a* **primary submachine** *of M if*

(i) $\exists q \in Q$ such that $P = < q >$;

(ii) $\forall s \in Q$ if $P \subseteq < s >$, then $P = < s >$.

Theorem 6.7.2 (Decomposition Theorem) *Let $M = (Q, X, \mu)$ be a ffsm. Let $\wp = \{P_1, P_2, \ldots, P_n\}$ be set of all distinct primary submachines of M. Then*

(i) $M = \cup_{i=1}^n P_i$;

(ii) $M \neq \cup_{i=1, i \neq j}^n P_i$ for any $j \in \{1, 2, \ldots, n\}$.

Proof. (i) Let $q_0 \in Q$. Now $\forall q_i \in Q$, either (a) $< q_i > \in \wp$ or (b) $\exists q_{i+1} \in Q \backslash S(q_i)$ such that $< q_i > \subset < q_{i+1} >$. Since Q is finite, either $< q_0 > \in \wp$ or there exists a positive integer k such that $< q_0 > \subset < q_k > \in \wp$. Thus $Q = \cup_{i=1}^n S(p_i)$ where $P_i = < p_i >$, $i = 1, 2, \ldots, n$. Hence $M = \cup_{i=1}^n P_i$.

(ii) Let $N = \cup_{i=1, i \neq j}^n P_i$ and let $P_j = < p_j >$. If $p_j \in \cup_{i=1, i \neq j}^n S(p_i)$, then $p_j \in S(p_i)$ for some $i \neq j$. Hence $P_j = < p_j > \subset P_i$. However this contradicts the maximality of P_j since $P_j \neq P_i$. Thus $p_j \notin \cup_{i=1, i \neq j}^n S(p_i)$. Hence $M \neq N$. ■

Corollary 6.7.3 *Let $M = (Q, X, \mu)$ be a ffsm. Then every singly generated submachine of $M \neq \phi$ is a submachine of a primary submachine of M.* ■

Corollary 6.7.4 *Let $M = (Q, X, \mu)$ be a ffsm. The $P_1, P_2, \ldots, P_n$ in Theorem 6.7.2 are unique.* ■

Definition 6.7.5 *Let $M = (Q, X, \mu)$ be a ffsm. Then* **rank** *of M, rank(M), is the number of distinct primary submachines of M.*

Theorem 6.7.6 *Let $M = (Q, X, \mu)$ be a ffsm. The following assertions are equivalent.*

(i) M is retrievable.

(ii) Every primary submachine of M is strongly connected.

Proof. (i)⇒(ii): Let P be a primary submachine of M. Then $P = < p >$ for some $p \in Q$. Then as in the proof of (iii)⇒(ii) of Theorem 6.6.6, $< p >$ is strongly connected.

(ii)⇒(i): Now $M = \cup_{i=1}^n P_i$ where P_i are primary submachines of M. Then the P_i are strongly connected. Thus M is the union of strongly connected submachines. By Theorem 6.6.6, (i) holds. ■

Lemma 6.7.7 *Let $M = (Q, X, \mu)$ be a ffsm. Then M has a strongly connected submachine.*

Proof. We prove the result by induction on $|Q| = n$. If $n = 1$, then the result is obvious. Suppose the result is true for all ffsms $N = (T, X, \nu)$ such that $|T| < n$, $n > 1$. Let $q \in Q$. Then $M' = (S(q), X, \mu|_{S(q)\times X\times S(q)})$ is a submachine of M. If M' is strongly connected, then the result follows. Suppose that M' is not strongly connected. Then $\exists p \in S(q)$ such that $q \notin S(p)$ and hence $S(p) \subset S(q)$. Now $|S(p)| < n$. Hence by the induction hypothesis the ffsm $M'' = (S(p), X, \mu|_{S(p)\times X\times S(p)})$ has a strongly connected submachine. Since M'' is a submachine of M, M has a strongly connected submachine. ∎

Theorem 6.7.8 *Let $M = (Q, X, \mu)$ be a ffsm. The following assertions are equivalent.*
(i) M is retrievable.
(ii) Every singly generated submachine of M is primary.
(iii) Every nonempty connected submachine of M is primary.

Proof. (i)⇒(ii): Now $M = \cup_{i=1}^{n} P_i$ where the P_i are primary submachines of M. By Theorem 6.5.11, the P_i are strongly connected. Let $N = < q >$ be a singly generated submachine of M. Then $< q > \subseteq P_i$ for some i. Hence $< q > = P_i$ by Theorem 6.5.11. Thus N is primary.

(ii)⇒(i): Since every singly generated submachine of M is primary, every singly generated submachine of M is strongly connected. Thus every primary submachine of M is strongly connected. By Theorem 6.7.6, (i) holds.

(ii)⇒(iii): Let $N = (T, X, \nu)$ be a nonempty connected submachine of M. Let $q \in T$. Suppose $S(q) \neq T$. Since N is connected, $S(T\backslash S(q)) \cap S(q) \neq \phi$. Let $r \in S(T\backslash S(q)) \cap S(q)$. Then $r \in S(t)$ for some $t \in T\backslash S(q)$ and $r \in S(q)$. Now $< r > \subseteq < t >$ and $< r > \subseteq < q >$. Since $< r >$ is primary, $< t > = < r > = < q >$. Hence $t \in S(q)$ which is a contradiction. Hence $N = < q >$ and so N is primary.

(iii)⇒(ii): Let $N = < s >$ be a singly generated submachine. By Lemma 6.7.7, N has a strongly connected submachine $B = < r >$, say. Then B is connected and hence primary. Thus $< r > = < s > = N$. Hence N is primary. ∎

Lemma 6.7.9 *Let $M = (Q, X, \mu)$ be a ffsm and let $N = (T, X, \nu)$ be a separated submachine of M. Then every primary submachine of N is also a primary submachine of M.*

Proof. Let $< q >$ be a primary submachine of N. Suppose $< q >$ is not a primary submachine of M. Then $\exists p \in Q\backslash S(q)$ such that $< q > \subset < p >$. Clearly $p \notin T$. Thus $p \in Q\backslash T$. Since $q \in S(p)$, $q \in S(Q\backslash T)$. Thus $q \in S(Q\backslash T) \cap T$, which a contradiction since N is separated. Hence $< q >$ is a primary submachine of M. ∎

Theorem 6.7.10 *Let $M = (Q, X, \mu)$ be a ffsm and let $N_i = (T_i, X, \nu_i), i = 1, 2, ..., n$ be the primary submachines of M. Then a proper submachine $N = (T, X, \nu)$ of M is separated if and only if for some $J \subseteq \{1, 2, \ldots, n\}$, $J \neq \phi$, $Q \backslash T = \cup_{i \in J} T_i$.*

Proof. Suppose $N = (T, X, \nu)$ be a proper separated submachine of M. Then $S(Q \backslash T) = Q \backslash T$. Since N is proper, the submachine $< Q \backslash T >$ is nonempty. Thus $< Q \backslash T >$ is the union of all its primary submachines. Since $< Q \backslash T >$ is separated every primary submachine of $< Q \backslash T >$ is a primary submachine of M. Thus $S(Q \backslash T) = \cup_{i \in J} T_i$ for some $J \subseteq \{1, 2, \ldots, n\}$, $J \neq \phi$. Since $Q \backslash T = S(Q \backslash T)$, $Q \backslash T = \cup_{i \in J} T_i$ for some $J \subseteq \{1, 2, \ldots, n\}$, $J \neq \phi$. Conversely, let $N = (T, X, \nu)$ be a proper submachine of M such that $Q \backslash T = \cup_{i \in J} T_i$ for some $J \subseteq \{1, 2, \ldots, n\}$, $J \neq \phi$. Then $S(Q \backslash T) = S(\cup_{i \in J} T_i) = \cup_{i \in J} S(T_i) = \cup_{i \in J} T_i = Q \backslash T$. Hence N is separated. ■

Corollary 6.7.11 *Let $M = (Q, X, \mu)$ be a ffsm. Then M is connected if and only if M has no proper submachine $N = (T, X, \nu)$ such that $Q \backslash T$ is the union of the sets of states of all primary submachines of M.* ■

Definition 6.7.12 *Let $M = (Q, X, \mu)$ be a ffsm and let $N = (T, X, \nu)$ be a submachine of M. A subset $R \subseteq Q$ is called a* ***generating set*** *of N, and is said to* ***generate*** *N if $N = < R >$.*

Lemma 6.7.13 *Let $M = (Q, X, \mu)$ be a ffsm and let $N_i = (T_i, X, \nu_i), i = 1, 2, ..., n$ be the primary submachines of M. Let $R \subseteq Q$. Then R generates M if and only if $\forall i, 1 \leq i \leq n, \exists r_i \in R$ such that $N_i = < r_i >$.*

Proof. Suppose that R generates M. Then $M = < R > = \cup_{r \in R} < r >$. Let $q_i \in T_i$ be such that $T_i = < q_i >$, $i = 1, 2, \ldots, n$. Then $q_i \in \cup_{r \in R} < r >$ and so $q_i \in < r >$ for some $r \in R$. Thus $< q_i > \subseteq < r >$. Since $< q_i >$ is primary, $< q_i > = < r >$. The converse is immediate. ■

Definition 6.7.14 *Let $M = (Q, X, \mu)$ be a ffsm. Let $R \subseteq Q$ be a generating set of M. Then R is said to be a* ***minimal generating set*** *of M if*

(i) $M = < R >$, and
(ii) $\forall r \in R$, $< R \backslash \{r\} > \neq M$.

Theorem 6.7.15 *Let $M = (Q, X, \mu)$ be a ffsm. Let $R \subseteq Q$ be a generating set of M. Then R is a minimal generating set of M if and only if $|R| =$ rank(M).*

Proof. Let $n =$ rank(M) and let $N_i = (T_i, X, \nu_i)$ be a primary submachine of M, $i = 1, 2, \ldots, n$. By Lemma 6.7.13, since R is a generating set of M, $\exists r_i \in R$ such that $< r_i > = T_i$, $i = 1, 2, \ldots, n$. Since the T_i are distinct, the r_i are distinct. Thus $|R| \geq$ rank(M). Now assume that R is a minimal generating set. Suppose $|R| >$ rank(M). Then $\exists r \in R$ such that

$r \notin \{r_1, r_2, \dots, r_n\}$. Thus $< R \backslash \{r\} > = Q$. Hence R is not minimal, a contradiction. Thus $|R| = \text{rank}(M)$. Conversely suppose that $|R| = \text{rank}(M)$. Then $R = \{r_1, r_2, \dots, r_n\}$. Hence $< R \backslash \{r_i\} > = \cup_{j \neq i} N_j \neq M$. Thus R is minimal. ■

Theorem 6.7.16 *Let $M = (Q, X, \mu)$ be a ffsm. Then M is not connected if and only if $\exists$ a generating set R of M with nonempty subsets R_1 and R_2 such that $< R_1 > \cap < R_2 > = \phi$ and $M = < R_1 > \cup < R_2 >$.*

Proof. Suppose that M is not connected. Then $\exists$ a proper submachine $N = (T, X, \nu)$ of M which is separated., i.e., $S(Q \backslash T) \cap T = \phi$. Let R_1 be a generating set of $< S(Q \backslash T) >$ and let R_2 be a generating set of N. Then $< R_1 > \cap < R_2 > = \phi$ and $M = < R_1 > \cup < R_2 >$. Conversely, suppose that R_1 and R_2 exists. Let $N = < R_2 >$. Then $< S(Q \backslash T) > = < R_1 >$. Hence $N = (T, X, \mu|_{T \times X \times T})$ is a proper submachine of M which is separated. ■

6.8 Admissible Partitions

In this section, we introduce the concept of a covering of a ffsm by another, admissible partitions and relations of a ffsm, μ-orthogonality of admissible partitions, irreducible ffsm, and the quotient of a ffsm induced by an admissible partition of the state set. We derive results concerning μ-orthogonality and covering, Theorems 6.8.13 and 6.8.14. We show that an admissible partition π of Q is maximal if and only if the quotient M/π is irreducible, Theorem 6.8.19. We also show that a ffsm can be covered by irreducible ffsms, Theorem 6.8.20. These results allow us to study fuzzy finite state machines via coverings of products of simpler machines. Irreducible finite state machines seems to arise naturally in some applications, e.g., biology. An example is given in [6] of a finite state machine arising from a metabolic pathway.

Definition 6.8.1 *Let $M = (Q, X, \mu)$ be a ffsm and $\mathcal{P} = \{Q_1, Q_2, \dots, Q_k\}$ be a partition of Q. Then $\mathcal{P}$ is called an* **admissible partition** *of Q if the following holds: Let $a \in X$, then $\forall i$, $\exists j$, $1 \leq i, j \leq k$ such that $\forall p_1, p_2 \in Q_i$, if $\mu(p_1, a, r) > 0$ for some $r \in Q$, then $\exists$ $t \in Q$ such that $\mu(p_2, a, t) \geq \mu(p_1, a, r)$ and $t, r \in Q_j$.*

Proposition 6.8.2 *Let $M = (Q, X, \mu)$ be a ffsm.*
(i) Let $1_Q = \{\{q\} \mid q \in Q\}$. Then 1_Q is an admissible partition of Q.
(ii) $\{Q\}$ is an admissible partition of Q.

Theorem 6.8.3 *Let $M = (Q, X, \mu)$ be a ffsm and $\mathcal{P} = \{Q_1, Q_2, \dots, Q_k\}$ be a partition of Q. The following are equivalent:*
(i) $\mathcal{P}$ is an admissible partition of Q.

(ii) Let $x \in X^*$. Then $\forall i$, $\exists j$, $1 \leq i, j \leq k$ such that $\forall p_1, p_2 \in Q_i$, if $\mu^*(p_1, x, r) > 0$ for some $r \in Q$, then $\exists\ t \in Q$ such that $\mu^*(p_2, x, t) \geq \mu^*(p_1, x, r)$ and $t, r \in Q_j$.

Proof. (i)$\Rightarrow$(ii) Let $x \in X^*$ and $len(x) = n$. Let $p_1, p_2 \in Q_i$ and $\mu^*(p_1, x, r) > 0$ for some $r \in Q$. If $n = 0$, then $x = \Lambda$ and $\mu^*(p_1, x, r) > 0$ implies that $p_1 = r$. Thus $\mu^*(p_2, x, p_2) = 1 = \mu^*(p_1, x, p_1)$. In this case $i = j$. Hence the result is true for $n = 0$. Suppose the result is true for all $y \in X^*$ such that $len(y) = n - 1$, where $n > 0$. Let $x = ya$, where $a \in X$. Now $\mu^*(p_1, x, r) = \mu^*(p_1, ya, r) = \vee\{\mu^*(p_1, y, s) \wedge \mu^*(s, a, r) \mid s \in Q\} > 0$. Since Q is finite, there exists $t \in Q$ such that $\mu^*(p_1, y, t) \wedge \mu^*(t, a, r) = \mu^*(p_1, ya, r)$. Thus $\mu^*(p_1, y, t) > 0$ and $\mu(t, a, r) = \mu^*(t, a, r) > 0$. By the induction hypothesis, there exists j and there exists $s \in Q$ such that $\mu^*(p_2, y, s) \geq \mu^*(p_1, y, t)$ and $s, t \in Q_j$. Now $s, t \in Q_j$ and $\mu(t, a, r) > 0$. Hence by (i), there exists l and there exists $q \in Q$ such that $\mu(s, a, q) \geq \mu(t, a, r)$ and $r, q \in Q_l$. Now $\mu^*(p_2, x, q) = \mu^*(p_2, ya, q) \geq \mu^*(p_2, y, s) \wedge \mu(s, a, q) \geq \mu^*(p_1, y, t) \wedge \mu(t, a, r) = \mu^*(p_1, ya, r) = \mu^*(p_1, x, r)$ and $q, r \in Q_l$. The result now follows by induction. ∎

Corollary 6.8.4 *Let $M = (Q, X, \mu)$ be a ffsm. Then every admissible partition $\mathcal{P}$ of Q induces an admissible relation $\sim$ on Q such that the set of all equivalence classes of $\sim$ is $\mathcal{P}$. Conversely, the set of all equivalence classes of an admissible relation on Q is an admissible partition of Q.* ∎

Lemma 6.8.5 *Let $M = (Q, X, \mu)$ be a ffsm and $\pi = \{H_i \mid i \in I\}$ be an admissible partition of Q. Let $i, j \in I$. Then $\forall q, q' \in H_i$, $\forall a \in X$,*

$$\vee\{\mu(q, a, r) \mid r \in H_j\} = \vee\{\mu(q', a, r) \mid r \in H_j\}.$$

Proof. Let $q, q' \in H_i$, $a \in X$, $A = \{\mu(q, a, r) \mid r \in H_j\}$, and $B = \{\mu(q', a, r) \mid r \in H_j\}$. Suppose $\mu(q, a, r) > 0$ for some $r \in H_j$. Since π is an admissible partition, there exists $r' \in Q$ such that $\mu(q', a, r') \geq \mu(q, a, r)$. Again by the admissibility of π, $r' \in H_j$. Similarly if $\mu(q', a, p) > 0$ for some $p \in H_j$, then there exists $p' \in H_j$ such that $\mu(q, a, p') \geq \mu(q', a, p)$. Also, by the admissibility of π, it follows that $\mu(q, a, r) = 0\ \forall r \in H_j$ if and only if $\mu(q', a, r) = 0\ \forall r \in H_j$. Hence

$$\vee\{\mu(q, a, r) \mid r \in H_j\} = \vee\{\mu(q', a, r) \mid r \in H_j\}.\ ∎$$

Theorem 6.8.6 *Let $M = (Q, X, \mu)$ be a ffsm. Let $\pi = \{H_i \mid i \in I\}$ be an admissible partition of Q. Define*

$$\mu^\pi : \pi \times X \times \pi \to [0, 1]$$

by

$$\mu^\pi(H_i, a, H_j) = \vee\{\mu(q, a, r) \mid r \in H_j\}$$

$\forall H_i, H_j \in \pi$ and $a \in X$, where $q \in H_i$. Then $M/\pi = (\pi, X, \mu^\pi)$ is a ffsm, called the ***quotient fuzzy finite state machine*** *with respect to π.*

Proof. By Lemma 6.8.5, μ^π is well-defined. ∎

Proposition 6.8.7 *Let $M = (Q, X, \mu)$ be a ffsm. Let $\pi = \{H_i \mid i \in I\}$ be an admissible partition of Q. Then for $q \in H_i$*

$$\mu^{\pi *}(H_i, x, H_j) \leq \vee\{\mu^*(q, x, r) \mid r \in H_j\}$$

$\forall H_i, H_j \in \pi$ and $x \in X^*$.

Proof. Let $H_i, H_j \in \pi$ and $x \in X^*$. Let $len(x) = n$. If $n = 0$, then $x = \Lambda$. If $H_i = H_j$ then $\mu^{\pi *}(H_i, x, H_j) = 1$ and $\vee\{\mu^*(q, x, r) \mid r \in H_j\} = \vee\{\mu^*(q, x, r) \mid r \in H_i\} = \mu^*(q, x, q) = 1$, where $q \in H_i$. If $H_i \neq H_j$, then $\mu^{\pi *}(H_i, x, H_j) = 0$ and $H_i \cap H_j = \phi$. Since $H_i \cap H_j = \phi$ and $q \in H_i$, $\vee\{\mu^*(q, x, r) \mid r \in H_j\} = 0$. Hence $\mu^{\pi *}(H_i, x, H_j) = \vee\{\mu^*(q, x, r) \mid r \in H_j\}$. Suppose that the result is true $\forall y \in X^*$ such that $len(y) = n - 1$, $n > 0$. Let $n > 0$ and $x = ya$, where $y \in X^*$, $a \in X$ and $len(y) = n - 1$. Now for $q \in H_i$ and $s \in H_k$

$$\begin{aligned}
\mu^{\pi *}(H_i, x, H_j) &= \mu^{\pi *}(H_i, ya, H_j) \\
&= \vee\{\mu^{\pi *}(H_i, y, H_k) \wedge \mu^{\pi *}(H_k, a, H_j) \mid H_k \in \pi\} \\
&= \vee\{(\vee\{\mu^*(q, y, r) \mid r \in H_k\}) \wedge \\
&\qquad (\vee\{\mu^*(s, a, p) \mid p \in H_j\}) \mid H_k \in \pi\} \\
&= \vee\{\vee\{\mu^*(q, y, r) \wedge \mu^*(s, a, p) \mid r \in H_k, p \in H_j\} \\
&\qquad \mid H_k \in \pi\} \\
&\leq \vee\{\vee\{\mu^*(q, y, r) \wedge \mu^*(r, a, p) \mid r \in H_k, p \in H_j\} \\
&\qquad \mid H_k \in \pi\} \\
&= \vee\{\vee\{\mu^*(q, y, r) \wedge \mu^*(r, a, p) \mid r \in H_k,\ H_k \in \pi\} \\
&\qquad \mid p \in H_j\} \\
&= \vee\{\vee\{\mu^*(q, y, r) \wedge \mu^*(r, a, p) \mid r \in Q\} \mid p \in H_j\} \\
&= \vee\{\mu^*(q, ya, p) \mid p \in H_j\} \\
&= \vee\{\mu^*(q, x, p) \mid p \in H_j\}. \blacksquare
\end{aligned}$$

Proposition 6.8.8 *Let $M = (Q, X, \mu)$ be a ffsm. Let $\pi = \{H_i \mid i \in I\}$ be an admissible partition of Q. Then for all $x \in X^*$*

$$\mu^*(q, x, p) \leq \mu^{\pi *}(H_i, x, H_j),$$

where $q \in H_i$, $p \in H_j$.

Proof. Let $q \in H_i$, $p \in H_j$. Let $x \in X^*$ and $len(x) = n$. If $n = 0$, then $x = \Lambda$. If $q = p$, then $H_i = H_j$ and $\mu^*(q, x, p) = 1 = \mu^{\pi *}(H_i, x, H_i)$. Suppose $q \neq p$. Then $\mu^*(q, x, p) = 0 \leq \mu^{\pi *}(H_i, x, H_j)$. Suppose $n = 1$. Then $x = a \in X$ and

$$\begin{aligned}
\mu^*(q, a, p) &= \mu(q, a, p) \\
&\leq \vee\{\mu(q, a, r) \mid r \in H_j\} \\
&= \mu^\pi(H_i, a, H_j) \\
&= \mu^{\pi *}(H_i, a, H_j).
\end{aligned}$$

Hence the result is true for $n = 0$ and $n = 1$. Suppose that the result is true $\forall y \in X^*$ such that $len(y) = n - 1$, $n > 0$. Let $n > 0$ and $x = ya$, where $y \in X^*$, $a \in X$ and $len(y) = n - 1$. Then

$$\begin{aligned}
\mu^*(q, x, p) &= \mu^*(q, ya, p) \\
&= \vee\{\mu^*(q, y, r) \wedge \mu(r, a, p) \mid r \in Q\} \\
&= \vee\{\vee\{\mu^*(q, y, r) \wedge \mu(r, a, p) \mid r \in H_k\} \mid H_k \in \pi\} \\
&\leq \vee\{\vee\{\mu^{\pi*}(H_i, y, H_k) \wedge \mu^{\pi}(H_k, a, H_j) \mid r \in H_k\} \mid H_k \in \pi\} \\
&\qquad \text{(by induction and } n = 1 \text{ case)} \\
&= \vee\{\mu^{\pi*}(H_i, y, H_k) \wedge \mu^{\pi}(H_k, a, H_j) \mid H_k \in \pi\} \\
&= \mu^{\pi*}(H_i, x, H_j).
\end{aligned}$$

The result now follows by induction. ■

Corollary 6.8.9 *Let $M = (Q, X, \mu)$ be a ffsm. Let $\pi = \{H_i \mid i \in I\}$ be an admissible partition of Q. Then for all $x \in X^*$, for all $H_i, H_j \in \pi$,*

$$\vee\{\mu^*(q, x, p) \mid p \in H_j\} \leq \mu^{\pi*}(H_i, x, H_j),$$

where $q \in H_i$. ■

Theorem 6.8.10 *Let $M = (Q, X, \mu)$ be a ffsm. Let $\pi = \{H_i \mid i \in I\}$ be an admissible partition of Q. Then for $q \in H_i$*

$$\mu^{\pi*}(H_i, x, H_j) = \vee\{\mu^*(q, x, r) \mid r \in H_j\}$$

$\forall H_i, H_j \in \pi$ and $x \in X^*$ ■

Definition 6.8.11 *Let $M = (Q, X, \mu)$ be a ffsm. Let π and τ be admissible partitions of Q. π and τ are called **μ-orthogonal** if*

(i) $\pi \cap \tau = 1_Q$, and

(ii) $\forall H_i, H_u \in \pi$, $\forall K_j, K_v \in \tau$, $\forall a \in X$, if $H_i \cap K_j = \{q_0\}$ and $H_u \cap K_v = \{p_0\}$, then

$$\mu(q_0, a, p_0) = \vee\{\mu(q_0, a, p) \wedge \mu(q_0, a, p') \mid p \in H_u, p' \in K_v\}.$$

Theorem 6.8.12 *Let $M = (Q, X, \mu)$ be a ffsm. Let π and τ be admissible partitions of Q. Then π and τ are μ-orthogonal if and only if*

(i) $\pi \cap \tau = 1_Q$, and

(ii) $\forall H_i, H_u \in \pi$, $\forall K_j, K_v \in \tau$, $\forall x \in X^*$, if $H_i \cap K_j = \{q_0\}$ and $H_u \cap K_v = \{p_0\}$, then

$$\mu^*(q_0, x, p_0) = \vee\{\mu^*(q_0, x, p) \wedge \mu^*(q_0, x, p') \mid p \in H_u, p' \in K_v\}.$$

Proof. Suppose π and τ are called μ-orthogonal. Then clearly (i) holds.

(ii) Let $H_i, H_u \in \pi$, $K_j, K_v \in \tau$, $x \in X^*$, $H_i \cap K_j = \{q_0\}$, and $H_u \cap K_v = \{p_0\}$. Suppose $len(x) = n$. If $n = 0$, then $x = \Lambda$. Now $\mu^*(q_0, x, p_0) = 0$ if $q_0 \neq p_0$ and $\mu^*(q_0, x, p_0) = 1$ if $q_0 = p_0$. Suppose $q_0 \neq p_0$. Then either

$H_i \cap H_u = \phi$ or $K_j \cap K_v = \phi$, say, $H_i \cap H_u = \phi$. Thus $q_0 \notin H_u$ and $q_0 \notin K_v$. Hence $\vee\{\mu^*(q_0, x, p) \wedge \mu^*(q_0, x, p') \mid p \in H_u, p' \in K_v\} = \vee\{0 \wedge \mu^*(q_0, x, p') \mid p \in H_u, p' \in K_v\} = 0$. Suppose $q_0 = p_0$. Then $H_i = H_u$ and $K_j = K_v$. Thus $\vee\{\mu^*(q_0, x, p) \wedge \mu^*(q_0, x, p') \mid p \in H_u, p' \in K_v\} = \mu^*(q_0, x, p_0) \wedge \mu^*(q_0, x, p_0) = 1$. Hence if $n = 0$, then $\mu^*(q_0, x, p_0) = \vee\{\mu^*(q_0, x, p) \wedge \mu^*(q_0, x, p') \mid p \in H_u, p' \in K_v\}$. Suppose that the result is true $\forall y \in X^*$ such that $len(y) = n - 1$, $n > 0$. Let $n > 0$ and $x = ya$, where $y \in X^*$, $a \in X$ and $len(y) = n - 1$. Then

$$\begin{aligned}
\mu^*(q_0, x, p_0) &= \mu^*(q_0, ya, p_0) \\
&= \vee\{\mu^*(q_0, y, r) \wedge \mu(r, a, p_0) \mid r \in Q\} \\
&= \vee\{\mu^*(q_0, y, r) \wedge (\vee\{\mu(r, a, p) \wedge \mu(r, a, p') \mid p \in H_u, \\
&\qquad p' \in K_v\}) \mid r \in Q\} \\
&= \vee\{(\vee(\mu^*(q_0, y, r) \wedge \mu(r, a, p))) \wedge (\vee(\mu^*(q_0, y, r) \\
&\qquad \wedge \mu(r, a, p'))) \mid p \in H_u, p' \in K_v,\ r \in Q\} \\
&= \vee\{(\vee\{\mu^*(q_0, y, r) \wedge \mu(r, a, p) \mid r \in Q\}) \wedge \\
&\qquad (\vee\{\mu^*(q_0, y, r) \wedge \mu(r, a, p') \mid r \in Q\}) \mid p \in H_u, p' \in K_v\} \\
&= \vee\{\mu^*(q_0, ya, p) \wedge \mu^*(q_0, ya, p') \mid p \in H_u, p' \in K_v\} \\
&= \vee\{\mu^*(q_0, x, p) \wedge \mu^*(q_0, x, p') \mid p \in H_u, p' \in K_v\}.
\end{aligned}$$

The result now follows by induction. The converse is trivial. ∎

Let $M = (Q, X, \mu)$ be a ffsm. Let π and τ be admissible partitions of Q. Consider the ffsms $M/\pi = (\pi, X, \mu^\pi)$ and $M/\tau = (\tau, X, \mu^\tau)$. Define $\mu^\wedge : (\pi \times \tau) \times X \times (\pi \times \tau) \to [0, 1]$ by

$$\mu^\wedge((H_i, K_j), a, (H_u, K_v)) = \mu^\pi(H_i, a, H_u) \wedge \mu^\tau(K_j, a, K_v)$$

$\forall H_i, H_u \in \pi, K_j, K_v \in \tau$ and $a \in X$. Then $M/\pi \wedge M/\tau = (\pi \times \tau, X, \mu^\wedge)$ is a fuzzy finite state machine. Note that

$$\mu^{\wedge *}((H_i, K_j), x, (H_u, K_v)) = \mu^{\pi *}(H_i, x, H_u) \wedge \mu^{\tau *}(K_j, x, K_v)$$

$\forall H_i, H_u \in \pi, K_j, K_v \in \tau$ and $x \in X^*$.

Theorem 6.8.13 *Let $M = (Q, X, \mu)$ be a ffsm. Let π and τ be admissible partitions of Q which are μ-orthogonal. Then $M \leq M/\pi \wedge M/\tau$.*

Proof. Define $\eta : \pi \times \tau \to Q$ by $\eta((H_i, K_j)) = q_0$, where $H_i \cap K_j = \{q_0\}$. Since π and τ are μ-orthogonal, η is one-to-one. Let ζ be the identity map on X. Let $H_i, H_u \in \pi, K_j, K_v \in \tau$ and $x \in X^*$. Suppose $H_i \cap K_j = \{q_0\}$ and $H_u \cap K_v = \{p_0\}$. Then

$$\mu^*(\eta((H_i, K_j)), x, \eta((H_u, K_v))) = \mu^*(q_0, x, p_0).$$

Also,

$$\begin{aligned}
\mu^{\wedge *}((H_i, K_j), x, (H_u, K_v)) &= \mu^{\pi *}(H_i, a, H_u) \wedge \mu^{\tau *}(K_j, x, K_v) \\
&= (\vee\{\mu^*(q_0, x, p) \mid p \in H_u\}) \wedge \\
&\quad (\vee\{\mu^*(q_0, x, p') \mid p' \in K_v\}) \\
&= \vee\{\mu^*(q_0, x, p) \wedge \mu^*(q_0, x, p') \\
&\quad \mid p \in H_u,\ p' \in K_v\} \\
&= \mu^*(q_0, x, p_0),
\end{aligned}$$

where the last inequality holds since π and τ are μ-orthogonal. Thus

$$\mu^*(\eta((H_i, K_j)), x, \eta((H_u, K_v))) = \mu^{\wedge *}((H_i, K_j), x, (H_u, K_v)).$$

Now

$$\begin{aligned}
\mu^{\wedge *}((H_i, K_j), x, (H_u, K_v)) &= \vee\{\mu^{\wedge *}((H_i, K_j), x, (H_r, K_s)) | \eta((H_r, K_s)) \\
&= \eta((H_u, K_v)), (H_r, K_s) \in \pi \times \tau\}
\end{aligned}$$

since η is one-to-one. Hence $\mu^*(\eta((H_i, K_j)), x, \eta((H_u, K_v))) = \vee\{\mu^{\wedge *}((H_i, K_j), x, (H_r, K_s)) \mid \eta((H_r, K_s)) = \eta((H_u, K_v)), (H_r, K_s) \in \pi \times \tau\}$. Consequently $M \leq M/\pi \wedge M/\tau$. ■

Theorem 6.8.14 *Let $M = (Q, X, \mu)$ be a ffsm. Let π be an admissible partitions of Q. If there exists a partition τ of Q such that π and τ are μ-orthogonal, then there exists a ffsm N such that $M \leq N \omega M/\pi$.*

Proof. Let $\pi = \{H_i\}_{i \in I}$ and $\tau = \{K_j\}_{j \in J}$ be μ-orthogonal partitions of Q. Let $N = (\tau, \pi \times X, \mu')$, where

$$\mu'(K_j, (H_i, a), K_v) = \vee\{\mu(q_0, a, p) \mid p \in K_v\},$$

and $\{q_0\} = H_i \cap K_j$. Since τ is admissible, μ' is well-defined. Define $\omega : \pi \times X \to \pi \times X$ to be the identity map. Define $\eta : \tau \times \pi \to Q$ by $\eta((K_j, H_i)) = q_0$, where $\{q_0\} = H_i \cap K_j$. Then η is one-to-one and onto. Let ζ be the identity map on X. Then for $\{p_0\} = H_u \cap K_v$,

$$\begin{aligned}
\mu(\eta((K_j, H_i)), a, \eta((K_v, H_u))) &= \mu(q_0, a, p_0) \\
&= \vee\{\mu(q_0, a, p') \wedge \mu(q_0, a, p) \\
&\quad \mid (p', p) \in K_v \times H_u\} \\
&= (\vee\{\mu(q_0, a, p') \mid p' \in K_v\}) \wedge \\
&\quad (\vee\{\mu(q_0, a, p) \mid p \in H_u\}) \\
&= \mu'(K_j, \omega(H_i, a), K_v) \wedge \mu^{\pi}(H_i, a, H_u) \\
&= \mu^{\omega}((K_j, H_i), a, (K_v, H_u)).
\end{aligned}$$

Thus

$$\mu^*(\eta((K_j, H_i)), x, \eta((K_v, H_u))) = \mu^{\omega *}((K_j, H_i), x, (K_v, H_u)) \qquad (6.1)$$

for $x \in X^*$ such that $len(x) = 1$. Suppose that (6.1) is true if $len(x) = n-1$, $n > 0$, where $x \in X^*$. Now $\mu^*(\eta((K_j, H_i)), xa, \eta((K_v, H_u))) = \vee\{\mu^*(\eta((K_j, H_i)), x, \eta((K, H))) \wedge \mu(\eta((K, H)), a, \eta((K_v, H_u))) \mid \eta(K, H) \in \eta(\tau\times\pi)\}$ (since η is onto) $= \vee\{\mu^{\omega *}((K_j, H_i), x, (K, H)) \wedge \mu^{\omega}((K, H)), a, (K_v, H_u)) \mid (K, H) \in (\tau\times\pi)\} = \mu^{\omega *}((K_j, H_i), xa, (K_v, H_u))$. Now $\mu^*(\eta((K_j, H_i)), \Lambda, \eta((K_v, H_u))) = 1$ if and only if $\eta((K_j, H_i)) = \eta((K_v, H_u))$ if and only if $(K_j, H_i) = (K_v, H_u)$ (since η is one-to-one) if and only if $\mu^{\omega *}((K_j, H_i), \Lambda, (K_v, H_u)) = 1$. From this, it follows it follows that (6.1) holds for $x = \Lambda$. Hence (6.1) holds $\forall x \in X^*$. Thus by induction, (η, ζ) is a covering of M by $N\omega M/\pi$. ∎

Definition 6.8.15 *Let Q be a nonempty set and π and τ be partitions of Q. Then $\pi \leq \tau$ if $\forall A \in \pi$, there exists $B \in \tau$ such that $A \subseteq B$.*

The proof of the following lemma is straightforward.

Lemma 6.8.16 *Let Q be a finite nonempty set. Let $\pi = \{H_i\}_{i=1}^n$ and $\tau = \{K_j\}_{j=1}^m$ be partitions of Q such that $\pi \leq \tau$. Then $\forall j$, $1 \leq j \leq m$, $K_j = H_{j_1} \cup H_{j_2} \cup \ldots \cup H_{j_{r_j}}$ for some $H_{j_1}, H_{j_2}, \ldots, H_{j_{r_j}} \in \pi$ and $m \leq n$. If $m = n$, the $\pi = \tau$.* ∎

Definition 6.8.17 *Let $M = (Q, X, \mu)$ be a ffsm. Let π be an admissible partition of Q. Then π is called* ***maximal*** *if π is nontrivial and if τ is any admissible partition of Q with $\pi \leq \tau \leq \{Q\}$, then either $\tau = \pi$ or $\tau = \{Q\}$.*

Definition 6.8.18 *Let $M = (Q, X, \mu)$ be a ffsm. Then M is called* ***irreducible*** *if $|Q| > 1$ and 1_Q and $\{Q\}$ are the only admissible partitions of Q.*

Theorem 6.8.19 *Let $M = (Q, X, \mu)$ be a ffsm. Let $\pi = \{H_i\}_{i=1}^n$ be an admissible partition of Q. Then π is maximal if and only if M/π is irreducible.*

Proof. Suppose π is maximal. Now $M/\pi = (\pi, X, \mu^\pi)$ where

$$\mu^\pi(H_i, a, H_j) = \vee\{\mu(q, a, r) \mid r \in H_j\}$$

$\forall H_i, H_j \in \pi$ and $a \in X$, where $q \in H_i$. Since π is maximal, $\pi \neq \{Q\}$. Thus $|\pi| > 1$. Let $\bar{\pi}$ be an admissible partition of π. Suppose $\bar{\pi} \neq 1_\pi$. Then there exists $\tau \subseteq \pi$ such that $\tau \in \bar{\pi}$ and $|\tau| > 1$. Suppose that $\tau \neq \pi$. Without loss of generality we may assume that $\tau = \{H_1, \ldots, H_m\}$, where $1 < m < n$. Let

$$\pi' = \{H_1 \cup \ldots \cup H_m, H_{m+1} \cup \ldots \cup H_n\}.$$

Then $\pi < \pi'$ and π' is a partition of Q. We now show that π' is admissible. It suffices to consider $q, p \in H_1 \cup \ldots \cup H_m$ with $q \in H_1$ and $p \in H_2$. Suppose $\mu(p, a, r) > 0$ where $r \in H_i$. Then

$$\mu^\pi(H_2, a, H_i) = \vee\{\mu(p, a, r') \mid r' \in H_i\} > 0.$$

Since $\bar{\pi}$ is admissible, there exists $K \in \vec{\pi}$ such that

$$\mu^{\pi}(H_1, a, K) \geq \mu^{\pi}(H_2, a, H_i)$$

and K and H_i belong to the same element of $\bar{\pi}$. Hence

$$\vee\{\mu(q, a, t') \mid t' \in K\} \geq \vee\{\mu(p, a, r') \mid r' \in H_i\}.$$

This implies that there exists $t \in K$ such that $\mu(q, a, t) \geq \mu(p, a, r) > 0$. Now $H_i \in \tau$ if and only $K \in \tau$ since K and H_i belong to the same element of $\bar{\pi}$. If $K, H_i \in \tau$, then $t, r \in H_1 \cup \ldots \cup H_m$ and if $K, H_i \notin \tau$, then $t, r \in H_{m+1} \cup \ldots \cup H_n$, i.e., t and r belong to the same element of π'. Hence π' is admissible. Since π is maximal, it follows that $\pi' = \{Q\}$ and so $\tau = \{Q\}$. This implies that $\bar{\pi} = \{\pi\}$. Thus M/π is irreducible.

Conversely, suppose that M/π is irreducible. Let τ be an admissible partition of Q such that $\pi \leq \tau \leq \{Q\}$. Suppose that $\pi \neq \tau$. By Lemma 6.8.16, there is no loss of generality in assuming that $\tau = \{H_1 \cup \ldots \cup H_m, H_{m+1}, \ldots, H_n\}$, $1 < m \leq n$. Since $\tau \neq \pi$,

$$\bar{\tau} = \{\{H_1, \ldots, H_m\}, \{H_{m+1}\}, \ldots, \{H_n\}\} \neq 1_{\pi}.$$

We now show that $\bar{\tau}$ is admissible. It suffices to consider H_1, H_2. Suppose $\mu^{\pi}(H_1, a, H_i) > 0$ for some $H_i \in \pi$. Then $\vee\{\mu(q, a, r') \mid r' \in H_i\} > 0$ where $q \in H_1$. Let $r \in H_i$ be such that $\mu(q, a, r) = \vee\{\mu(q, a, r') \mid r' \in H_i\} > 0$. Since τ is admissible, $\forall p \in H_2$, there exists $t_p \in Q$ such that $\mu(p, a, t_p) \geq \mu(q, a, r)$ and r, t_p are in the same element of τ $\forall p \in H_2$. Now if $H_i \notin \{H_1, \ldots, H_m\}$, then $r, t_p \in H_i$ $\forall p \in H_2$ and $\mu^{\pi}(H_2, a, H_i) \geq \mu(p, a, t_p) \geq \mu(q, a, r) = \mu^{\pi}(H_1, a, H_i)$. Suppose $H_i \in \{H_1, \ldots, H_m\}$. Then $r, t_p \in H_1 \cup \ldots \cup H_m$ for all $p \in H_2$. Let $\mu(p, a, t_{p'}) = \vee\{\mu(p, a, t_p) \mid p \in H_2\}$. Since $t_{p'} \in H_1 \cup \ldots \cup H_m$, $t_{p'} \in H_k$ for some k, $1 \leq k \leq m$. Hence $\mu^{\pi}(H_2, a, H_k) = \vee\{\mu(q, a, r') \mid r' \in H_k\} \geq \mu(p, a, t_{p'}) \geq \mu(q, a, r) = \mu^{\pi}(H_1, a, H_i)$ and H_k, $H_i \in \{H_1, \ldots, H_m\}$. Consequently $\bar{\tau}$ is an admissible partition of π. Since M/π is irreducible, $\bar{\tau} = \{\pi\}$ and so $\tau = \{Q\}$. Hence π is maximal. ■

Theorem 6.8.20 *Let $M = (Q, X, \mu)$ be a ffsm and $|Q| = n \geq 2$. Then*

$$M \leq N_1\omega_1 N_2\omega_2 \ldots \omega_{m-1} N_m,$$

where $N_1, N_2, \ldots, N_m$ are irreducible ffsms and the state sets Q_i of N_i are such that $|Q_i| < n$.

Proof. Since $|Q| \geq 2$, we can choose a maximal admissible partition π of Q. Clearly $|\pi| < |Q|$. By Theorem 6.8.14, there exists a ffsm N such that $M \leq N\omega M/\pi$ for some suitable ω. Since π is maximal, M/π is irreducible. Also π is the state set of M/π and $|\pi| < |Q|$. If R is the state set of N, then $|R| < |Q|$ by the construction of N. We can now apply Theorem 6.8.14 to

N, to obtain $N \leq N'\omega' N/\pi'$ such that N/π' is irreducible and the number of states in N' is less that the number of states in N. Thus, we have

$$M \leq N'\omega' N/\pi'\omega M/\pi.$$

Continue this process and apply Theorem 6.8.14 to N'. Since $|Q|$ is finite, this process must stop after a finite number of steps. Hence $M \leq N_1\omega_1 N_2\omega_2 \ldots \omega_{m-1} N_m$, where $N_1, N_2, \ldots, N_m$ are irreducible ffsms and the state sets Q_i of N_i are such that $|Q_i| < n$. ■

6.9 On Fuzzy Recognizers

In this section, we define and examine the concept of a fuzzy recognizer. If $L(M)$ is the language recognized by an incomplete fuzzy recognizer M, we show that there is a completion $\overline{M}$ of M such that $L(\overline{M}) = L(M)$, Theorem 6.9.14. We also show that if A is a recognizable set of words, then there is a complete accessible fuzzy recognizer M_A such that $L(M_A) = A$, Theorem 6.9.20. Our long term goal is to determine rational decompositions of recognizable sets. In fact, we wish to determine a decomposition which gives a constructive characterization of a recognizable set. We lay ground work for this determination by proving Theorems 6.9.28 and 6.9.29.

Let $M = (Q, X, \mu)$ be a ffsm. Let $A, B \subseteq X^*$. Then $AB = \{uv \mid u \in A, v \in B\}$. For $x \in X^*$, $xA = \{x\}A$ and $Ax = A\{x\}$.

Definition 6.9.1 *Let $M = (Q, X, \mu)$ be a ffsm. Let $q \in Q, x \in X^*$ and $A \subseteq X^*$. Define $q * x : Q \to [0,1]$ and $q * A : Q \to [0,1]$ by*

$$(q * x)(p) = \mu^*(q, x, p)$$

$\forall p \in Q$ and

$$q * A = \cup_{x \in A} q * x.$$

Definition 6.9.2 *Let $M = (Q, X, \mu)$ be a ffsm. Let $\alpha : Q \to [0,1]$, $x \in X^*$ and $A \subseteq X^*$. Define*

(i) $\alpha * x : Q \to [0,1]$ by

$$(\alpha * x)(p) = \vee\{\alpha(q) \wedge \mu^*(q, x, p) \mid q \in Q\}$$

$\forall p \in Q$,

(ii) $\alpha * A : Q \to [0,1]$

$$\alpha * A = \cup_{x \in A} \alpha * x,$$

(iii) $\alpha * x^{-1} : Q \to [0,1]$ by

$$(\alpha * x^{-1})(p) = \vee\{\alpha(q) \wedge \mu^*(p, x, q) \mid q \in Q\}$$

$\forall p \in Q$,

(iv) $\alpha * A^{-1} : Q \to [0,1]$

$$\alpha * A^{-1} = \cup_{x \in A} \alpha * x^{-1}.$$

Lemma 6.9.3 *Let $M = (Q, X, \mu)$ be a ffsm. Let α be a fuzzy subset of Q and $A \subseteq X^*$. Then*

(i)

$$(\alpha * A)(p) = \vee\{\vee\{\alpha(q) \wedge \mu^*(q, x, p) \mid q \in Q\} \mid x \in A\},$$

(ii)

$$(\alpha * A^{-1})(p) = \vee\{\vee\{\alpha(q) \wedge \mu^*(p, x, q) \mid q \in Q\} \mid x \in A\}$$

$\forall p \in Q$. ■

Lemma 6.9.4 *Let $M = (Q, X, \mu)$ be a ffsm and let $\alpha : Q \to [0,1]$. Let $S = \{\alpha * x \mid x \in X^*\}$. Then S is a finite set.*

Proof. Let $x \in X^*$. Then $(\alpha * x)(p) = \vee\{\alpha(q) \wedge \mu^*(q, x, p) \mid q \in Q\}$. Since μ is finite valued, it follows that μ^* is finite valued. Also, α is finite valued. It now follows that the number of mappings of the form $\alpha * x : Q \to [0,1]$ is finite. Hence S is finite. ■

Theorem 6.9.5 *Let α be a fuzzy subset of Q, $x, y \in X^*$ and $A, B \subseteq X^*$. Then*

(i) $(\alpha * x) * y = \alpha * (xy)$,

(ii) $(\alpha * A) * y = \alpha * (Ay)$

(iii) $(\alpha * A) * B = \alpha * (AB)$,

(iv) $(\alpha * x^{-1}) * y^{-1} = \alpha * (yx)^{-1}$

(v) $(\alpha * A^{-1}) * y^{-1} = \alpha * (yA)^{-1}$

(ii) $(\alpha * A^{-1}) * B^{-1} = \alpha * (BA)^{-1}$.

Proof. (i) Let $p \in Q$. Then

$$\begin{aligned}
((\alpha * x) * y)(p) &= \vee\{(\alpha * x)(q) \wedge \mu^*(q, y, p) \mid q \in Q\} \\
&= \vee\{\vee(\{\alpha(r) \wedge \mu^*(r, x, q) \mid r \in Q\}) \wedge \mu^*(q, y, p) \\
&\quad \mid q \in Q\} \\
&= \vee\{\alpha(r) \wedge (\vee\{\mu^*(r, x, q) \wedge \mu^*(q, y, p) \mid q \in Q\}) \\
&\quad \mid r \in Q\} \\
&= \vee\{\alpha(r) \wedge \mu^*(r, xy, p) \mid r \in Q\} \\
&= (\alpha * (xy))(p).
\end{aligned}$$

Hence $(\alpha * x) * y = \alpha * (xy)$.

(ii) Let $p \in Q$. Then

$$\begin{aligned}((\alpha * A) * y)(p) &= \vee\{(\alpha * A)(q) \wedge \mu^*(q,y,p) \mid q \in Q\} \\ &= \vee\{(\vee\{(\alpha * x)(q) \mid x \in A\}) \wedge \mu^*(q,y,p) \mid q \in Q\} \\ &= \vee\{\vee\{(\alpha * x)(q) \wedge \mu^*(q,y,p) \mid q \in Q\} \mid x \in A\} \\ &= \vee\{((\alpha * x) * y)(p) \mid x \in A\} \\ &= \vee\{(\alpha * (xy))(p) \mid x \in A\} \\ &= (\cup_{x\in A}\alpha * (xy))(p) \\ &= (\alpha * (Ay))(p).\end{aligned}$$

Hence $(\alpha * A) * y = \alpha * (Ay)$.

(iii) $(\alpha * A) * B = \cup_{y\in B}((\alpha * A) * y) = \cup_{y\in B}\alpha * (Ay) = \alpha * (AB)$.

(iv) Let $p \in Q$. Then

$$\begin{aligned}((\alpha * x^{-1}) * y^{-1})(p) &= \vee\{(\alpha * x^{-1})(q) \wedge \mu^*(p,y,q) \mid q \in Q\} \\ &= \vee\{(\vee\{\alpha(r) \wedge \mu^*(q,x,r) \mid r \in Q\}) \wedge \mu^*(p,y,q) \\ &\quad \mid q \in Q\} \\ &= \vee\{\alpha(r) \wedge (\vee\{\mu^*(q,x,r) \wedge \mu^*(p,y,q) \mid q \in Q\}) \\ &\quad \mid r \in Q\} \\ &= \vee\{\alpha(r) \wedge (\vee\{\mu^*(p,y,q) \wedge \mu^*(q,x,r) \mid q \in Q\}) \\ &\quad \mid r \in Q\} \\ &= \vee\{\alpha(r) \wedge \mu^*(p,yx,r) \mid r \in Q\} \\ &= (\alpha * (yx)^{-1})(p).\end{aligned}$$

Hence $(\alpha * x^{-1}) * y^{-1} = \alpha * (yx)^{-1}$.

(v) Let $p \in Q$. Then

$$\begin{aligned}((\alpha * A^{-1}) * y^{-1})(p) &= \vee\{(\alpha * A^{-1})(q) \wedge \mu^*(p,y,q) \mid q \in Q\} \\ &= \vee\{(\vee\{(\alpha * x^{-1})(q) \mid x \in A\}) \wedge \mu^*(p,y,q) \\ &\quad \mid q \in Q\} \\ &= \vee\{\vee\{(\alpha * x^{-1})(q) \wedge \mu^*(p,y,q) \mid q \in Q\} \mid x \in A\} \\ &= \vee\{((\alpha * x^{-1}) * y^{-1})(p) \mid x \in A\} \\ &= \vee\{(\alpha * (yx)^{-1}(p) \mid x \in A\} \\ &= (\cup_{x\in A}\alpha * (yx)^{-1})(p) \\ &= (\alpha * (yA)^{-1})(p).\end{aligned}$$

Hence $(\alpha * A^{-1}) * y^{-1} = \alpha * (yA)^{-1}$.

(vi) $(\alpha * A^{-1}) * B^{-1} = \cup_{y\in B}((\alpha * A^{-1}) * y^{-1}) = \cup_{y\in B}(\alpha * (yA)^{-1}) = \alpha * (BA)^{-1}$. ■

Definition 6.9.6 *$M = (Q, X, \mu)$ be a ffsm. Let $\alpha : Q \to [0,1]$ and $T : Q \to [0,1]$. Define*

$$\alpha^{-1} \circ T = \{x \in X^* | \alpha(q) \wedge \mu^*(q,x,p) \wedge T(p) > 0 \textit{ for some } q,p \in Q\}.$$

Definition 6.9.7 *Let Q and X be finite subsets. A* ***fuzzy recognizer*** *is a five tuple $M = (Q, X, \mu, \overline{q_0}, F)$, where*

Q is a finite nonempty set of states,

X is a finite nonempty set of input symbols,

$\mu : Q \times X \times Q \to [0,1]$ is a function, called the ***fuzzy transition function,***

$\overline{q_0}$ is a fuzzy subset of Q, i.e., $\overline{q_0} : Q \to [0,1]$, called the ***initial fuzzy state,*** *and*

F is a fuzzy subset of Q, i.e., $F : Q \to [0,1]$, called the ***fuzzy subset of final states.***

Clearly, if $M = (Q, X, \mu, \overline{q_0}, F)$ is a fuzzy recognizer, then $N = (Q, X, \mu)$ is a fuzzy finite state machine. We call N the fuzzy finite state machine associated with the fuzzy recognizer M.

Definition 6.9.8 *Let $M = (Q, X, \mu, \overline{q_0}, F)$ be a fuzzy recognizer. Let $x \in X^*$. Then x is said to be* ***recognized*** *by M if*

$$\vee_{q\in Q}(\overline{q_0}(q) \wedge (\vee_{p\in Q}\{\mu^*(q,x,p) \wedge F(p)\})) > 0.$$

Lemma 6.9.9 *Let $M = (Q, X, \mu, \overline{q_0}, F)$ be a fuzzy recognizer. Let $x \in X^*$. Then x is recognized if and only if there exists $p, q \in Q$ such that $\overline{q_0}(q) \wedge \mu^*(q,x,p) \wedge F(p) > 0$.* ■

Definition 6.9.10 *Let $M = (Q, X, \mu, \overline{q_0}, F)$ be a fuzzy recognizer. Let*

$$L(M) = \{x \in X^* \mid x \text{ is recognized by } M\}.$$

$L(M)$ is called the ***language recognized*** *by the fuzzy recognizer M.*

Lemma 6.9.11 *Let $M = (Q, X, \mu, \overline{q_0}, F)$ be a fuzzy recognizer. Then*

$$L(M) = \{x \in X^* \mid \overline{q_0}(q) \wedge \mu^*(q,x,p) \wedge F(p) > 0 \text{ for some } q, p \in Q\}. \blacksquare$$

Definition 6.9.12 *Let $M = (Q, X, \mu)$ be a fuzzy finite state machine. M is called* ***complete*** *if for all $q \in Q$, $a \in X$, there exists $p \in Q$ such that $\mu(q,a,p) > 0$.*

Definition 6.9.13 *Let $M = (Q, X, \mu, \overline{q_0}, F)$ be a fuzzy recognizer. Then M is called* ***complete*** *if the fuzzy finite state machine associated with M is complete.*

Let $M = (Q, X, \mu, \overline{q_0}, F)$ be a fuzzy recognizer such that M is not complete. Let $\bar{Q} = Q \cup \{t\}$, where t is an element such that $t \notin Q$. For all $q \in Q$, let $0 < m_q \leq 1$. Let $0 < m \leq 1$. Define $\bar{\mu} : \bar{Q} \times X \times \bar{Q} \to [0,1]$ by for all $p, q \in Q$, $a \in X$,

$$\bar{\mu}(p,a,q) = \mu(p,a,q),$$

for all $p \in Q$, $a \in X$

$$\bar{\mu}(p,a,t) = \begin{cases} m_p & \text{if } \vee\{\mu(p,a,q) \mid q \in Q\} = 0 \\ 0 & \text{if } \vee\{\mu(p,a,q) \mid q \in Q\} > 0 \end{cases}$$

and

$$\bar{\mu}(t,a,p)=\begin{cases} m & \text{if } t=p \\ 0 & \text{if } t\neq p. \end{cases}$$

Define $\overline{\overline{q_0}} : \bar{Q} \to [0,1]$ by

$$\overline{\overline{q_0}}(p)=\begin{cases} \overline{q_0}(p) & \text{if } p\neq t \\ 0 & \text{if } p=t. \end{cases}$$

Define $\overline{F} : \bar{Q} \to [0,1]$ by

$$\overline{F}(p)=\begin{cases} F(p) & \text{if } p\in Q \\ 0 & \text{otherwise.} \end{cases}$$

It is easy to see that the fuzzy recognizer $\bar{M} = (\bar{Q}, X, \bar{\mu}, \overline{\overline{q_0}}, \overline{F})$ is complete. $\bar{M}$ is called a **completion** of M.

Theorem 6.9.14 *Let $M = (Q, X, \mu, \overline{q_0}, F)$ be an incomplete fuzzy recognizer. Let $\bar{M}$ be a completion of M. Then $L(M) = L(\bar{M})$.*

Proof. Let $x \in L(M)$. Then $\exists q, p \in Q$ such that $\overline{q_0}(p) \wedge \mu^*(q,x,p) \wedge F(p) > 0$. This implies that $\overline{\overline{q_0}}(p) \wedge \bar{\mu}^*(q,x,p) \wedge F(p) > 0$. Hence $x \in L(\bar{M})$. Thus $L(M) \subseteq L(\bar{M})$. Now let $x \in L(\bar{M})$. There exists $q, p \in \bar{Q}$ such that $\overline{\overline{q_0}}(p) \wedge \bar{\mu}^*(q,x,p) \wedge \overline{F}(p) > 0$. This implies that $\overline{\overline{q_0}}(p) > 0$. Thus $p \in Q$ and so $\overline{q_0}(p) = \overline{\overline{q_0}}(p)$. Suppose $\mu^*(q,x,p) = 0$. Since $\bar{\mu}^*(q,x,p) > 0$ and $\mu^*(q,x,p) = 0$, we must have $p = t$. This is a contradiction since $p \in Q$. Hence $\mu^*(q,x,p) > 0$. Since $p \in Q$, $F(p) = \overline{F}(p) > 0$. Thus $\overline{q_0}(p) \wedge \mu^*(q,x,p) \wedge F(p) > 0$ and so $x \in L(M)$. Consequently, $L(M) = L(\bar{M})$. ■

Definition 6.9.15 *Let $M = (Q, X, \mu, \overline{q_0}, F)$ be a fuzzy recognizer. Let*

$$S = \{q \in Q \mid \vee\{\vee\{\overline{q_0}(p) \wedge \mu^*(p,x,q) \mid p \in Q\} \mid x \in X^*\} > 0\}.$$

Definition 6.9.16 *Let $M = (Q, X, \mu, \overline{q_0}, F)$ be a fuzzy recognizer. Then M is called **accessible** if $S = Q$.*

Theorem 6.9.17 *Let $M = (Q, X, \mu, \overline{q_0}, F)$ be a fuzzy recognizer. Then M is accessible if and only if $(\overline{q_0} * X^*)(q) > 0 \ \forall q \in Q$.*

Theorem 6.9.18 *Let $M = (Q, X, \mu, \overline{q_0}, F)$ be a fuzzy recognizer. Then*
(i) $L(M) = \overline{q_0}^{-1} \circ F$.
(ii) Let $x \in X^*$. Let $A = L(M)$. Then

$$x^{-1}A = (\overline{q_0} * x)^{-1} \circ F$$

where $x^{-1}A = \{y \in X^* \mid xy \in A\}$.

Proof. (i) Straightforward.

(ii) Let $y \in (\overline{q_0} * x)^{-1} \circ F$. Then $(\overline{q_0} * x)(q) \wedge \mu^*(q,y,p) \wedge F(p) > 0$ for some $q, p \in Q$. This implies that $(\overline{q_0} * x)(q) > 0$ and $\mu^*(q,y,p) \wedge F(p) > 0$. Now $(\overline{q_0} * x)(q) = \vee_{r \in Q}\{\overline{q_0}(r) \wedge \mu^*(r,x,q)\} > 0$ and so $\overline{q_0}(r) \wedge \mu^*(r,x,q) > 0$ for some $r \in Q$. Thus we have $\overline{q_0}(r) \wedge \mu^*(r,x,q) \wedge \mu^*(q,y,p) \wedge F(p) > 0$ for some $q,p,r \in Q$. This implies that $\overline{q_0}(r) \wedge \mu^*(r,xy,p) \wedge F(p) > 0$ for some $p, r \in Q$. Thus $xy \in A$ and so $y \in x^{-1}A$. Now let $y \in x^{-1}A$. Then $xy \in A$ and so there exists $p, r \in Q$ such that $\overline{q_0}(r) \wedge \mu^*(r,xy,p) \wedge F(p) > 0$. This implies that $\mu^*(r,xy,p) > 0$. Now $\mu^*(r,xy,p) = \vee_{q \in Q}\{\mu^*(r,x,q) \wedge \mu^*(q,y,p)\} > 0$. Thus there exists $q \in Q$ such that $\mu^*(r,x,q) \wedge \mu^*(q,y,p) > 0$, i.e., $\mu^*(r,x,q) > 0$ and $\mu^*(q,y,p) > 0$. Now $(\overline{q_0} * x)(q) = \vee_{s \in Q}\{\overline{q_0}(s) \wedge \mu^*(s,x,q)\} \geq \overline{q_0}(r) \wedge \mu^*(r,x,q) > 0$. Hence $(\overline{q_0} * x)(q) \wedge \mu^*(q,y,p) \wedge F(p) > 0$ for some $q, p \in Q$. Thus $y \in (\overline{q_0} * x)^{-1} \circ F$. Hence $x^{-1}A = (\overline{q_0} * x)^{-1} \circ F$. ■

Definition 6.9.19 *Let $A \subseteq X^*$. Then A is called* ***recognizable*** *if $\exists$ a fuzzy recognizer M such that $A = L(M)$.*

Theorem 6.9.20 *Let A be a recognizable subset of X^*. Then there exists a complete accessible fuzzy recognizer M_A such that $L(M_A) = A$.*

Proof. There exists a fuzzy recognizer $M = (Q, X, \mu, \overline{q_0}, T)$ such that $L(M) = A$. Let $Q_A = \{x^{-1}A \mid x \in X^*\}$. Let $0 < m_A,\ n_A,\ t_A \leq 1$. Define $\mu_A : Q_A \times X \times Q_A \to [0,1]$ by

$$\mu_A(x^{-1}A, a, y^{-1}A) = \begin{cases} m_A & \text{if } (xa)^{-1}A = y^{-1}A \\ 0 & \text{otherwise} \end{cases}$$

$\forall x^{-1}A, y^{-1}A \in Q_A$, $a \in X$. Define $\overline{q_0}_A : Q_A \to [0,1]$ by

$$\overline{q_0}_A(x^{-1}A) = \begin{cases} n_A & \text{if } x^{-1}A = A \\ 0 & \text{otherwise} \end{cases}$$

$\forall x^{-1}A \in Q_A$. Define $F : Q_A \to [0,1]$ by

$$F(x^{-1}A) = \begin{cases} t_A & \text{if } x \in A \\ 0 & \text{otherwise} \end{cases}$$

Set $M_A = (Q_A, X, \mu_A, \overline{q_0}_A, F)$. Next we show that M_A is a complete fuzzy recognizer such that $L(M_A) = A$. Let $x^{-1}A, y^{-1}A, u^{-1}A, v^{-1}A \in Q_A$, $a, b \in X$. Let $(x^{-1}A, a, y^{-1}A) = (u^{-1}A, b, v^{-1}A)$. Then $x^{-1}A = u^{-1}A$, $y^{-1}A = v^{-1}A$ and $a = b$. Now $(xa)^{-1}A = a^{-1}(x^{-1}A) = b^{-1}(u^{-1}A) = (ub)^{-1}A$. Hence $(xa)^{-1}A = y^{-1}A$ if and only if $(ua)^{-1}A = v^{-1}A$. This shows that μ_A is well-defined. By Theorem 6.9.18, $x^{-1}A = (\overline{q_0} * x)^{-1} \circ T$ for all $x \in X^*$. By Lemma 6.9.4, $S = \{\overline{q_0} * x \mid x \in X^*\}$ is finite. It now follows that $D = \{(\overline{q_0} * x)^{-1} \circ T \mid x \in X^*\}$ is finite. Hence Q_A is a finite set. Clearly M_A is a complete accessible recognizer.

Let $x \in A$. Then $F(x^{-1}A) = t_A > 0$. Let $x = a_1 a_2 \ldots a_n$, $a_i \in X$ for all i. Now $\mu_A^*(A, x, x^{-1}A) \geq \mu_A(A, a_1, a_1^{-1}A) \wedge\ \mu_A(a_1^{-1}A,\ a_2, (a_1a_2)^{-1}A) \wedge$

$\mu_A((a_1a_2)^{-1}A, a_3, (a_1a_2a_3)^{-1}A) \wedge \ldots \wedge \mu_A((a_1a_2\ldots a_{n-1})^{-1}A, a_n, (a_1a_2\ldots a_n)^{-1}A) = m_A \wedge m_A \wedge \ldots \wedge m_A = m_A > 0$. Thus $\overline{q_0}_A(A) \wedge \mu_A^*(A, x, x^{-1}A) \wedge F(x^{-1}A) > 0$. This implies that $x \in L(M_A)$. Now let $x \in L(M_A)$. There exist $y^{-1}A, z^{-1}A \in Q_A$ such that $\overline{q_0}_A(y^{-1}A) \wedge \mu_A^*(y^{-1}A, x, z^{-1}A) \wedge F(z^{-1}A) > 0$. Thus $\overline{q_0}_A(y^{-1}A) > 0$, $\mu_A^*(y^{-1}A, x, z^{-1}A) > 0$, and $F(z^{-1}A) > 0$. By the definition of $\overline{q_0}_A$, it follows that $y^{-1}A = A$. Thus $\mu_A^*(A, x, z^{-1}A) = \mu_A^*(y^{-1}A, x, z^{-1}A) > 0$ and so $x^{-1}A = z^{-1}A$ by the definition of μ_A^*. Hence $F(x^{-1}A) = F(z^{-1}A) > 0$ and so $x \in A$. Consequently, $A = L(M_A)$. ■

Let $M = (Q_A, X, \mu, \overline{q_0}, F)$. The relation $\sim$ defined on X^* by for all $x, y \in X^*$, $x \sim y$ if and only if ($\mu^*(q, x, p) > 0$ if and only if $\mu^*(q, y, p) > 0$ for all $q, p \in Q$). Then $\sim$ is a congruence relation by Theorem 6.1.5. Also, if $x \sim y$, then $x \in L(M)$ if and only if $y \in L(M)$. Let $A \subseteq X^*$. Define a relation $\sim_A$ on X^* by for all $x, y \in X^*$, $x \sim_A y$ if and only if ($uxv \in A$ if and only if $uyv \in A$ for all $u, v \in X^*$). It is easily seen that $\sim_A$ is a congruence relation on X^*. Clearly, for all $x, y \in X^*$, if $x \sim y$, then $x \sim_A y$. However, in general, the converse may not be true, i.e., if $x \sim_A y$, then $x \sim y$.

Theorem 6.9.21 *Let A be a recognizable subset of X^*. Let M_A be the fuzzy recognizer constructed in the proof of Theorem 6.9.20. Define $\sim$ on X^* by for all $x, y \in X^*$, $x \sim y$ if and only if ($\mu_A^*(u^{-1}A, x, v^{-1}A) > 0$ if and only if $\mu_A^*(u^{-1}A, y, v^{-1}A) > 0$ for all $u^{-1}A, v^{-1}A \in Q_A$). Define $\sim_A$ on X^* by for all $x, y \in X^*$, $x \sim_A y$ if and only if ($uxv \in A$ if and only if $uyv \in A$ for all $u, v \in X^*$). Then $\forall x, y \in X^*$, $x \sim y$ if and only if $x \sim_A y$.*

Proof. Let $x, y \in X^*$ and $x \sim_A y$. Then $\forall u, v \in X^*$, $uxv \in A$ if and only if $uyv \in A$. Thus $(ux)^{-1}A = (uy)^{-1}A$ $\forall u \in X^*$. Hence for all $u^{-1}A, v^{-1}A \in Q_A$, $(ux)^{-1}A = v^{-1}A$ if and only if $(uy)^{-1}A = v^{-1}A$. Thus by the definition of μ_A, $\mu_A^*(u^{-1}A, x, v^{-1}A) > 0$ if and only if $\mu_A^*(u^{-1}A, y, v^{-1}A) > 0$. Hence $x \sim y$. The converse is trivial. ■

Theorem 6.9.22 *Let $A \subseteq X^*$. The following assertions are equivalent.*

(i) A is recognizable.

(ii) $X^*/\sim_A$ is finite, where $X^*/\sim_A$ denotes the set of all congruence classes with respect to $\sim_A$.

(iii) A is the union of congruence classes of a congruence relation of X^* of finite index.

Proof. (i)⇒(ii): Since A is recognizable, for the fuzzy recognizer M_A, the relations $\sim$ and $\sim_A$ are equivalent. Thus $X^*/\sim_A = X^*/\sim$. Since $X^*/\sim$ is finite by Theorem 6.1.6, $X^*/\sim_A$ is finite.

(ii)⇒(iii): Since $X^*/\sim_A$ is finite, $\sim_A$ is a congruence relation of finite index. Now $A = \cup\{[x] \mid x \in A\}$, where $[x]$ is the equivalence class with respect to the relation $\sim_A$.

(iii)⇒(i): Suppose $\sim$ is a congruence of finite index on X^*. For $x \in X^*$, let $[x]$ denote the equivalence class with respect to $\sim$. Let $A = \cup\{[\alpha_i] \mid \alpha_i \in X^*,\ i = 1, 2, \ldots, n\}$. Let $Q = X^*/\sim = \{[x] \mid x \in X^*\}$. Since $\sim$ is of finite index, Q is a finite set. Let $0 < m, n, t \leq 1$. Define $\mu : Q \times X \times Q \to [0, 1]$ by

$$\mu([x], a, [y]) = \begin{cases} m & \text{if } [xa] = [y] \\ 0 & \text{otherwise} \end{cases}$$

$\forall [x], [y] \in Q,\ a \in X$. Define $\overline{q_0} : Q \to [0, 1]$ by

$$\overline{q_0}([x]) = \begin{cases} n & \text{if } [x] = [\Lambda] \\ 0 & \text{otherwise} \end{cases}$$

$\forall [x] \in Q$. Define $F : Q \to [0, 1]$ by

$$F([x]) = \begin{cases} t & \text{if } x \in A \\ 0 & \text{otherwise} \end{cases}$$

$\forall [x] \in Q$. Set $M = (Q, X, \mu, \overline{q_0}, F)$. Then M is a fuzzy recognizer. Let $x \in A$. Now $\overline{q_0}([\Lambda]) = n > 0$ and $F([x]) = t > 0$. Let $x = a_1a_2 \ldots a_n$, $a_i \in X$ for all i. Now $\mu^*([\Lambda], x, [x]) \geq \mu([\Lambda], a_1, [a_1]) \wedge \mu([a_1],\ a_2, [a_1a_2]) \wedge \mu([a_1a_2], a_3, [a_1a_2a_3]) \wedge \ldots \wedge \mu([a_1a_2 \ldots a_{n-1}], a_n, [a_1a_2 \ldots a_{n-1}a_n]) = m \wedge m \wedge \ldots \wedge m = m > 0$. Thus $\overline{q_0}([\Lambda]) \wedge \mu^*([\Lambda], x, [x]) \wedge F([x]) > 0$. This implies that $x \in L(M)$. Now let $x \in L(M)$. There exist $[y], [z] \in Q$ such that $\overline{q_0}([y]) \wedge \mu^*([y], x, [z]) \wedge F([z]) > 0$. Thus $\overline{q_0}([y]) > 0$, $\mu^*([y], x, [z] > 0$, and $F([z]) > 0$. By the definition of $\overline{q_0}$, it follows that $[y] = [\Lambda]$. Thus $\mu^*([\Lambda], x, [z]) = \mu^*([y], x, [z]) > 0$ and so $[x] = [z]$ by the definition of μ^*. Hence $F([x]) = F([z]) > 0$ and so $x \in A$. Consequently, $A = L(M)$. ■

Definition 6.9.23 *Let $M_1 = (Q_1, X, \mu_1, \overline{q_{01}}, F_1)$ and $M_2 = (Q_2, X, \mu_2, \overline{q_{02}}, F_2)$ be fuzzy recognizers. Then $M = M_1 \cup M_2 = (Q_1 \times Q_2, X, \mu_1 \wedge \mu_2, \overline{q_{01}} \wedge \overline{q_{02}}, F_1 \vee F_2)$.*

Theorem 6.9.24 *Let $M_1 = (Q_1, X, \mu_1, \overline{q_{01}}, F_1)$ and $M_2 = (Q_2, X, \mu_2, \overline{q_{02}}, F_2)$ be fuzzy recognizers. Then*

$$(\mu_1 \wedge \mu_2)^*((q_1, q_2), x, (p_1, p_2)) = \mu_1^*(q_1, x, p_1) \wedge \mu_2^*(q_2, x, p_2)$$

$\forall ((q_1, q_2), x, (p_1, p_2)) \in (Q_1 \times Q_2) \times X \times (Q_1 \times Q_2)$.

Proof. Let $((q_1, q_2), x, (p_1, p_2)) \in (Q_1 \times Q_2) \times X \times (Q_1 \times Q_2)$ and $len(x) = n$. If $n = 0$ or $n = 1$, then the result is true by definition. Assume it is true for $len(x) = n$. Let $a \in X$. Now

$$\begin{aligned}(\mu_1 \wedge \mu_2)^*((q_1, q_2), xa, (p_1, p_2)) &= \vee\{(\mu_1 \wedge \mu_2)^*((q_1, q_2), x, (r_1, r_2)) \wedge \\ &\quad (\mu_1 \wedge \mu_2)((r_1, r_2), a, (p_1, p_2)) \\ &\quad \mid (r_1, r_2) \in Q_1 \times Q_2\} \\ &= \vee\{\mu_1^*(q_1, x, r_1) \wedge \mu_2^*(q_2, x, r_2) \wedge \\ &\quad \mu_1(r_1, a, p_1) \wedge \mu_2(r_2, a, p_2) \\ &\quad \mid (r_1, r_2) \in Q_1 \times Q_2\} \\ &= \vee\{\mu_1^*(q_1, x, r_1) \wedge \mu_1(r_1, a, p_1) \mid \\ &\quad r_1 \in Q_1\} \wedge \{\mu_2^*(q_2, x, r_2) \wedge \\ &\quad \mu_1(r_2, a, p_2) \mid r_2 \in Q_1\} \\ &= \mu_1^*(q_1, xa, p_1) \wedge \mu_2^*(q_2, xa, p_2). \blacksquare\end{aligned}$$

Theorem 6.9.25 *Let* $M_1 = (Q_1, X, \mu_1, \overline{q_{01}}, F_1)$ *and* $M_2 = (Q_2, X, \mu_2, \overline{q_{02}}, F_2)$ *be complete fuzzy recognizers. Then*

$$L(M_1 \cup M_2) = L(M_1) \cup L(M_2).$$

Proof. Let $x \in L(M_1 \cup M_2)$. Then $\exists (q_1, q_2), (p_1, p_2) \in Q_1 \times Q_2$ such that

$$(\overline{q_{01}} \wedge \overline{q_{02}})(q_1, q_2) \wedge (\mu_1 \wedge \mu_2)^*((q_1, q_2), x, (p_1, p_2)) \wedge (F_1 \vee F_2)(p_1, p_2) > 0.$$

Thus $\exists (q_1, q_2), (p_1, p_2) \in Q_1 \times Q_2$ such that

$$\overline{q_{01}}(q_1) \wedge \overline{q_{02}}(q_2) \wedge \mu_1^*(q_1, x, p_1) \wedge \mu_2^*(q_2, x, p_2) \wedge (F_1(p_1) \vee F_2(p_2)) > 0.$$

This implies that $\exists (q_1, q_2), (p_1, p_2) \in Q_1 \times Q_2$ such that

$$(\overline{q_{01}}(q_1) \wedge \mu_1^*(q_1, x, p_1) \wedge F_1(p_1)) \vee (\overline{q_{02}}(q_2) \wedge \mu_2^*(q_2, x, p_2) \wedge F_2(p_2)) > 0.$$

Hence $\exists (q_1, q_2), (p_1, p_2) \in Q_1 \times Q_2$ such that either $(\overline{q_{01}}(q_1) \wedge \mu_1^*(q_1, x, p_1) \wedge F_1(p_1)) > 0$ or $(\overline{q_{02}}(q_2) \wedge \mu_2^*(q_2, x, p_2) \wedge F_2(p_2)) > 0$. Thus $x \in L(M_1) \cup L(M_2)$. Hence $L(M_1 \cup M_2) \subseteq L(M_1) \cup L(M_2)$. Now let $x \in L(M_1) \cup L(M_2)$. Then $x \in L(M_1)$ or $x \in L(M_2)$. Suppose $x \in L(M_1)$. Then $\exists q_1, p_1 \in Q_1$ such that $\overline{q_{01}}(q_1) \wedge \mu_1^*(q_1, x, p_1) \wedge F_1(p_1) > 0$. Now $\exists q_2 \in Q_2$ such that $\overline{q_{02}}(q_2) > 0$. Since M_2 is complete, $\exists p_2 \in Q_2$ such that $\mu_2^*(q_2, x, p_2) > 0$. Thus we have $\overline{q_{01}}(q_1) \wedge \overline{q_{02}}(q_2) > 0$, $\mu_1^*(q_1, x, p_1) \wedge \mu_2^*(q_2, x, p_2) > 0$ and $F_1(p_1) \vee F_2(p_2) > 0$. Hence

$$(\overline{q_{01}} \wedge \overline{q_{02}})(q_1, q_2) \wedge (\mu_1 \wedge \mu_2)^*((q_1, q_2), x, (p_1, p_2)) \wedge (F_1 \vee F_2)(p_1, p_2) > 0$$

and so $x \in L(M_1 \cup M_2)$. Thus $L(M_1) \cup L(M_2) \subseteq L(M_1 \cup M_2)$. Consequently, $L(M_1 \cup M_2) = L(M_1) \cup L(M_2)$. ■

Definition 6.9.26 *Let* $M_1 = (Q_1, X, \mu_1, \overline{q_{01}}, F_1)$ *and* $M_2 = (Q_2, X, \mu_2, \overline{q_{02}}, F_2)$ *be fuzzy recognizers. Then* $M = M_1 \cap M_2 = (Q_1 \times Q_2, X, \mu_1 \wedge \mu_2, \overline{q_{01}} \wedge \overline{q_{02}}, F_1 \wedge F_2)$.

Theorem 6.9.27 *Let* $M_1 = (Q_1, X, \mu_1, \overline{q_{01}}, F_1)$ *and* $M_2 = (Q_2, X, \mu_2, \overline{q_{02}}, F_2)$ *be fuzzy recognizers. Then*

$$L(M_1 \cap M_2) = L(M_1) \cap L(M_2).$$

Proof. Now $x \in L(M_1 \cap M_2)$ if and only if $\exists (q_1, q_2), (p_1, p_2) \in Q_1 \times Q_2$ such that

$$(\overline{q_{01}} \wedge \overline{q_{02}})(q_1, q_2) \wedge (\mu_1 \wedge \mu_2)^*((q_1, q_2), x, (p_1, p_2)) \wedge (F_1 \wedge F_2)(p_1, p_2) > 0.$$

if and only if $\exists (q_1, q_2), (p_1, p_2) \in Q_1 \times Q_2$ such that

$$\overline{q_{01}}(q_1) \wedge \overline{q_{02}}(q_2) \wedge \mu_1^*(q_1, x, p_1) \wedge \mu_2^*(q_2, x, p_2) \wedge (F_1(p_1) \wedge F_2(p_2)) > 0.$$

if and only if $\exists (q_1, q_2), (p_1, p_2) \in Q_1 \times Q_2$ such that

$$(\overline{q_{01}}(q_1) \wedge \mu_1^*(q_1, x, p_1) \wedge F_1(p_1)) \wedge (\overline{q_{02}}(q_2) \wedge \mu_2^*(q_2, x, p_2) \wedge F_2(p_2)) > 0.$$

if and only if $x \in L(M_1) \cap L(M_2)$. Hence $L(M_1 \cap M_2) = L(M_1) \cap L(M_2)$. ■

In the proofs of the next two theorems, we use the fact that if $M = (Q, X, \mu)$ is a ffsm, $q, p \in Q$, and $x = x_1 x_2 \ldots x_n$, where $x_i \in X$, $i = 1, 2, \ldots n$, then

$$\begin{aligned} \mu^*(q, x, p) &= \vee\{\mu(q, x_1, r_1) \wedge \mu(r_1, x_2, r_2) \wedge \ldots \\ &\quad \wedge \mu(r_{n-1}, x_n, p) \mid r_i \in Q,\ i = 1, 2, \ldots n-1\}. \end{aligned}$$

The above fact follows from Lemma 6.1.2.

Theorem 6.9.28 *Let* $A, B \subseteq X^*$. *If* A *and* B *are recognizable, then* $A \cdot B$ *is recognizable.*

Proof. Let $M_1 = (Q_1, X, \mu_1, \overline{q_{01}}, F_1)$ and $M_2 = (Q_2, X, \mu_2, \overline{q_{02}}, F_2)$ be recognizers of A and B, respectively. Let

$$I_i = \{q \in Q_i \mid \overline{q_{0i}}(q) > 0\}$$

and

$$T_i = \{q \in Q_i \mid F_i(q) > 0\}$$

for $i = 1, 2$. Define $\overline{q_0^{\Delta}} : Q_1 \times \mathcal{P}(Q_2) \to [0, 1]$ by $\forall (q, P) \in Q_1 \times \mathcal{P}(Q_2)$,

$$\overline{q_0^{\Delta}}(q, P) = \begin{cases} 0 & \text{if } P \neq \phi \\ \overline{q_{01}}(q) & \text{if } P = \phi. \end{cases}$$

Define $F^{\Delta} : Q_1 \times \mathcal{P}(Q_2) \to [0, 1]$ by $\forall (q, P) \in Q_1 \times \mathcal{P}(Q_2)$

$$F^{\Delta}(q, P) = \vee\{F_2(p) \mid p \in P\}.$$

Define $\mu_1 \Delta \mu_2 : (Q_1 \times \mathcal{P}(Q_2)) \times X \times (Q_1 \times \mathcal{P}(Q_2)) \to [0,1]$ as follows:

$$\mu_1 \Delta \mu_2((q,P), c, (q', s_2^c(P)) = \begin{cases} \mu_1(q,c,q') \wedge \vee\{\mu_2(p,c,p') \mid p \in P, \\ \qquad\qquad p' \in s_2^c(P)\} \text{ if } q' \notin T_1,\ P \neq \phi \\ \mu_1(q,c,q') \qquad\qquad \text{if } q' \notin T_1,\ P = \phi; \end{cases}$$

$$\begin{aligned} \mu_1 \Delta \mu_2((q,P), c, (q', s_2^c(P) \cup I_2) &= \mu_1(q,c,q') \wedge \vee\{\mu_2(p,c,p') \\ &\qquad \mid p \in P, p' \in s_2^c(P) \cup I_2\} \end{aligned}$$

if $q \in T_1$, where $s_2^c(P) = \{p' \in Q_2 \mid \mu_2(p,c,p') > 0, p \in P\}$; and

$$\mu_1 \Delta \mu_2 \text{ is 0 elsewhere.}$$

Let $a \in A$ and $b \in B$. Then $\exists q \in I_1$, $q' \in T_1$, $p \in I_2$, $p' \in T_2$ such that $\mu_1^*(q,a,q') > 0$ and $\mu_2^*(p,b,p') > 0$. Now

$$ab = u_1 \ldots u_{i_1} u_{i_1+1} \ldots u_{i_2} \ldots u_{i_j} u_{i_j+1} \ldots u_{i_k} v$$

where $u_l \in X$, $v \in X^*$, i_j is the smallest such that $i_j > \ldots > i_1 \geq i_0$ and $u_1 \ldots u_{i_j} \in A$ for $j = 1, \ldots, k$, $i_0 = 1$ and there does not exist i_{k+1} such that $u_1 \ldots u_{i_{k+1}} \in A$. Let $\widehat{u_0} = u_1 \ldots u_{i_1}$ and $\widehat{u_j} = u_{i_j+1} \ldots u_{i_{j+1}}$, $j = 1, \ldots, k-1$. For all $x \in X^*$ and $\forall P \in \mathcal{P}(Q_2)$, let

$$s^x(P) = \{p' \in Q_2 \mid \mu_2^*(p,x,p') > 0, p \in P\}.$$

Let

$$\begin{aligned} R_{ab} &= s_2^v(s_2^{\widehat{u_{k-1}}}(s_2^{\widehat{u_{k-2}}}(\ldots s_2^{\widehat{u_0}}(I_2)\ldots)))\cup \\ &\quad s_2^v(s_2^{\widehat{u_{k-1}}}(s_2^{\widehat{u_{k-2}}}(\ldots s_2^{\widehat{u_1}}(I_2)\ldots))) \cup \ldots \cup s_2^v(I_2). \end{aligned} \tag{6.2}$$

Let $q'' \in Q_1$ be such that $\mu_1^*(q', b, q'') > 0$. Now $a = u_1 \ldots u_{i_j}$ for some i_j. Let

$$(I_2)_a = s_2^{\widehat{u_{j-1}}}(s_2^{\widehat{u_{j-2}}}(\ldots s_2^{\widehat{u_0}}(I_2)\ldots)).$$

Then

$$\begin{aligned} (\mu_1 \Delta \mu_2)^*((q,\phi), ab, (q'', R_{ab})) &= \vee\{(\mu_1 \Delta \mu_2)^*((q,\phi), a, (v,R)) \\ &\qquad \wedge(\mu_1 \Delta \mu_2)^*((v,R), b, (q'', R_{ab})) \mid \\ &\qquad (v,R) \in Q_1 \times \mathcal{P}(Q_2)\} \\ &\geq (\mu_1 \Delta \mu_2)^*((q,\phi), a, (q', (I_2)_a)) \\ &\qquad \wedge(\mu_1 \Delta \mu_2)^*((q', (I_2)_a), b, (q'', R_{ab})) \\ &> 0 \end{aligned}$$

by the definition of $(I_2)_a$ and since R_{ab} is a continuation of $(I_2)_a$, i.e., $R_{ab} = ((I_2)_a)_b$.

Conversely, let $x \in L(M_1 \Delta M_2)$. Then

$$\overline{q_0^{\Delta}}(q,\phi) \wedge (\mu_1 \Delta \mu_2)^*((q,\phi), x, (q'', (I_2)_x)) > 0,$$

where $\mu_1^*(q,x,q'') > 0$ and $(I_2)_x$ is defined as follows: we have that $x = uv$, where $u = \widehat{u_0} \dots \widehat{u_{k-1}}$ and where the $\widehat{u_j}$ and v are defined as above (k exists since $x \in L(M_1 \Delta M_2)$). Then $(I_2)_x$ is the right hand side of equation (6.2). Since $x \in L(M_1 \Delta M_2)$, $(\mu_1 \Delta \mu_2)^*((q,\phi), x, (q'', (I_2)_x)) > 0$ and so $\exists j$, $\exists p \in I_2$,

$$s_2^{u_{i_j+1} \dots u_{i_{j+1}} \widehat{u_{i_{j+1}}} \dots \widehat{u_k} v}(p) \in (I_2)_x \cap T_2.$$

Thus $u_{i_j+1} \dots u_{i_{j+1}} \widehat{u_{i_{j+1}}} \dots \widehat{u_k} v \in B$. Since $u_1 \dots u_{i_j} \in A$, $x \in A \cdot B$. ■

Theorem 6.9.29 *Let $A \subseteq X^*$ be recognizable. Then A^* is recognizable.*

Proof. Let $M = (Q, X, \mu, \overline{q_0}, F)$ be a recognizer of A. Define $\mu' : \mathcal{P}(Q) \times X \times \mathcal{P}(Q) \to [0,1]$ as follows:

$$\mu'(P, c, s^c(P)) = \vee\{\mu(p,c,q) \mid p \in P, q \in s^c(P)\} \text{ if } s^c(P) \cap T = \phi$$

$$\mu'(P, c, s^c(P) \cup I) = \vee\{\mu(p,c,q) \mid p \in P, q \in s^c(P) \cup I\} \text{ if } s^c(P) \cap T \neq \phi$$

$$\mu'(P, c, P') = 0 \text{ otherwise,}$$

where $c \in X$, $P, P' \in \mathcal{P}(Q)$, $T = \{q \in Q \mid F(q) > 0\}$, $I = \{q \in Q \mid \overline{q_0}(q) > 0\}$, and $s^c(P) = \{q \in Q \mid \mu(p,c,q) > 0, p \in P\}$. Define $F' : \mathcal{P}(Q) \to [0,1]$ by $F'(P) = \vee\{F(p) \mid p \in P\}$ and $\overline{q_0'} : \mathcal{P}(Q) \to [0,1]$ by $\overline{q_0'}(I) = \vee\{\overline{q_0}(q) \mid q \in I\}$, $\overline{q_0'}$ is 0 otherwise, $T' = \{P \in \mathcal{P}(Q) \mid F'(P) > 0\} = \{P \in \mathcal{P}(Q) \mid P \cap T \neq \phi\}$. Let $M' = (\mathcal{P}(Q), X, \mu', \overline{q_0'}, F')$. Let $\alpha \in A^*$. Then $\alpha = a_1 a_2 \dots a_n$, where $a_i \in A$ and if $a_i = a_{i_1} \dots a_{i_{n_i}}$ for $a_{i_1}, \dots, a_{i_{n_i}} \in X$, $\nexists m_i < n_i$ such that $a_{i_1} \dots a_{i_{n_i}} \in A$, $i = 1, 2, \dots, n$. Let $I_\alpha = s^{a_n}(\dots s^{a_2}(s^{a_1}(I) \cup I) \cup \dots) \cup I$. Now $\mu'^*(I, \alpha, I_\alpha) \geq \mu'^*(I, a_1, s^{a_1}(I) \cup I) \wedge \mu'^*(s^{a_1}(I) \cup I, a_2, s^{a_2}(s^{a_1}(I) \cup I) \cup I) \wedge \dots \wedge \mu'^*(s^{a_{n-1}}(\dots s^{a_2}(s^{a_1}(I) \cup I) \cup \dots) \cup I, a_n, I_\alpha) > 0$ since $\exists i \in I$, $\exists p \in s^{a_1}(I) \cap T$ such that $\mu^*(i, a_1, p) > 0$ (because $a_1 \in L(M)$), $\exists i' \in I$, $\exists p' \in s^{a_2}(s^{a_1}(I) \cup I) \cap T$ such that $\mu^*(i', a_2, p') > 0$ (because $a_2 \in L(M)$) and so on. Thus $\alpha \in L(M')$. Hence $A^* \subseteq L(M')$. Let $\beta \in L(M')$. Then $\exists P \in T'$ such that $\mu'^*(I, \beta, P) > 0$. Now suppose $\exists k$ such that

$$\beta = u_1 \dots u_{i_1} u_{i_1+1} \dots u_{i_j} u_{i_j+1} \dots u_{i_k} v,$$

where $u_1 \dots u_{i_1} \in A$ with i_1 smallest and the i_j are the smallest such that $u_{i_{j-1}+1} \dots u_{i_j} \in A$, $j = 2, \dots, k$. Let $\widehat{u_1} = u_1 \dots u_{i_1}$ and $\widehat{u_j} = u_{i_{j-1}+1} \dots u_{i_j}$, $j = 2, \dots, k$. Then $u_1 \dots u_{i_{j-1}} \in A^*$ and $P = (I)_\beta$ where

$$(I)_\beta = s^v(s^{\widehat{u_k}}(\dots s^{\widehat{u_2}}(s^{\widehat{u_1}}(I) \cup I) \cup \dots I) \cup I) \cup I.$$

There exists j, $\exists q \in I$ such that $\mu^*(q, \widehat{u_j} \ldots \widehat{u_k}v, p) > 0$ for some $p \in (I)_\beta \cap T$. Thus $\mu'^*((I)_{\widehat{u_1}\ldots\widehat{u_{j-1}}}, \widehat{u_j} \ldots \widehat{u_k}v, (I)_\beta) > 0$. Hence $\widehat{u_j} \ldots \widehat{u_k}v \in A^*$. If no such k exists, then a similar argument shows that $\beta \in A$. Thus $\beta \in A^*$. Hence $L(M') \subseteq A^*$. Consequently $L(M') = A^*$. ∎

If $L(M)$ is the language recognized by an incomplete fuzzy recognizer, we showed that there is a completion $\overline{M}$ of M such that $L(\overline{M}) = L(M)$. We also showed that if A is a recognizable set of words, then there is a complete accessible fuzzy recognizer M_A such that $L(M_A) = A$. If A and B are recognizable sets of words, then $A \cdot B$ and A^* are recognizable. These results are significant in that they lay the ground work for determining methods of decomposing recognizable sets and thus for giving a constructive characterization of recognizable sets. In particular, we hope to show an analog of Kleene's result for fuzzy finite state machines, namely, that the class of recognizable subsets of X^* equals the class of all regular subsets of X^*.

6.10 Minimal Fuzzy Recognizers

In this section, we show that for any fuzzy recognizer M_n there is a deterministic fuzzy recognizer M_d with the same behavior, Theorem 6.10.5. Then we show that there is complete accessible deterministic fuzzy recognizer M_{dA} with the same behavior as M_n and M_d and which is minimal, Theorem 6.10.8. The long term goal is to develop methods of decomposing a recognizable set of a fuzzy finite state machine. One method would be to follow along the lines of Kleene to give a constructive characterization of a recognizable set. In this section, we lay the foundation for the accomplishment of our goal.

Let $M = (Q, X, \mu, \overline{q_0}, F)$ be a fuzzy recognizer and $N = (Q, X, \mu)$. Then N is a fuzzy finite state machine. We call N the fuzzy finite state machine associated with the fuzzy recognizer M.

Definition 6.10.1 *Let Q and X be finite nonempty sets and let $\mu : Q \times X \times Q \to [0, 1]$. Then μ is called a **fuzzy function** of $Q \times X$ into Q if for all $q \in Q$, $a \in X$, if $\mu(q, a, p) > 0$ and $\mu(q, a, p') > 0$ for some $p, p' \in Q$, then $p = p'$.*

Theorem 6.10.2 *Let $M = (Q, X, \mu, \overline{q_0}, T)$ be a fuzzy recognizer. Then μ is a fuzzy function of $Q \times X$ into Q if and only if μ^* is a fuzzy function of $Q \times X^*$ into Q.*

Proof. Suppose μ is a fuzzy function. Let $q \in Q$ and $x \in X^*$. Suppose $\mu^*(q, x, p) > 0$ and $\mu^*(q, x, p') > 0$ for some $p, p' \in Q$. If $x = \Lambda$, then $p = q = p'$. Suppose $x \neq \Lambda$. Let $x = a_1 a_2 \cdots a_n \in X^*$, $a_i \in X$. There exists

$q_1, q_2, \ldots, q_{n-1}, q'_1, q'_2, \ldots, q'_{n-1} \in Q$ such that $\mu(q, a_1, q_1) \wedge \mu(q_1, a_2, q_2) \wedge \ldots \wedge \mu(q_{n-1}, a_n, p) > 0$ and $\mu(q, a_1, q'_1) \wedge \mu(q'_1, a_2, q'_2) \wedge \ldots \wedge \mu(q'_{n-1}, a_n, p') > 0$. This implies that $\mu(q, a_1, q_1) > 0$, $\mu(q_i, a_{i+1}, q_{i+1}) > 0$, $i = 1, 2, \ldots, n-2$, $\mu(q_{n-1}, a_n, p) > 0$, $\mu(q, a_1, q'_1) > 0$, $\mu(q'_i, a_{i+1}, q'_{i+1}) > 0$, $i = 1, 2, \ldots, n-2$, $\mu(q'_{n-1}, a_n, p') > 0$. Now $\mu(q, a_1, q_1) > 0$ and $\mu(q, a_1, q'_1) > 0$. Since μ is a fuzzy function, $q_1 = q'_1$. Suppose $q_j = q'_j$, $j = 1, 2, \ldots, i$, $i < n-2$. Now $\mu(q_i, a_{i+1}, q_{i+1}) > 0$ and $\mu(q_i, a_{i+1}, q'_{i+1}) > 0$ implies that $q_{i+1} = q'_{i+1}$ since μ is a fuzzy function. Hence by induction $q_j = q'_j$, $j = 1, 2, \ldots, n-1$. Hence $\mu(q_{n-1}, a_n, p) > 0$ and $\mu(q_{n-1}, a_n, p') > 0$ implies that $p = p'$. Thus μ^* is a fuzzy function. The converse is trivial. ■

Definition 6.10.3 *A* ***deterministic fuzzy recognizer*** *is a fuzzy recognizer* $M_d = (Q_d, X, \mu, \overline{q_0}, T)$ *such that*

(i) there exists a unique $s_0 \in Q_d$ *such that* $\overline{q_0}(s_0) > 0$; s_0 *is called the* ***initial state,***

(ii) μ *is a fuzzy function of* $Q \times X$ *into* Q, *and*

(iii) for all $x \in X^*$, *there exists a unique* $q_x \in Q_d$ *such that* $\mu^*(s_0, x, q_x) > 0$.

Let $M_d = (Q_d, X, \mu, \overline{q_0}, T)$ be a deterministic fuzzy recognizer. Let $F_d = \{q \in Q_d \mid T(d) > 0\}$. F_d is called the set of **final states** of M_d.

Theorem 6.10.4 *Let* $M = (Q, X, \mu, \overline{q_0}, T)$ *be a fuzzy recognizer. Suppose* M *is complete and* μ *is a fuzzy function of* $Q \times X$ *into* Q. *Let* $s_0 \in Q$. *Then the following are equivalent.*

(i) For all $a \in X$, *there exists a unique* $q_a \in Q$ *such that* $\mu(s_0, a, q_a) > 0$.

(ii) For all $x \in X^*$, *there exists a unique* $q_x \in Q$ *such that* $\mu^*(s_0, x, q_x) > 0$.

Proof. (i)⇒(ii): Let $x \in X^*$ and $len(x) = n$. If $x = \Lambda$, i.e., $n = 0$, then $\mu^*(s_0, \Lambda, s_0) = 1 > 0$ and if $\mu^*(s_0, \Lambda, p) > 0$, then by the definition of μ^*, $s_0 = p$. Suppose the result is true for all $y \in X^*$ such that $len(y) < len(x)$, where $len(x) = n \geq 1$. Let $x = ya$, where $y \in X^*$, $a \in X$, $len(y) = n - 1$. By the induction hypothesis, there exists a unique $q_y \in Q$ such that $\mu^*(s_0, y, q_y) > 0$. Since M is complete, there exists $p \in Q$ such that $\mu(q_y, a, p) > 0$. Thus $\mu^*(s_0, x, p) \geq \mu^*(s_0, y, q_y) \wedge \mu(q_y, a, p) > 0$. Since μ is a fuzzy function of $Q \times X$ into Q, μ^* is a fuzzy function of $Q \times X^*$ into Q by Theorem 6.10.2. Thus if $\mu^*(s_0, x, p) > 0$ and $\mu^*(s_0, x, p') > 0$ for some $p, p' \in Q$, then $p = p'$. It now follows that there exists a unique $q_x \in Q$ such that $\mu^*(s_0, x, q_x) > 0$.

(ii)⇒(i): Immediate. ■

Theorem 6.10.5 *For each fuzzy recognizer* $M_n = (Q, X, \mu, \overline{q_0}, F)$, *one can construct a deterministic fuzzy recognizer* $M_d = (Q_d, X, \mu, \overline{q_0}, F)$ *such that* $L(M_d) = L(M_n)$.

Proof. For all $x \in X^*$, set

$$Q_x = \{q' \in Q \mid \exists q \in Q \text{ such that } \overline{q_0}(q) \wedge \mu^*(q, x, q') > 0\}.$$

Then

$$Q_\Lambda = \{q' \in Q \mid \overline{q_0}(q') > 0\}.$$

Let $Q_d = \{Q_x \mid x \in X^*\}$. Define $i : Q_d \to [0,1]$ by $\forall Q_x \in Q_d$,

$$i(Q_x) = \begin{cases} \vee\{\overline{q_0}(q) \mid q \in Q_\Lambda\} & \text{if } x = \Lambda \\ 0 & \text{if } x \neq \Lambda. \end{cases}$$

Let $F_d = \{Q_x \in Q_d \mid F(q) > 0 \text{ for some } q \in Q_x\}$. Define $T : F_d \to [0,1]$ by $\forall Q_x \in F_d$, $T(Q_x) = \vee\{F(q) \mid q \in Q_x\}$. Define $\nu : Q_d \times X \times Q_d \to [0,1]$ by $\forall (Q_y, a, Q_x) \in Q_d \times X \times Q_d$

$$\nu(Q_y, a, Q_x) = \begin{cases} \vee\{\mu^*(q, y, q') \wedge \mu(q', a, r) \mid q \in Q_y,\ q' \in Q,\ r \in Q_{ya}\} \\ \qquad\qquad \text{if } x = ya \\ 0 \qquad\qquad \text{otherwise.} \end{cases}$$

Let $M_d = (Q_d, X, \nu, i, T)$. We now show that $L(M_n) = L(M_d)$.

Now $x \in L(M_n)$ if and only if $\overline{q_0}(q) \wedge \mu^*(q, x, q') \wedge F(q') > 0$ for some $q, q' \in Q$ if and only if $F(q') > 0$ for some $q' \in Q_x$ if and only if $T(Q_x) > 0$. It suffices to show that $\nu^*(Q_\Lambda, x, Q_x) > 0$ for then $x \in L(M_d)$ if and only if $i(Q_\Lambda) \wedge \nu^*(Q_\Lambda, x, Q_x) \wedge T(Q_x) > 0$ if and only if $T(Q_x) > 0$ (since $i(Q_\Lambda) > 0$ and $\nu^*(Q_\Lambda, x, Q_x) > 0$) if and only if $x \in L(M_n)$. We show $\nu^*(Q_\Lambda, x, Q_x) > 0$ by induction on $len(x)$. Suppose $len(x) = 0$. Then $x = \Lambda$ and $\nu^*(Q_\Lambda, \Lambda, Q_\Lambda) = 1 > 0$. Suppose $len(x) \geq 1$ and the result is true for all $y \in X^*$ such that $len(y) < len(x)$. Let $x = ya$, where $a \in X$. Then $\nu^*(Q_\Lambda, x, Q_x) = \vee\{\nu^*(Q_\Lambda, y, r) \wedge \nu(r, a, Q_x) \mid r \in Q_d\}$ and $\nu^*(Q_\Lambda, y, Q_y) > 0$ by the induction hypothesis. Hence it suffices to show that $\nu(Q_y, a, Q_x) > 0$, but the later inequality is true by the definition of ν since $x = ya$. ■

Let $M_d = (Q_d, X, \mu, \overline{q_0}, T)$ be a complete accessible deterministic fuzzy recognizer and let s_0 denote the initial state of M_d. For $x \in X^*$, we let $q_x \in Q_d$ denote the unique state such that $\mu^*(s_0, x, q_x) > 0$.

Theorem 6.10.6 *Let $M_d = (Q_d, X, \mu, \overline{q_0}, T)$ be a complete accessible deterministic fuzzy recognizer. Let s_0 be the initial state of M_d. For all $q \in Q_d$, let $q^{-1} \circ T = \{y \in X^* \mid \mu^*(q, y, p) \wedge T(p) > 0 \text{ for some } p \in Q_d\}$. Let $A = L(M_d)$.*

(i) For all $x \in X^*$, $x^{-1}A = q_x^{-1} \circ T$.

(ii) Let $q \in Q_d$. Then there exists $x \in X^*$ such that $q^{-1} \circ T = x^{-1}A$ and $q = q_x$.

(iii) Let $q \in Q_d$ be such that $T(q) > 0$. Then $q^{-1} \circ T = x^{-1}A$ for some $x \in A$.

(iv) $A = \Lambda^{-1}A = s_0^{-1} \circ T$.

(v) Let $x = a_1a_2 \cdots a_n \in X^*$, *where* $a_i \in X$, $i = 1, 2, \ldots, n$. *Then*

$$\mu^*(s_0, x, q_x) = \mu(s_0, a_1, q_{a_1}) \wedge \mu(q_{a_1}, a_2, q_{a_1a_2}) \wedge \ldots \wedge \mu(q_{a_1 \ldots a_{n-1}}, a_n, q_x).$$

Proof. (i) Let $y \in x^{-1}A$. Then $xy \in A$ and so $\overline{q_0}(s_0) \wedge \mu^*(s_0, xy, p) \wedge T(p) > 0$ for some $p \in Q_d$. This implies that $\mu^*(s_0, xy, p) > 0$. Thus there exists $q \in Q_d$ such that $\mu^*(s_0, x, q) \wedge \mu^*(q, y, p) > 0$. Hence $\mu^*(s_0, x, q) > 0$. Since M_d is deterministic, $q = q_x$. Hence $\overline{q_0}(s_0) \wedge \mu^*(s_0, x, q_x) \wedge \mu^*(q_x, y, p) \wedge T(p) > 0$. Thus $y \in q_x^{-1} \circ T$. Hence $x^{-1}A \subseteq q_x^{-1} \circ T$. Now let $y \in q_x^{-1} \circ T$. Then $\mu^*(q_x, y, p) \wedge T(p) > 0$ for some $p \in Q_d$. This implies that $\overline{q_0}(s_0) \wedge \mu^*(q_x, y, p) \wedge T(p) > 0$. Now $\mu^*(s_0, x, q_x) > 0$. Hence $\overline{q_0}(s_0) \wedge \mu^*(s_0, x, q_x) \wedge \mu^*(q_x, y, p) \wedge T(p) > 0$ for some $p \in Q_d$. Thus $\overline{q_0}(s_0) \wedge \mu^*(s_0, xy, p) \wedge T(p) > 0$ for some $p \in Q_d$. Hence $xy \in A$ or $y \in x^{-1}A$. It now follows that $x^{-1}A = q_x^{-1} \circ T$.

(ii) Let $q \in Q_d$. Since M_d is accessible, there exists $x \in X^*$ such that $\mu^*(s_0, x, q) > 0$. Since M_d is deterministic, it follows that $q = q_x$. By (i), $x^{-1}A = q_x^{-1} \circ T = q^{-1} \circ T$.

(iii) By (ii), $q^{-1} \circ T = x^{-1}A$ for some $x \in X^*$. Since $T(q) > 0$, $\mu^*(q, \Lambda, q) \wedge T(q) > 0$ and so $\Lambda \in q^{-1} \circ T$. Thus $\Lambda \in x^{-1}A$ and so $x \in A$.

(iv) Let $y \in A$. Then $\overline{q_0}(s_0) \wedge \mu^*(s_0, y, q) \wedge T(q) > 0$ for some $q \in Q_d$. This implies that $\mu^*(s_0, y, q) \wedge T(q) > 0$ for some $q \in Q_d$ and so $y \in s_0^{-1} \circ T$. Thus $A \subseteq s_0^{-1} \circ T$. Now let $y \in s_0^{-1} \circ T$. Then $\mu^*(s_0, y, q) \wedge T(q) > 0$ for some $q \in Q_d$. Thus $\overline{q_0}(s_0) \wedge \mu^*(s_0, y, q) \wedge T(q) > 0$ for some $q \in Q_d$ and so $y \in A$. It now follows that $A = s_0^{-1} \circ T$.

(v) Now $\mu^*(s_0, x, q_x) = \vee\{\mu(s_0, a_1, q_1) \wedge \mu(q_1, a_2, q_2) \ldots \wedge \mu(q_{n-1}, a_n, q_x) \mid q_1, q_2, \cdots, q_{n-1} \in Q_d\}$. Since Q_d is finite, there exists $q_1, q_2, \cdots, q_{n-1} \in Q_d$ such that $\mu^*(s_0, x, q_x) = \mu(s_0, a_1, q_1) \wedge \mu(q_1, a_2, q_2) \ldots \wedge \mu(q_{n-1}, a_n, q_x)$. Since $\mu^*(s_0, x, q_x) > 0$, $\mu(s_0, a_1, q_1) \wedge \mu(q_1, a_2, q_2) \wedge \mu(q_2, a_3, q_3) \wedge \ldots \wedge \mu(q_{n-1}, a_n, q_x) > 0$. This implies that $\mu(s_0, a_1, q_1) > 0$ and so $q_1 = q_{a_1}$ since M_d is deterministic. Now $\mu(s_0, a_1, q_{a_1}) \wedge \mu(q_{a_1}, a_2, q_2) > 0$ implies that $\mu^*(s_0, a_1a_2, q_2) > 0$ and since M_d is deterministic $q_2 = q_{a_1a_2}$. We see that an argument by induction will yield $q_i = q_{a_1 \ldots a_i}$ for all $i = 1, 2, \cdots, n-1$. ■

Theorem 6.10.7 *Let A be a recognizable subset of X^*. Then there exists a complete accessible deterministic fuzzy recognizer M_{dA} such that $L(M_{dA}) = A$.*

Proof. Let $M_d = (Q_d, X, \mu, \overline{q_0}, T)$ be a deterministic fuzzy recognizer such that $A = L(M_d)$. Since M_d is deterministic, there exists a unique $s_0 \in Q_d$ such that $\overline{q_0}(s_0) > 0$.

Let $Q_{dA} = \{x^{-1}A \mid x \in X^*\}$. Let $m_A = \vee\{\mu(q, a, p) \mid q, p \in Q, a \in X\}$. Then $1 \geq m_A > 0$. Let $0 < t_A \leq 1$. Define $\mu_{dA} : Q_{dA} \times X \times Q_{dA} \to [0, 1]$ by

$$\mu_{dA}(x^{-1}A, a, y^{-1}A) = \begin{cases} m_A & \text{if } (xa)^{-1}A = y^{-1}A \\ 0 & \text{otherwise} \end{cases}$$

$\forall x^{-1}A, y^{-1}A \in Q_{dA}$, $a \in X$. Define $\overline{q_0}_{dA} : Q_A \to [0,1]$ by

$$\overline{q_0}_{dA}(x^{-1}A) = \begin{cases} \overline{q_0}(s_0) & \text{if } x^{-1}A = A \\ 0 & \text{otherwise} \end{cases}$$

$\forall x^{-1}A \in Q_A$. Define $T_{dA} : Q_A \to [0,1]$ by

$$T_{dA}(x^{-1}A) = \begin{cases} t_A & \text{if } x \in A \\ 0 & \text{otherwise.} \end{cases}$$

Set $M_{dA} = (Q_{dA}, X, \mu_{dA}, \overline{q_0}_{dA}, T_{dA})$. Next we show that M_{dA} is a complete accessible fuzzy deterministic recognizer such that $L(M_{dA}) = A$. Clearly $\overline{q_0}_{dA}(A) = \overline{q_0}(s_0) > 0$ and $\overline{q_0}_{dA}(x^{-1}A) = 0$ if $A \neq x^{-1}A$. Thus M_{dA} has a unique initial state.

Let $x^{-1}A, y^{-1}A, u^{-1}A, v^{-1}A \in Q_{dA}$, $a, b \in X$. Let $(x^{-1}A, a, y^{-1}A) = (u^{-1}A, b, v^{-1}A)$. Then $x^{-1}A = u^{-1}A$, $y^{-1}A = v^{-1}A$ and $a = b$. Now $(xa)^{-1}A = a^{-1}(x^{-1}A) = b^{-1}(u^{-1}A) = (ub)^{-1}A$. Hence $(xa)^{-1}A = y^{-1}A$ if and only if $(ua)^{-1}A = v^{-1}A$. This shows that μ_{dA} is well-defined.

Let $x^{-1}A, y^{-1}A, z^{-1}A \in Q_{dA}$ and

$$\mu_{dA}(x^{-1}A, a, y^{-1}A) > 0$$

and

$$\mu_{dA}(x^{-1}A, a, z^{-1}A) > 0.$$

Then $y^{-1}A = (xa)^{-1}A = z^{-1}A$. Hence M_{dA} is deterministic.

By Theorem 6.10.6, Q_{dA} is a finite set. Clearly M_{dA} is a complete accessible fuzzy recognizer.

Let $x \in A$. Then $T_{dA}(x^{-1}A) = t_A > 0$. Let $x = a_1 a_2 \ldots a_n$, $a_i \in X$ for all i. Now $\mu^*_{dA}(A, x, x^{-1}A) \geq \mu_{dA}(A, a_1, a_1^{-1}A) \wedge \mu_{dA}(a_1^{-1}A, a_2, (a_1a_2)^{-1}A) \wedge \mu_{dA}((a_1a_2)^{-1}A, a_2, (a_1a_2a_3)^{-1}A) \wedge \ldots \wedge \mu_{dA}((a_1a_2 \ldots a_{n-1})^{-1}A, a_n, (a_1a_2 \ldots a_n)^{-1}A) = m_A \wedge m_A \wedge \ldots \wedge m_A = m_A > 0$. Thus

$$\overline{q_0}_{dA}(A) \wedge \mu^*_{dA}(A, x, x^{-1}A) \wedge T_{dA}(x^{-1}A) > 0.$$

This implies that $x \in L(M_{dA})$. Now let $x \in L(M_{dA})$. There exists $z^{-1}A \in Q_{dA}$ such that $\overline{q_0}_{dA}(A) \wedge \mu^*_{dA}(A, x, z^{-1}A) \wedge T_{dA}(z^{-1}A) > 0$. Thus $\overline{q_0}_{dA}(A) > 0$, $\mu^*_{dA}(A, x, z^{-1}A) > 0$, and $T_{dA}(z^{-1}A) > 0$. Now $\mu^*_{dA}(A, x, z^{-1}A) > 0$ implies that $x^{-1}A = z^{-1}A$ by the definition of μ^*_{dA}. Hence $T_{dA}(x^{-1}A) = T_{dA}(z^{-1}A) > 0$ and so $x \in A$. Consequently, $A = L(M_{dA})$. ∎

Theorem 6.10.8 *Let $A \subseteq X^*$ be recognizable. Suppose $M_d = (Q_d, X, \mu, \overline{q_0}, T)$ is a complete accessible deterministic fuzzy recognizer with behavior A. Let M_{dA} be the complete accessible deterministic fuzzy recognizer as constructed in the proof of Theorem 6.9.14. Then $\exists$ a function $f : Q_d \to Q_{dA}$ such that*

(i) $f(s_0) = A$;

(ii) $f^{-1}(F_{dA}) = F_d$, where $F_d = \{q \in Q_d \mid T(q) > 0\}$ and $F_{dA} = \{x^{-1}A \in Q_{dA} \mid T_{dA}(x^{-1}A) > 0\}$

(iii) $\mu_{dA}(f(q), a, f(p)) \geq \mu(q, a, p)$ for all $p, q \in Q_d$, $a \in X$;

(iv) f is surjective.

Proof. Define $f : Q_d \to Q_{dA}$ by $\forall q \in Q_d$, $f(q) = q^{-1} \circ T$. By Theorem 6.10.6(iii), f maps Q_d into Q_{dA}.

(i) $f(s_0) = s_0^{-1} \circ T = A$ by Theorem 6.10.6(iv).

(ii) Let $q \in F_d$. Then $T(q) > 0$. There exists $t \in A$ such that $q^{-1} \circ T = t^{-1}A$ by Theorem 6.10.6(iii). Now $t \in A = L(M_{dA})$. Hence $\mu^*_{dA}(A, t, y^{-1}A) \wedge T_{dA}(y^{-1}A) > 0$ for some $y^{-1}A$. This implies that $\mu^*_{dA}(A, t, y^{-1}A) > 0$ and $T_{dA}(y^{-1}A) > 0$. Now $\mu^*_{dA}(A, t, y^{-1}A) > 0$ implies that $t^{-1}A = y^{-1}A$ and $T_{dA}(y^{-1}A) > 0$ implies that $y^{-1}A \in F_{dA}$. Hence $f(q) = q^{-1} \circ T = t^{-1}A = y^{-1}A \in F_{dA}$, i.e., $q \in f^{-1}(F_{dA})$. Suppose $q \in f^{-1}(F_{dA})$. Then $T_{dA}(q^{-1} \circ T) = T_{dA}(f(q)) > 0$. Now $q^{-1} \circ T = t^{-1}A$ for some $t \in X^*$. Thus $T_{dA}(t^{-1}A) > 0$ and by the definition of T_{dA} it follows that $t \in A$. Since $t \in A$, $\Lambda \in t^{-1}A = q^{-1} \circ T$. Hence $\mu^*(q, \Lambda, r) \wedge T(r) > 0$ for some $r \in Q_d$. This implies that $q = r$. Thus $T(q) > 0$ and so $q \in F_d$. Consequently $f^{-1}(F_{dA}) = F_d$.

(iii) Let $q, p \in Q_d$. Now $f(q) = q^{-1} \circ T$ and $f(q) = p^{-1} \circ T$. By Theorem 6.10.6, there exists $x, y \in X^*$ such that $q^{-1} \circ T = x^{-1}A$ and $p^{-1} \circ T = y^{-1}A$. Let $a \in X$. Suppose $\mu_{dA}(f(q), a, f(p)) = 0$. Then $\mu_{dA}(x^{-1}A, a, y^{-1}A) = 0$ and so $(xa)^{-1}A \neq y^{-1}A$. We claim that $\mu(q, a, p) = 0$.

Suppose $\mu(q, a, p) > 0$. Either there exists $t \in (xa)^{-1}A$ such that $t \notin y^{-1}A$ or there exists $t \in y^{-1}A$ such that $t \notin (xa)^{-1}A$. First suppose that there exists $t \in (xa)^{-1}A$ such that $t \notin y^{-1}A$. Since $xat \in A$, $\mu^*(s_0, xat, r) \wedge T(r) > 0$ for some $r \in Q_d$. Thus $\mu^*(s_0, xat, r) > 0$. This implies that there exists $q', p' \in Q_d$ such that $\mu^*(s_0, x, q') \wedge \mu(q', a, p') \wedge \mu^*(p', t, r) > 0$. Since M_d is deterministic, $q' = q$ (note this follows from the proof of Theorem 6.10.6 by the choice of x). Thus $\mu(q, a, p') > 0$ and since M_d is deterministic, it follows that $p = p'$. Hence $\mu^*(p, t, r) > 0$. By the choice of y (as in the proof of Theorem 6.10.6), $\mu^*(s_0, y, p) > 0$. It now follows that $\mu^*(s_0, y, p) \wedge \mu^*(p, t, r) \wedge T(r) > 0$ and so $t \in y^{-1}A$, a contradiction. Now suppose there exists $t \in y^{-1}A$ such that $t \notin (xa)^{-1}A$. Then $xat \notin A$. Now $yt \in A$. Thus $\mu^*(s_0, yt, r) \wedge T(r) > 0$ for some $r \in Q_d$. Thus there exists $p'' \in Q_d$ such that $\mu^*(s_0, y, p'') \wedge \mu^*(p'', t, r) > 0$. From this it follows that $p = p''$. Thus $\mu^*(p, t, r) > 0$. Hence $\mu^*(s_0, x, q) \wedge \mu(q, a, p) \wedge \mu^*(p, t, r) \wedge T(r) > 0$ and so $xat \in A$, a contradiction. Hence $\mu(q, a, p) = 0$. From the definition of μ_{dA}, it now follows that $\mu_{dA}(f(q), a, f(p)) \geq \mu(q, a, p)$.

(iv) Let $x^{-1}A \in Q_{dA}$. By Theorem 6.10.6, $x^{-1}A = q_x^{-1} \circ T$. Thus f is surjective. ∎

We regard the recognizer M_{dA} as being **minimal complete recognizer** of the recognizable subset A, where the term 'minimal' refers to the properties described in Theorem 6.10.8, in particular (iv) implies that $|Q_{dA}| \leq |Q_d|$.

6.11 References

1. Arbib, M.A., *Algebraic Theory of Machines, Languages, and Semigroups*, Academic Press, New York, 1968.

2. Bavel, Z., *Introduction to the Theory of Automata*, Reston Publishing Co., Inc., Reston Virginia, A Prentice Hall Company, 1983.

3. Comer, S. D., Polygroups derived from cogroups, *J. Alg.*, 89 (1984), 397 - 405.

4. Dörfler, W., The cartesian composition of automata, *Math. Systems Theory*, 11 (1978) 239-257.

5. Eilenberg, S., *Automata, Languages and Machines,* Vol. A, B, Academic Press, New York, 1974, 1976.

6. Holcombe, W.M.L., *Algebraic Automata Theory*, Cambridge University Press, New York, 1982.

7. Jantosciak, J., Homomorphisms, Equivalences and reductions in hypergroups, *Rivista Di Matematica Pura Ed Applicata* 9 (1991) 23 - 47.

8. Kandel, A. and Lee, S. C., *Fuzzy Switching and Automata: Theory and Applications*, Crane Russak, 1980.

9. Malik, D.S. and Mordeson, J.N., On fuzzy recognizers, *Kybernets*, 28 (1999) 47-60.

10. Malik, D.S. and Mordeson, J.N., Minimal fuzzy recognizers, *J. Fuzzy Math.*, 7 (1999) 381-389.

11. Malik, D. S., Mordeson, J. N. and Sen, M. K., Semigroups of fuzzy finite state machines, Advances in Fuzzy Theory and Techonology, Edited by P.P. Wang, Vol II, 1994, 87-98.

12. Malik, D. S., Mordeson, J. N. and Sen, M. K., Submachines of fuzzy finite state machines, *J. Fuzzy Math*, 2 (1994) 781-792.

13. Malik, D. S., Mordeson, J. N. and Sen, M. K., On subsystems of a fuzzy finite state machine, *Fuzzy Sets and Systems*, 68 (1994) 83-92.

14. Malik, D. S., Mordeson, J. N. and Sen, M. K., The cartesian composition of fuzzy finite state machines, *Kybernets*, 24 (1995).

15. Malik, D. S., Mordeson, J. N. and Sen, M. K., On fuzzy regular languages, *Inform. Sci.,* 88 (1996) 263-273.

16. Malik, D. S., Mordeson, J. N. and Sen, M. K., Products of fuzzy finite state machines, *Fuzzy Sets and Systems,* 92 (1997) 95-102.

17. Steimann, F. and Adlassnig, K., Clinical monotoring with fuzzy automata, *Fuzzy Sets and Systems,* 61 (1994) 37-42.

18. Wee, W. G., On generalizations of adaptive algorithm and application of the fuzzy sets concept to pattern classification, Ph, D. Thesis, Purdue University, June, 1967.

19. Zadeh, L. A., Fuzzy sets, *Inform. Control,* 8 (1965) 338 - 353.

20. Zadeh, L. A., Fuzzy sets and systems, Proc. Symp. System Theory, Polytechnic Institute of Brooklyn, 29 - 37 (1965).

21. Zariski, O. and Samuel, P., *Commutative Algebra* Vol I, D. Van Nostrand Company, Inc 1958.

Appendix

BBB

In this section, we describe the BBB algorithm (see Chapter 2) in C++. At the end of this section, we also illlustrate how to enter data for this program.

```
// BBB program
#include <iostream.h>
#include <fstream.h>
#include <iomanip.h>
struct nodeType
{
    float dv;   // decision value L(x)
    float bl;   // branch label l
    int pt;     // parent
    int ec;     // eldest child
    int ys;     // younger sibling
    int nc;     // number of children
    int *ch;    // pointer to children
};
int BBB(nodeType graph[], int vertex, float DV);
void printPath(nodeType graph[], int vertex);
int main()
{
   nodeType *graph;
```

```
int numberOfVertices;
int j,k;
int BBBLeaf;
char filename[50];
ifstream infile;
cout<<"Processing "<<endl;
cout<<"Enter input file name: ";
cin>>filename;
infile.open(filename,ios::nocreate);
if(!infile)
{
  cout<<"Input file does not exist."<<endl;
  return 1;
}
cout.setf(ios::fixed,ios::floatfield);
cout.setf(ios::showpoint);
cout<<setprecision(3);
infile>>numberOfVertices;
cout<<"Number of nodes = "<<numberOfVertices<<endl;
graph = new nodeType[numberOfVertices];
graph[0].bl = 0;
graph[0].pt = -1;
graph[0].ys = -1;
graph[0].dv = 1;
infile>>graph[0].nc;
if(graph[0].nc != 0)
{
  graph[0].ch = new int[graph[0].nc];
  for(k = 0; k < graph[0].nc; k++)
  infile>>graph[0].ch[k];
  graph[0].ec = 0;
}
for(j = 1; j < numberOfVertices; j++)
{
  infile>>graph[j].bl;
  infile>>graph[j].nc;
  if(graph[j].nc != 0)
  {
    graph[j].ch = new int[graph[j].nc];
    for(k = 0; k < graph[j].nc; k++)
    infile>>graph[j].ch[k];
  }
  graph[j].dv = 0;
  graph[j].pt = -1;
  graph[j].ec = -1;
```

```
      graph[j].ys = -1;
   }//end for
   BBBLeaf = BBB(graph,0,0);
   cout<<"BBBLeaf = "<<BBBLeaf<<endl;
   cout<<"Path = 0";
   printPath(graph, BBBLeaf);
   cout<<endl;
   cout<<"Decision Value = "<<graph[BBBLeaf].dv<<endl;
   return 0;
}

int BBB(nodeType graph[], int vertex, float DV)
{
   int x, y, p;
   float l;
   int child, nchild;
   bool step2, step4, step5;
   bool step6, step7, step8;
   bool found;
   if(graph[vertex].nc == 0)
      return vertex;
   x = vertex;
   step2 = true;
   while(step2)
   {
      step4 = true;
      step5 = true;
      step6 = true;
      //step3
      child = graph[x].ch[0];
      l = (graph[x].dv) * (graph[child].bl);
      if(l <= DV)
      {
         graph[x].dv = 0;
         graph[child].pt = x;
         x = child;
         y = -1;
         step2 = step4 = step5 = false;
      }
      else
      {
         graph[x].ec = child;
        graph[child].pt = x;
         graph[child].dv = l;
         if(graph[child].nc == 0)
```

```
      step2 = false;
      else
      {
        found = false;
        for(int j = 1; j < graph[x].nc; j++)
        {
          nchild = graph[x].ch[j];
          l = (graph[x].dv)*(graph[nchild].bl);
          if(l <= DV)
          {
            graph[graph[x].ch[j-1]].ys = -1;
            step2 = false;
            found = true;
            break; // exit for; go to step4
          }
          graph[nchild].pt = x;
          graph[graph[x].ch[j-1]].ys = nchild;
          graph[nchild].dv = l;
          if(graph[nchild].nc == 0)
          {
            step2 = false;
            graph[nchild].ys = -1;
            found = true;
            break; //exit for go to step4
          }
        }//end for
        if(!found)
        {
          graph[graph[x].ch[graph[x].nc - 1]].ys = -1;
        }
      }
      //step4
      if(step4)
      {
        x = graph[x].ec;
        if(graph[x].nc != 0)
          step2 = true;
        else
          step2 = false;
      }
    }//end else
  } //end while step2
        //step 5
  if(step5)
    y = x;
```

```
  // step6
while(step6)
{
  step7 = true;
  step8 = true;
  if(graph[x].pt == 0)
  {
      return y;
  }
else
{
  x = graph[x].pt;
  step7 = true;
}
while(step7)
{
  if(graph[x].ys == -1)
  {
    step7 = false;
    step8 = false;
    step6 = true;
    break;
  }
  else
  {
    step6 = false;
    x = graph[x].ys;
    if( y != -1)
    {
      if(graph[x].dv <= graph[y].dv)
      {
        step7 = false;
        step8 = false;
        step6 = true;
        break; //exit while step7; go to 6
      }
    }
    else
      if(graph[x].dv <= DV)
      {
        step7 = false;
        step8 = false;
        step6 = true;
        break;
      }
```

```
        if(step8)
        {
         if(y != -1)
         {
           cout<<"Calling BBB: x = "<<x
                <<", L(y) = "<<graph[y].dv<<endl;
           p = BBB(graph,x,graph[y].dv);
         }
         else
         {
           cout<<"Calling BBB: x = "<<x
           <<", L(y) = "<<DV<<endl;
           p = BBB(graph,x,DV);
         }
         if(p != -1)
           y = p;
         step7 = true;
         graph[x].ys = -1;
         }
        }
      }//end while step7
    }//end while step6
}//end main
void printPath(nodeType graph[], int vertex)
{
      if(vertex != 0)
      {
        printPath(graph, graph[vertex].pt);
        cout<<"-"<<vertex;
      }
}//end printPath
```

We now illustrate how to enter data for a particular program. We assume that the input data is stored in a file. Consider the following decision tree.

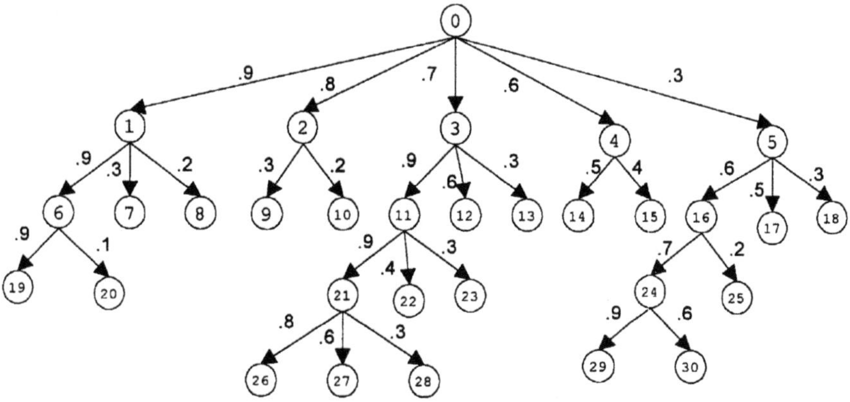

This decision tree has 31 nodes. In the input file the data must be stored in the following form.

31
5 1 2 3 4 5
.9 3 6 7 8
.8 2 9 10
.7 3 11 12 13
.6 2 14 15
.3 3 16 17 18
.9 2 19 20
.3 0
.2 0
.3 0
.2 0
.9 3 21 21 23
.6 0
.3 0
.5 0
.4 0
.6 2 24 25
.5 0
.3 0
.9 0
.1 0
.9 3 26 27 28
.4 0
.3 0
.7 2 29 30
.2 0

.8 0
.6 0
.3 0
.9 0
.6 0

The first line indicates the number of nodes in the decision tree. The first number in the second line indicates that the root node has 5 children. The children of the root node are 1, 2, 3, 4, and 5. The data in the remaining rows is in the following form: The first number indicates the l value of the node, the second number indicates the number of children this node has. If the second entry is nonzero, we alos list the children of that node.

The output of the program for this data is: (We assume that data is stored in the file bbbinp6.txt and this file is on the floppy drive A.)

Processing
Enter input file name: a:bbbinp6.txt
Number of nodes = 31
Calling BBB: x = 2, L(y) = 0.729
BBBLeaf = 19
Path = 0-1-6-19
Decision Value = 0.729

Index

List of Symbols

$\mathbb{N}$	the set of positive integers, p. 1
$\mathbb{Z}$	the set integers, p. 1
$\mathbb{Q}$	the set of rational numbers, p. 1
$\mathbb{R}$	the set of real numbers, p. 1
$\mathbb{R}^+$	the set of positive real numbers, p. 1
$\mathbb{N}_n$	the set of first n positive integers, p. 1
$\|S\|$	the cardinality of a set S, p. 1
$\in$	belongs to, p. 1
$\notin$	does not belong to, p. 1
$\subseteq$	subset, p. 1
$\subset$	proper subset, p. 1
$\supseteq$	contains, p. 1
$\supset$	properly contains, p. 1
$\emptyset$	the empty set, p. 1
$\cup$	the union of sets, p. 2
$\cap$	the intersection of sets, p. 2
$A\backslash B$	relative complement of B in A, p. 2
$\overline{A}$	relative complement of A in its universe, p. 2
$A \times B$	Cartesian cross product of A and B, p. 2
$\mathcal{P}(X)$	power set of X, p. 2
$\mathrm{Dom}(R)$	domain of a relation R, p. 3
$\mathrm{Im}(R)$	image of a relation R, p. 3
$[x]$	equivalence class of an element x, p. 3
$\circ$	composition of fuzzy and crisp relations, p. 5
μ_α	α-cut of the fuzzy subset μ, p. 7
$\mathrm{Supp}(\mu)$	support of the fuzzy subset μ, p. 7
$\mathcal{FP}(X)$	the fuzzy power set of X, p. 7

$\overline{\mu}$	complement of the fuzzy subset μ, p. 8
$\wedge$	infimum, p. 8
$\vee$	supremum, p. 8
FPI	fuzzy prime implicant, p. 21
$R\psi Q$	fuzzy consensus of the phrases R and Q, p.23
EFPI	essential fuzzy prime implicant, p. 30
$(P,\overline{P})$	a cut in the network, p.65
$C(P,\overline{P})$	capacity of a cut, p.65
$\overrightarrow{(X,\overline{X})}$	set of arcs of the cut $(X,\overline{X})$ directed from X to $\overline{X}$, p.68
$\overleftarrow{(X,\overline{X})}$	set of arcs of the cut $(X,\overline{X})$ directed from $\overline{X}$ to X, p.68
$\tilde{\wedge}$	extended min, p. 109
$\tilde{+}$	extended sum, p.112
$\odot$	max-product binary operation, p.139
$C(X)$	the convex MP-span of X, p.140
MPSM	max-product sequential-like machine, p. 142
$len(x)$	length of the tape x, p.142
Λ	empty string, empty tape, or identity of the semi group, p. 142
i.d.	initial distribution, p. 143
IMPSM	initialized max-product sequential-like machine, p. 143
$\rho(A)$	collection of all rows of a matrix A, p.144
$\rho(M)$	p.147, 162
$\rho(\overline{M})$	p.147, 162
$\otimes$	max-min binary operation, p. 148
$C(X)$	convex span of X, p. 149
MSLM	maximin sequential-like machine, p. 155
sd	state distribution, p. 157
IMSLM	initialized maximin sequential-like machine, p. 157
NSLM	nondeterministic sequential-like machine, p. 163
DSLM	deterministic sequential-like machine, p. 163
$\mathbb{R}^{\geq 0}$	the set of all nonnegative real numbers, p. 165
$\mathbb{R}^{\geq 0}_{\infty}$	$\mathbb{R}^{\geq 0}_{\infty} = \mathbb{R}^{\geq 0} \cup \{\infty\}$, p. 165
CMG	context-free max-product grammar, p. 165
CFFL	context-free fuzzy language, p. 166
$L(f,r,>)$	$L(f,r,>) = \{x \in T^* \mid f(x) > r\}$, p. 168
$L(f,r,\geq)$	$L(f,r,\geq) = \{x \in T^* \mid f(x) \geq r\}$, p. 169
$L(f,r,=)$	$L(f,r,=) = \{x \in T^* \mid f(x) = r\}$, p. 169
$\mathcal{L}_r$	family of all languages of the form $L(f, r, >)$, where f is a $CFFL$, p. 171
$\mathcal{L}$	$\mathcal{L} = \cup_{r \in \mathbb{R}^{\geq 0}} \mathcal{L}_r$, p. 171
$\mathcal{C}$	the family of all CFL, p. 171
$\mathcal{R}$	the family of all regular languages, p. 171
RFG	regular fuzzy grammer, p. 174
FFA	finite fuzzy automaton, p. 174
FL	fuzzy language, p. 174

L(RFG)	regular fuzzy language, p. 174
$L(\theta)$	$L(\theta) = \{x \mid x \in \sum^*, \mu(x) \geq \theta\}$, p. 175
$\mathbf{FR}(\Sigma^*)$	family of F-regular languages, p.180
$\widetilde{E(M)}$	p.199
M/π	quotient fuzzy finite state machine, p. 220
■	end of proof

Studies in Fuzziness and Soft Computing

Vol. 26. A. Yazici and R. George
Fuzzy Database Modeling, 1999
ISBN 3-7908-1171-8

Vol. 27. M. Zaus
Crisp and Soft Computing with Hypercubical Calculus, 1999
ISBN 3-7908-1172-6

Vol. 28. R.A. Ribeiro, H.-J. Zimmermann, R.R. Yager and J. Kacprzyk (Eds.)
Soft Computing in Financial Engineering, 1999
ISBN 3-7908-1173-4

Vol. 29. H. Tanaka and P. Guo
Possibilistic Data Analysis for Operations Research, 1999
ISBN 3-7908-1183-1

Vol. 30. N. Kasabov and R. Kozma (Eds.)
Neuro-Fuzzy Techniques for Intelligent Information Systems, 1999
ISBN 3-7908-1187-4

Vol. 31. B. Kostek
Soft Computing in Acoustics, 1999
ISBN 3-7908-1190-4

Vol. 32. K. Hirota and T. Fukuda
Soft Computing in Mechatronics, 1999
ISBN 3-7908-1212-9

Vol. 33. L.A. Zadeh and J. Kacprzyk (Eds.)
Computing with Words in Information/ Intelligent Systems 1, 1999
ISBN 3-7908-1217-X

Vol. 34. L.A. Zadeh and J. Kacprzyk (Eds.)
Computing with Words in Information/ Intelligent Systems 2, 1999
ISBN 3-7908-1218-8

Vol. 35. K.T. Atanassov
Intuitionistic Fuzzy Sets, 1999
ISBN 3-7908-1228-5

Vol. 36. L.C. Jain (Ed.)
Innovative Teaching and Learning, 2000
ISBN 3-7908-1246-3

Vol. 37. R. Słowiński and M. Hapke (Eds.)
Scheduling Under Fuzziness, 2000
ISBN 3-7908-1249-8

Vol. 38. D. Ruan (Ed.)
Fuzzy Systems and Soft Computing in Nuclear Engineering, 2000
ISBN 3-7908-1251-X

Vol. 39. O. Pons, M.A. Vila and J. Kacprzyk (Eds.)
Knowledge Management in Fuzzy Databases, 2000
ISBN 3-7908-1255-2

Vol. 40. M. Grabisch, T. Murofushi and M. Sugeno (Eds.)
Fuzzy Measures and Integrals, 2000
ISBN 3-7908-1255-2

Vol. 41. P. Szczepaniak, P. Lisboa and J. Kacprzyk (Eds.)
Fuzzy Systems in Medicine, 2000
ISBN 3-7908-1263-4

Vol. 42. S. Pal, G. Ashish and M. Kundu (Eds.)
Soft Computing for Image Processing, 2000
ISBN 3-7908-1217-X

Vol. 43. L.C. Jain, B. Lazzerini and U. Halici (Eds.)
Innovations in ART Neural Networks, 2000
ISBN 3-7908-1270-6

Vol. 44. J. Aracil and F. Gordillo (Eds.)
Stability Issues in Fuzzy Control, 2000
ISBN 3-7908-1277-3

Vol. 45. N. Kasabov (Ed.)
Future Directions for Intelligent Information Systems on Information Sciences, 2000
ISBN 3-7908-1276-5

Vol. 46. J.N. Mordeson and P.S. Nair
Fuzzy Graphs and Fuzzy Hypergraphs, 2000
ISBN 3-7908-1286-2

Vol. 47. E. Czogała† and J. Łęski
Fuzzy and Neuro-Fuzzy Intelligent Systems, 2000
ISBN 3-7908-1289-7

Vol. 48. M. Sakawa
Large Scale Interactive Fuzzy Multiobjective Programming, 2000
ISBN 3-7908-1293-5

Vol. 49. L.I. Kuncheva
Fuzzy Classifier Design, 2000
ISBN 3-7908-1298-6

Vol. 50. F. Crestani and G. Pasi (Eds.)
Soft Computing in Information Retrieval, 2000
ISBN 3-7908-1299-4

Vol. 51. J. Fodor, B. De Baets and P. Perny (Eds.)
Preferences and Decisions under Incomplete Knowledge, 2000
ISBN 3-7908-1303-6

Vol. 52. E.E. Kerre and M. Nachtegael (Eds.)
Fuzzy Techniques in Image Processing, 2000
ISBN 3-7908-1304-4

Vol. 53. G. Bordogna and G. Pasi (Eds.)
Recent Issues on Fuzzy Databases, 2001
ISBN 3-7908-1319-2

Studies in Fuzziness and Soft Computing

Vol. 54. P. Sinčák and J. Vaščák (Eds.)
Quo Vadis Computational Intelligence?, 2000
ISBN 3-7908-1324-9

Vol. 55. J. N. Mordeson, D. S. Malik
and S.-C. Cheng
Fuzzy Mathematics in Medicine, 2001
ISBN 3-7908-1325-7

Vol. 56. L. Polkowski, S. Tsumoto and T. Y. Lin
(Eds.)
Rough Set Methods and Applications, 2001
ISBN 3-7908-1328-1

Vol. 57. V. Novák and I. Perfilieva (Eds.)
Discovering the World with Fuzzy Logic, 2001
ISBN 3-7908-1330-3

GPSR Compliance
The European Union's (EU) General Product Safety Regulation (GPSR) is a set of rules that requires consumer products to be safe and our obligations to ensure this.

If you have any concerns about our products, you can contact us on

ProductSafety@springernature.com

In case Publisher is established outside the EU, the EU authorized representative is:

Springer Nature Customer Service Center GmbH
Europaplatz 3
69115 Heidelberg, Germany

www.ingramcontent.com/pod-product-compliance
Ingram Content Group UK Ltd.
Pitfield, Milton Keynes, MK11 3LW, UK
UKHW020136290926
15471UKWH00003B/525
9783662003749